Metal-Catalyzed Polymerization

Metal-Catalyzed Polymerization

Fundamentals to Applications

Edited by
Samir H. Chikkali

CRC Press
Taylor & Francis Group
Boca Raton London New York

CRC Press is an imprint of the
Taylor & Francis Group, an **informa** business

CRC Press
Taylor & Francis Group
6000 Broken Sound Parkway NW, Suite 300
Boca Raton, FL 33487-2742

First issued in paperback 2020

© 2018 by Taylor & Francis Group, LLC
CRC Press is an imprint of Taylor & Francis Group, an Informa business

No claim to original U.S. Government works

ISBN-13: 978-0-367-57314-0 (pbk)
ISBN-13: 978-1-4987-6757-6 (hbk)

Library of Congress Cataloging-in-Publication Data

Names: Chikkali, Samir.
Title: Metal-catalyzed polymerization : fundamentals to applications / [edited by] Samir Chikkali [and three others].
Description: Boca Raton : CRC Press, 2018. | Includes bibliographical references.
Identifiers: LCCN 2017008378 | ISBN 9781498767576 (hardback : alk. paper)
Subjects: LCSH: Polymerization. | Metal catalysts.
Classification: LCC QD281.P6 M3955 2018 | DDC 547/.28--dc23
LC record available at https://lccn.loc.gov/2017008378

Visit the Taylor & Francis Web site at
http://www.taylorandfrancis.com

and the CRC Press Web site at
http://www.crcpress.com

Contents

Preface

As polymer science progresses, the scientific world would like to have more and better control over polymerization reactions. Gone are the days when radical polymerization was the subject of many investigations in academic laboratories and industry. Today the scientific community prefers to have controlled reactions and precise arrangement of desired monomer sequences in a polymer chain. This is exactly what a metal-catalyzed/mediated polymerization offers. A survey of recent trends in polymer science shows that a large amount of recent literature deals with metal-catalyzed polymerization methods.

However, there is hardly any book on metal-catalyzed polymerization reactions that can serve as a single reference to organometallic and polymer chemistry community. Metal-catalyzed C–C coupling reactions are well established in organic and organometallic chemistry, and the understanding has been compiled in various books at the graduate, postgraduate, and higher levels. On one hand the chemistry of metal-catalyzed C–C coupling reactions—such as Suzuki–Miyaura, Stille, Heck, Negishi, Sonogashira, and Kumada coupling—has been extended to polymerization, on the other hand there are no books that can provide an overview of this interdisciplinary area. Therefore, we believe a dedicated book on metal-catalyzed polymerization will bridge the gap between organometallic chemistry and traditional polymer chemistry and, in the process, guide the field toward its full potential. The ultimate aim of this book is to

- Present a modern textbook/reference book for undergraduates, postgraduates, and advanced students.
- Provide an overview of the state-of-the-art development in metal-catalyzed polymerization reactions.
- Extract the best of both worlds (organometallic and polymer) for the younger generation polymer chemistry community.
- Provide a common platform for organometallic and polymer chemistry readers.
- Develop a disruptive understanding of the subject.
- Serve as a single reference for the academic and industrial communities.

To achieve these goals, this book presents an overview of seven highly sought-after topics in polymer chemistry. Although the focus is on the fundamentals, the state-of-the-art industrial development in these seven topics is also presented with real-world processes and technologies. Chapter 1 describes the essentials of organometallic chemistry and its relevance to metal-catalyzed polymerization. Subsequent chapters present a great deal of understanding in metal-catalyzed Ziegler–Natta/insertion/coordination polymerization that stands out as one of the most explored and industrially applied metal-catalyzed polymerization reactions. Although a comprehensive treatment is given to classical Ziegler–Natta process, new developments in metallocene-grade polyolefins, postmetallocenes, waxy polyolefin, and ultra-high-molecular-weight polyethylene are discussed in detail in Chapter 2. Chapter 3 presents the new polymerization technique called C1 polymerization, which offers a better alternative to prepare functional polyolefins than classical Ziegler–Natta polymerization.

The uniqueness of olefin metathesis in metal-catalyzed polymerization is summarized in Chapter 4, along with ring-opening polymerization. Striking a right balance between concepts, mechanisms, significance, and implications for sustainable chemistry is the hallmark of this chapter. Taming radicals to obtain control is the subject of many discussions since the discovery of controlled radical polymerization (CRP) by Professor K. Matyjaszewski in the mid-1990s. The approach in this chapter is to introduce the concept, present the mechanisms in various CRP techniques, and demonstrate how these can be adapted to produce various polymer architectures and advanced materials of industrial significance. The difference between a CRP technique and organometallic-mediated radical polymerization is presented in Chapter 5. The fundamental principles of organometallic-mediated radical polymerization, recent developments, and advantages over existing metal-catalyzed polymerization methods are outlined in Chapter 6. Chapter 7 is designed to present the relevance of C–C coupling reactions in classical organic chemistry and how these can be extended to metal-catalyzed polymerization methods. This relatively young field is dealt with by a fitting explanation and appropriate examples.

Although the book covers a wide spectrum of metal-catalyzed polymerizations, it is comprehensive and provides a clear overview of the field. It is, however, almost impossible to discuss every aspect of ligand designing, catalyst preparation, and polymerization, and to cite all original references. The references cited reflect the personal choice of the authors and serve to direct readers for the sake of completeness.

It is inevitable for a book of this kind to have errors, and this book is no exception. We will be pleased to hear from readers and will make corrections as needed. We hope you enjoy reading this book and that it will inspire you to contribute to the field of metal-catalyzed polymerization. We would add that a book on any topic is not the end of the matter but rather the beginning of a new chapter in the field.

Acknowledgments

Many individuals have contributed directly and indirectly to this book, and without their help, we would not have been able to complete this book.

Dr. Chikkali gratefully acknowledges his indebtedness to his mentors, peers, and students who presented their observations on various occasions and shared their thoughts. Many of these are crystallized in this book. He is grateful to his divisional colleagues for an inspiring environment and encouraging discussions on the courses he offers. This book is a direct outcome of one such discussion. He is particularly thankful for the creative atmosphere of CSIR-National Chemical Laboratory, Pune, Maharashtra, India, and the long legacy of the institute's support to the best science. He also thanks CRC Press editor Renu Upadhyay for the encouragement and freedom he received in writing this book.

Dr. Chikkali sincerely thanks his parents for their encouragement and his patient and supportive wife.

Contributors

Ashootosh V. Ambade
Polymer Science and Engineering
  Division
CSIR-National Chemical Laboratory
Pune, India

Bas de Bruin
Homogeneous and Supramolecular
  Catalysis Group
Van't Hoff Institute for Molecular
  Sciences (HIMS)
Faculty of Science
University of Amsterdam (UvA)
Amsterdam, the Netherlands

Samir H. Chikkali
Polymer Science and Engineering
  Division
CSIR-National Chemical Laboratory
Pune, India

Daniel L. Coward
EaStCHEM School of Chemistry
University of Edinburgh
Edinburgh, United Kingdom

Benjamin R. M. Lake
EaStCHEM School of Chemistry
University of Edinburgh
Edinburgh, United Kingdom

Sandeep Netalkar
Polymer Science and Engineering
  Division
CSIR-National Chemical Laboratory
Pune, India

Ketan Patel
Polymer Science and Engineering
  Division
CSIR-National Chemical Laboratory
Pune, India

Michael Shaver
EaStCHEM School of Chemistry
University of Edinburgh
Edinburgh, United Kingdom

1 Introduction to Organometallics

Samir H. Chikkali and Sandeep Netalkar

CONTENTS

1.1 WHY IS IT IMPORTANT?

Polymers are gigantic macromolecules that are constructed by linking together a large number of smaller molecules called *monomers* through the process of *polymerization.* Polymer chemistry has evolved from a primitive, pragmatic discipline in the early 1920s to a well-established academic course today. The early efforts by Hermann Staudinger laid the foundation of rational polymer science and today it is nearly well understood as any other contemporary science. The fundamentals of chemistry and physics of polymers are detailed in many text books and we refer the reader to dedicated main stream polymer science books.[1] Nevertheless, the subject is forging ahead and is conquering new territories, and understanding of these new frontiers is equally important.

To project the future of polymer chemistry, we examined the research topics that appeared most frequently in one of the standard polymer science journals *Macromolecules*. We screened the first issue of 1990, 2000, and 2016, and Table 1.1 summarizes the polymerization techniques reported in these issues. In 1990s the conventional research areas such as radical polymerization, radical initiator development, living cationic polymerization, photopolymerization, and condensation polymerization were still dominating. By the turn of the century (2000), these research topics tapered down and metal-catalyzed polymerization such as controlled radical polymerization, ring-opening metathesis polymerization (ROMP), and so on occupied the driver seat. The year 2016 saw enhanced academic activity in the metal-catalyzed polymerization methods. These trends in the evolution of polymer chemistry presented in Table 1.1 will convince you that there is relatively less academic activity in the conventional areas of radical polymerization, or ionic polymerization or condensation polymerization. In contrast, metal-catalyzed polymerization

---

TABLE 1.1

Evolution of Research Topics in Polymer Chemistry as Appeared in the First Issue of Macromolecules in January 1990, 2000, and 2016

January 1990	January 2000	January 2016
Condensation polymerization	Controlled radical polymerization	Metallocenes
Photopolymerization	Ring-opening polymerization	Living cationic polymerization: Metal catalyst
Solution polymerization	Ring-opening metathesis polymerization	Organometallic polymers
Radical polymerization	Metal-catalyzed polymerization	Sequential polymerization by metal catalyst
Radical initiators	Organometallic polymers	Self-assembly of polymers
Cyclopolymerization by metals	Solid-state polymerization	Polycondensation by metals
Living cationic polymerization		Thiol-ene chemistry
		Reversible addition fragmentation chain transfer

Publications

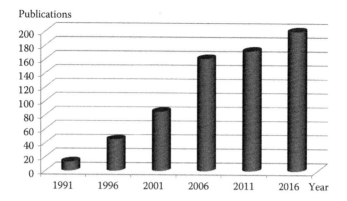

FIGURE 1.1 Ring-opening metathesis polymerization (ROMP) was chosen as one of the representative metal-catalyzed/mediated polymerization methods and the graph depicts the evolution of publications in ROMP between 1991 and 2016. The above-mentioned data were obtained by a search in Scifinder using the keyword *ROMP*.

methods are being extensively investigated as they provide better control over the polymerization process and can tune the polymer properties.

Number of publications in a research theme could be one of the useful indicators to foresee the unseen and predict the future. We chose ROMP as a representative metal-catalyzed polymerization method and accessed number of publications that appeared in ROMP every five years from 1991. Figure 1.1 summarizes the search results and it is obvious from this figure that the number of publications grew from merely 20 in 1991 to around 200 in 2016.[2] This representative sample indicates an upward trend in ROMP, and careful analysis of the two metrics (Table 1.1 and Figure 1.1) allows to forecast that polymer science is going to witness enhanced activity in metal-catalyzed or metal-mediated polymerization methods in the near future. Therefore, it is in the greater interest of the polymer community to venture with the organometallic or inorganic chemistry and vice versa. An interdisciplinary field requires adequate exchange between various stakeholders, and metal-catalyzed polymerization will make an impact if the organometallic chemist and polymer scientist can learn from each other and speak the same language. In our attempts to enhance this interaction between polymer scientist and inorganic/organometallic chemist, we provide a comprehensive overview of this interdisciplinary field of metal-catalyzed polymerization methods. This chapter will serve as an anchoring point: we will begin with brief history of coordination complexes, followed by rise of organometallic chemistry, introduction to various metal-catalyzed methods in polymerization, and this chapter will present elementary steps in metal-catalyzed polymerization methods. At the end of this chapter, a brief discussion about the subsequent chapters will be initiated.

An in-depth understanding of the field of metal-catalyzed polymerization will require at least basic knowledge of coordination chemistry, organometallic chemistry, and polymer chemistry. The subsequent sections in this chapter will initiate a brief discussion on coordination, organometallic, and polymer chemistry.

1.2 COORDINATION CHEMISTRY AND WERNER COMPLEXES[3]

The word coordination means *the organization of the different elements of a complex body or activity so as to enable them to work together smoothly and effectively*. So it is not a surprise that coordination chemistry implies the organization of ligands around a central metal into a complex by dative or covalent interaction that enables it to function as one entity. Metal complexes are also known as coordination compounds. $[Co(NH_3)_6]Cl_3$, hexaamminecobalt (III) chloride is an example of coordination complex where the central Co^{3+} metal ion is surrounded by six ammonia molecules and three chloride anions.

Although the coordination compounds such as Prussian blue $[Fe_7(CN)_{18} \cdot xH_2O]$ that was used as pigment were known since the early eighteenth century, there was no reasonable rationalization to explain the structures and their properties. Danish scientist, Sophus Mads Jorgensen, attempted to explain the properties of these compounds but he was not very successful. However his work laid the ground for Alfred Werner whose pioneering work on coordination compounds of cobalt and ammonia led the field of coordination chemistry to rise in prominence which won Alfred Werner the Nobel prize in 1913.

While working with cobalt and ammonium complexes, Werner realized that several chemically different forms of cobalt ammonia chloride $[CoCl_3(NH_3)_x]$ might exist which could be responsible for different colors and properties of these compounds. These complexes have three chloride ions in its empirical formula but the number of Cl^- ions that precipitated upon reaction with Ag^+ ions was not always the same. Based on the number of chloride ions precipitated, Werner predicted the structure of complex and Table 1.2 summarizes their structural formula along with the color of the compound.

Based on these observations, Werner postulated that the metal ions in coordination compounds possess two types of valencies', namely ionizable or primary valency and non-ionizable or secondary valency.

Primary valency is the number of charges (negative or positive) needed to satisfy the charge on the metal ion. Primary valency is what we call oxidation state today. For example in $[Co(NH_3)_6]Cl_3$ the +3 charge on cobalt ion is satisfied by three chloride anions. Therefore, the primary valency for the abovementioned complex is three. Primary valency is nondirectional.

Secondary valency is the number of molecules that are directly coordinated to the metal ions. The secondary valency of the metal ion in a complex may be fulfilled by negative ions or neutral ligands having a lone pair of electron (e.g.,

---

TABLE 1.2
Predicted Structure of Cobalt Ammonia Chloride Complexes from the Number of Ionized Chloride Ions

Compound	Color	Precipitated Cl⁻	Structural Formula
$[CoCl_3 6NH_3]$	Yellow	3	$[Co(NH_3)_6]Cl_3$
$[CoCl_3 5NH_3]$	Purple	2	$[Co(NH_3)_5Cl]Cl_2$
$[CoCl_3 4NH_3]$	Green	1	*trans*-$[Co(NH_3)_4Cl_2]Cl$
$[CoCl_3 4NH_3]$	Violet	1	*cis*-$[Co(NH_3)_4Cl_2]Cl$

H_2O, NH_3) or even by some cationic ligands. In the above-mentioned example of $[Co(NH_3)_6]Cl_3$, six ammonia molecules, directly coordinated to Co^{3+}, satisfy the secondary valency. Therefore the secondary valency of cobalt is six. Due to the direct coordination, the secondary valencies are directional in nature.

Similarly, in the complex $[Pt(NH_3)_6]Cl_4$, the primary valency of Pt is four and its secondary valency is six.

The concept of secondary valency, introduced by Werner is called as coordination number. Although the primary valencies are ionic in nature, the secondary valencies are oriented in space around the central metal ion in definite geometrical arrangement.

Detailed description of Werner's theory and structure of coordination complexes are beyond the scope of this chapter and it is advised to refer standard inorganic text books for further reference.[4]

1.2.1 NOMENCLATURE OF COORDINATION COMPLEXES

There are certain conventions that are followed in naming coordination compounds which are listed as follows:

1. Nomenclature of ligands: The ligands in the complex are named first and then the metals are named.
   a. The ligands are named first in an alphabetical order if more than one ligand is present in the complex.
   b. For anionic ligands the suffix "o" is used. For example, chloride ligands are named as chloro, sulphate as sulphato, nitrite as nitrito, and so on.
   c. Neutral ligands are named as it is with few exceptions such as water is termed *aqua*, ammonia as *ammine*, carbon monoxide as *carbonyl*, NO as nitrosyl and the N_2 and O_2 are called *dinitrogen* and *dioxygen*.
   d. The same ligands which are present in two or more numbers are indicated by a Greek prefix such as di- (or bis), tri-, tetra-, penta-, hexa-, hepta-, octa-, nona-, (ennea-), deca-, and so on for 1, 2, 3,... 10, respectively. If the polydentate ligands already contain Greek prefixes the names are placed in parentheses. The prefixes for the number of ligands become bis-, tris-, tetrakis-, pentakis-, and so on. For example, $[Ni(PPh_3)_2Cl_2]$ is named dichlorobis(triphenylphosphine)nickel(II).
   e. The symbol μ- is used to indicate a bridging ligand and this symbol must be used before every bridging ligand. For example, in $[(H_3N)_3Co(OH)_3Co(NH_3)_3]$, since the three OH groups bridge the two cobalt ions, the symbol μ- is placed before its name. It is named as triamminecobalt(III)-μ-trihydroxotriamminecobalt(III).
2. After the ligands the metal ions in the complex are named, following the conventions as given below:
   a. After the ligands the central metal ion is named. If the complex ion is cation, the metal is named after the element itself. If the complex is an anion then the name of the complex ends with the suffix -ate. For example, Fe in a complex anion is called ferrate and Pt is called palatinate.

b. The oxidation state of the metal ion is indicated as a roman numeral in parentheses. For example, $[Cr(NH_3)_3(H_2O)_3]Cl_3$ is named as triamminetriaquachromium(III) chloride, the amine ligand is put before aqua (alphabetical order), chromium in the complex ion is cationic hence retains the name chromium, since the chloride ion lies outside the complex sphere it is named as chloride.

Few representative examples for practice would be as follows:

1. $[Pt(NH_3)_5Cl]Br_3$ is named as pentaamminechloroplatinum(IV) bromide.
2. $[Pt(H_2NCH_2CH_2NH_2)_2Cl_2]Cl_2$ is named as dichlorobis(ethylenediamine) platinum(IV) chloride.
3. $[(H_3N)_3Co(OH)_3Co(NH_3)_3]$ is named as triamminecobalt(III)-μ-trihydroxo triamminecobalt(III).

As witnessed in this section, ligands play an important role in defining a metal complex and also tunes the properties of a metal complex.

1.3 CONCEPT OF LIGANDS

The concept of ligand is closely associated with or is discussed alongside the concept of complexes and vice versa. The word *ligand* originates from the latin word *ligare*, which means to bind/tie. The term ligand is used in coordination chemistry to mean an atom or a molecule that donates one or more of its available electrons to the central metal atoms or ions, which are collectively called as a complex (alternatively known as a coordination entity). The ligands may be neutral, negatively charged, and in rare case may be positively charged. Based on this, the ligands have been classified as L-, X-, or Z-type of ligands.

1.3.1 L-TYPE LIGAND

Ligands derived from neutral precursors are classified as L-type ligands. Examples for L-type ligands are water (H_2O), ammonia (NH_3), triphenylphosphine (PPh_3), carbon monoxide (CO), and so on. These types of ligands contribute two electrons from their available lone pair of electrons toward the metals forming dative coordinate bond and hence do not affect the valency of the metal. These ligands can therefore be called as Lewis bases. The contribution of these ligands to the electron count of the complex is two electrons each. Figure 1.2 depicts the L-type ligands with available lone pairs (marked by an arrow).

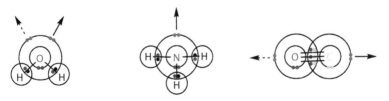

FIGURE 1.2   Representation of L-type ligands along with electron lone pairs.

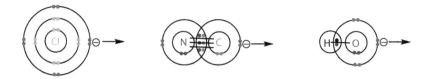

FIGURE 1.3 Representation of X-type ligands along with negative charge.

1.3.2 X-Type Ligand

The X-type of ligand is derived from anionic precursors. Examples include chloride (Cl^-), hydroxide (OH^-), cyanide (CN^-), and so on, which contribute only one electron in forming a covalent bond with the metal center in which the metal contributes the other electron thereby increasing the valency of the metal by one unit. The contribution of these ligands to the electron count of the complex is one electron each. Typical examples of X-type ligands are presented in Figure 1.3.

1.3.3 Z-Type Ligands

Occasionally, ligands such as NO^+, N_2H_5, BF_3, BR_3, and AlR_3 function as electron pair acceptors and are classified as Z-type ligands. In the case of such Z-type ligands, both the electrons are contributed by the metal toward forming the dative covalent bond increasing the valency count of the metal by two units. These ligands therefore function as Lewis acids. The contribution of these ligands to the electron count of the complex is zero electrons.

In a single ligand, there can be a number of sites (donor atoms) that can bind to the same central metal atom, which is described by the term called *denticity*. If the coordination of a single ligand to the central metal atom is through a single site, then the ligands are referred as monodentate ligands. Similarly, bidentate ligand binds to metal through its two different donor sites: tridentate ligand through three sites and polydentate ligand through many sites. Table 1.3 lists types of ligands based on their coordination ability with examples of each type.

TABLE 1.3

Types of Ligands Based on Denticity and Coordination Modes

Monodentate	Bidentate	Tridentate	Polydentate
Aqua, OH_2	Ethylenediamine,	Diethylenetriamine,	Ethylenediamine
Ammine, NH_3	$H_2NCH_2CH_2NH_2$	$NH(CH_2CH_2NH_2)_2$	Tetraacetic acid (EDTA^{4-}),
hydroxo, OH^-	Bipyridine Oxalato, $C_2O_4^{2-}$		($^-OOCCH_2)_2NCH_2-$
Chloro, Cl^-	Acetylacetonato,		$CH_2(NCH_2COO^-)_2$
	$(CH_3COCHCOCH_3)^-$		

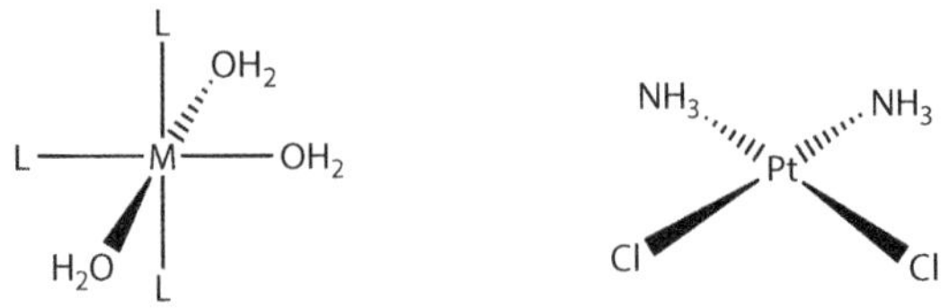

FIGURE 1.4 Typical monodentate ligand coordination.

1.3.3.1 Monodentate Ligands

Monodentate ligands can be of L-, X-, or Z-type ligands that coordinate to a metal through a single atom. Even though the oxygen in water (H_2O) has two electron lone pairs (see Figure 1.2) and Cl^- (Figure 1.3) has four electron lone pairs (which for geometrical constraints it cannot donate to same metal), it coordinates to metal through only one atom. Hence H_2O cannot be a bidentate ligand and Cl^- cannot be a polydentate ligand (see Figure 1.4). Monodentate ligands such as triphenyl phosphine are quite often encountered in C–C coupling reaction. The practical application of these ligands will be elucidated in Chapters 2, 3, and 7.

1.3.3.2 Bidentate Ligands

As explained earlier, even though H_2O has two lone pairs of electrons, it cannot function as a bidentate ligand. Hypothetically, if we consider that H_2O donates both of its lone pair to single central metal ions, it would still be bound to the metal through a single atom and hence would be considered as a monodentate ligand. For a ligand to function as a bidentate system, two valence pairs of electrons on two separate atom sites are required. For example, ethylene diamine and oxalate dianion bind to the metal through two separate atoms and hence is considered as bidentate ligand (Figure 1.5). Bidentate ligands form one of the most important classes of ligands in metal-catalyzed polymerization. The significance of bidentate ligands is presented in Chapters 2, 3, 4, 6, and 7. Similarly, a ligand with three lone pairs of electrons would serve as tridentate ligand (Figure 1.5) and so on.

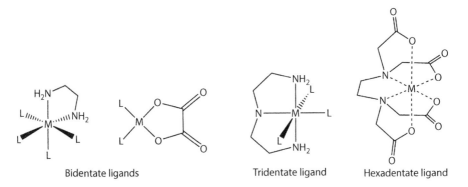

Bidentate ligands   Tridentate ligand   Hexadentate ligand

FIGURE 1.5 Typical bidentate, tridentate, and hexadentate ligand coordination.

FIGURE 1.6 Pictorial representation of bidentate ligand chelation.

1.3.3.3 Chelating Ligands

Ligand with denticity of more than one tends to form ring-like structure through its more than one site of coordination with the same central metal atom. Such ligands are referred as chelating ligands (*Chelate* is derived from the Greek word *chele* meaning *claw* or *to grab on*). The chelation of bidentate (or tri-, tetra-, multidentate) ligands to metals can be imagined as the claws of the lobster holding the metal (Figure 1.6).

The presence of two or more coordinate bonds between a bi/polydentate ligand and the central metal atom has a stabilizing effect than two monodentate ligands separately coordinating to the metal center. If there would be a competition between a chelating ligand and a similar nonchelating monodentate ligand, the metal will prefer former over the latter to achieve stability. The symbol *Kappa* is used to specify which atom of a ligand is bonding to the central metal and in cases where two or more donor atoms of a ligand are involved, a superscript numerical is used to represent the number of such ligations to same central metal atom. For example, a bidentate ethylene diamine (en) that is bound to same metal with an electron pair on two different nitrogen atoms may be represented by κ^2. Another example of the *Kappa* coordination would be dichlorido[(ethane-1,2-diyldinitrilo-$\kappa^2 N,N'$)tetraacetato]platinate(4⁻). Figure 1.7 depicts the structure of the platinum complex with -κ^2 coordination mode.

1.3.3.3.1 Bridging Ligands

In case of some ligands that have more than one available lone pairs of electrons such as Cl⁻, H_2O can donate the additional pair of electrons to one or more different metals thereby holding the two metals together like a bridge. This does not change the denticity of Cl⁻ or H_2O ligand since only one donor site is attached to two or more

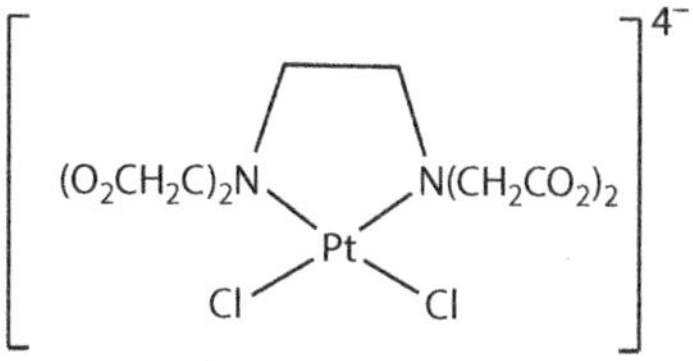

$$\left[(O_2CH_2C)_2N \diagdown \diagup N(CH_2CO_2)_2 \atop Cl \diagup \underset{Pt}{} \diagdown Cl \right]^{4-}$$

FIGURE 1.7 Pictorial representation of *Kappa* chelation by a bidentate ligand.

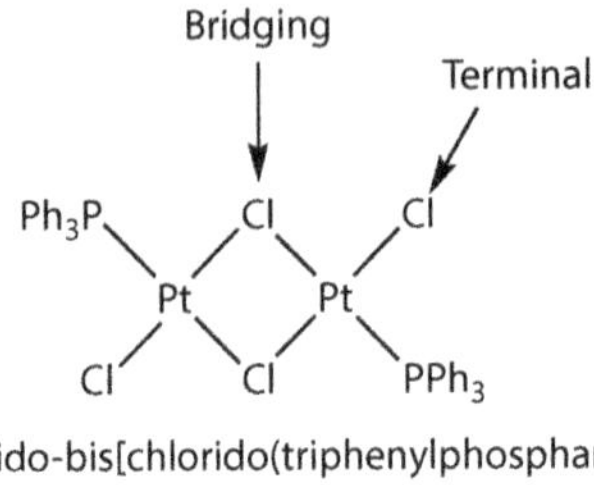

di-μ-chlorido-bis[chlorido(triphenylphosphane)platinum]

FIGURE 1.8   Pictorial representation of *Mu* (μ) bridging by a chlorido-ligand.

different metals. To represent a bridging ligand, the Greek character "μ" is used as prefix to the ligand, spelled as *mu*, with a superscript indicating the number of metals the single ligand bridges. To differentiate the bridging ligands from nonbridging, the latter are referred to as terminal ligands when both the types are present in the single complex. In the simplest case, a ligand may bridge two metals, which are represented as μ^2, sometimes the superscript in this case is omitted just to signify the least bridging that is possible. An example of this type would be di-μ-chlorido-bis [chloride(triphenylphoaphane)platinum] with two bridging chlorides and two terminal chlorides (Figure 1.8).

Sometimes, the various symbols used to denote various modes and the types of ligand coordinations (κ, μ, η) may appear confusing. However, to avoid such confusions, IUPAC has set definitive guide lines and the chemistry community follows these conventions.[5]

1.3.3.3.2   Ambidentate Ligands
Monodentate ligands that have more than one prospective coordinating site but due to geometrical constraints can bind to metal with only one site at a time are termed as ambidentate ligands. Examples of ambidentate ligands are thiocyanate anion SCN^-, (which can bind either through S or N atoms), NO_2^- (N or O), SO_3^{2-} (S or O), CN^- (C or N), and so on.

1.3.3.3.3   Hapticity
Hapticity in coordination chemistry is a concept denoting the number of adjacent or proximate atoms of a single ligand bonded to the single central metal atom. The Greek letter η spelled as *eta* is used to represent hapticity of a ligand with a superscript indicating the number of adjacent atoms. Hapticity should not be confused with denticity. If a multiple number of atoms are bound to the central metal that are not adjacent it is multidentate ligand.

In ferrocene molecule the central Fe^{2+} is bounded by two cyclopentadienyl rings. Since five adjacent carbon atoms of each cyclopentadienyl ring on either side are bounded to Fe^{2+} ion, the hapticity of each cyclopentadienyl ring is 5 represented as η^5 (Figure 1.9). The molecular formula is written as $Fe(\eta^5\text{-}(C_5H_5)_2)$.

In Zeise's salt since two adjacent carbon atoms of ethylene are coordinated to Pt through its π(pi) electron cloud, the hapticity of ethylene in Zeise's salt is denoted as η^2 ($K[PtCl_3(\eta^2\text{-}C_2H_4]\cdot H_2O)$ (Figure 1.9).

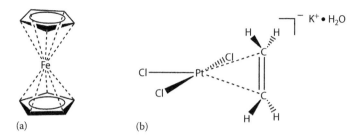

FIGURE 1.9 Typical examples of hapticity in metal complexes, $Fe(\eta^5\text{-}(C_5H_5)_2)$ (a) and $(K[PtCl_3(\eta^2\text{-}C_2H_4]H_2O)$ (b).

1.4 COORDINATION CHEMISTRY TO ORGANOMETALLIC CHEMISTRY[6]

As the name indicates, interaction between inorganic and organic chemistry community led to the development of a new branch of chemistry called organometallic chemistry. To put it forthrightly, organometallic chemistry is a specialized branch of coordination chemistry dealing with the study of compounds possessing metal-carbon bonds and the reactions involving either the formation or breaking of M-C bonds that may be transient or enduring in nature, whereas the coordination chemistry covers the remaining aspect of metal complex other than carbon ligands. Although this meticulous definition of organometallic chemistry emphasizes on the importance of the M-C bond and its chemistry (by chemistry we mean its synthesis, characterization, reactions, and applications in catalysis or in biological system), it is largely superficial and there exists varied perspectives so as to include the compounds possessing bonds between metal and other common elements encountered in organic chemistry such as metal-nitrogen, metal-oxygen, metal-halogen, and also metal-hydrogen-bonded compounds. Hence it would be appropriate in the context of generalization to define organometallic chemistry as the chemistry of metal-organyls rather than that of only metal-carbon. Similarly, not only all the transition metals, lanthanides, actinides, alkali, and alkaline earth metals but also the metalloids including the elements in group 13 and the heavier members of the group 14–16, that is, borderline *under the stairs* elements from groups 13–16 such as B, Si, P, As, Se, and Te come squarely under the domain of organometallic chemistry. However the expression *organoelement* is used frequently to describe the organic chemistry of these non- and semi-metals. Some examples to draw a distinction between the metals and the metalloids are organoboron, organosilicon, organophosphorous, organoarsenic, and so on. Some of the common concepts that are useful in organometallic and coordination chemistry are briefly defined in Section 1.4.1.

1.4.1 CONE ANGLE

Ligands play a central role in organometallic chemistry and subsequent catalysis. Therefore, any ligand designing needs to be based on certain rationales. To assist rational ligand designing, various general concepts have been developed in the recent past. In his efforts to quantify ligand properties, Tolman introduced the concept of cone

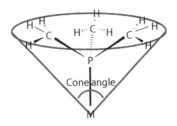

FIGURE 1.10 Pictorial representation of cone angle of trimethyl phosphine.

angle (Θ).[7] For monodentate ligands, this angle is considered as ligand cone angle or Tolman cone angle that can be extended to any monodentate ligand. It is simply the measure of the steric size of ligand. If we consider that the monodentate phosphine ligand (such as trimethyl phosphine) is coordinated to the metal at the axial position, the cone angle may be defined as an angle subtended by the cone with the metal at the vertex and the hydrogen (or any other atom for other ligands) at the perimeter of the cone (Figure 1.10).

Apart from the cone angle, Tolman attempted quantification of electronic properties of ligands and introduced the χ-parameter.[8] The χ-parameter quantifies electronic donor properties of ligands and was calculated from CO-frequency in a metal complex of type $[M(L)(CO)_x]$, where L could be a phosphine ligand.

1.4.2 Natural Bite Angle

The significance of bite angle was realized much later and the concept of *Natural Bite Angle* (β_n) was introduced to describe the properties of chelating bidentate ligands. The natural bite angle can be simply defined as an angle with which the two donor atoms of a bidentate ligand bite into the metal.[9,10] The bite angle is a geometric factor associated with the bi- and higher dentate ligands used to describe the angle of the ring made at the metal center by the chelating ligand (as depicted in Figure 1.11). This angle is a measure of distortion from a perfect geometry due to electronic and steric factors present in the ligand itself or the overall complex including the nature of the metal in the complex. Thus, ligands with denticity of more 1.3.8 than one can

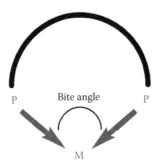

FIGURE 1.11 Illustration of $\angle$P-M-P bite angle when a diphosphine is coordinated to the metal M.

potentially coordinate to a metal with certain bite angle. The ligand bite angle can be calculated by molecular modeling of the backbone.

In idealized octahedral and square planar complexes, the preferred bite angle is close to 90° and in tetrahedral complexes it is closely around 109°. Depending on the ligand and metal pair, the bite angle can deviate from these standard angles. The ligand–metal–ligand bite angle also has an influence on the properties of the overall complex with respect to stability. Whereas the ideal bite angle range would have stabilizing effect on the complexes; the strain in the ligand–metal–ligand angle would have a destabilizing effect that would also reflect in the properties (application) of such system either in catalysis or in biology. Such an effect of bite angle can be assumed to be steric in origin.

1.4.3 Ligand Types in Organometallic Chemistry

The ligands discussed earlier (the L-type, X-type, and Z-type) that occupy the coordination sites on the metal do not actually directly participate in the reactions of the metal as such but their electronic and steric nature does influence the reactivity pattern of the complex which they are a part of. These are referred as spectator ligands. In contrast, some ligands can alter the functioning of a metal complex and can actively take part in the reactivity of a metal complex. Therefore, it is appropriate to classify organometallic compounds according to their ligands. Thus, the ligands in organometallic chemistry can be broadly classified into: (1) σ-bonding ligands; (2) σ-donor and π-acceptor ligands; and (3) σ/π-donor and π-acceptor ligands.

1.4.3.1 σ-Bonding Ligands

Metal alkyls such as diethyl zinc ($ZnEt_2$) were the first molecular compounds with a metal-carbon bond and Frankland is often considered as the father of organometallic chemistry. However, the major impact of organometallic chemistry was realized only after the discovery of Grignard reagent, organolithium reagent by Schlenk, or organoaluminum reagents by Ziegler. In all these organometallic compounds the ligand and the metal share one electron each to form a σ-bond to produce a metal alkyl complex.

1.4.3.2 σ-Donor and π-Acceptor Ligands

Ligands capable of donating σ-electrons and accepting π-electrons come under this class. A large body of ligands such as carbonyls, phosphines, carbenes, isocyanides, and so on fall under this category. A classic example of σ-donor and π-acceptor ligand complex in organometallic chemistry would be a metal carbonyl complex of type $[M(CO)_x]$. Figure 1.12 depicts the M-CO bonding in a metal complex.

The sp-hybrid orbital on the carbonyl donates 2 electrons to the metal to form a σ-bond and the metal reciprocates by back donating the electrons from its filled dπ-orbitals to empty π*-orbitals on the carbonyl. Similarly, phosphines carry a lone pair of electron on the central atom, which is donated to the metal to form a metal-phosphine σ-bond. But phosphines are also π-acids and are capable of accepting (π* P-R bonds can accept the electrons from filled metal d-orbitals) electrons from metal.

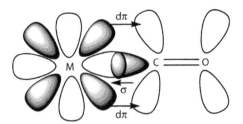

FIGURE 1.12 Electronic structure of a metal-carbonyl complex depicting σ-donation π-acceptor behavior of carbonyl ligands.

1.4.3.3 σ/π-Donor and π-Acceptor Ligands

The π-system ligands participate in the functioning and reactivity of the metal that they are bound to. Hence, the name *actor* ligand correctly justifies their role. In addition to influencing the electronic and steric properties of the metal, these actor ligands directly participate in the functioning of the metal. These interactions are of two types. The π-electron density on a ligand is donated to the metal orbitals and in turn the metal-filled orbitals reciprocate the favor by donating back the electron density from the filled metal orbitals to empty π* orbitals on the ligand. This is a synergic process. Broadly, the π-systems can be divided into linear π-systems and cyclic π-systems.

1.4.3.3.1 Linear π-Systems

Alkenes or olefin ligands are the most common type of π-systems in organometallic chemistry. The simplest among the alkene π-donor ligands would be ethylene. An example of an organometallic complex with π-donor ligands would be Zeise's salt ($K[PtCl_3(C_2H_4)]\cdot H_2O$) with a η^2-ethylene ligation (Figure 1.13).

The metal-alkene π-interaction is best explained by Dewar–Chatt–Duncanson model named after Michael J. S. Dewar, Joseph Chatt, and L. A. Duncanson. The π-bonding interaction is initiated by a sigma-type electron donation from the electron rich C=C pi orbital to the empty metal d-orbitals. This is the typical ligand HOMO

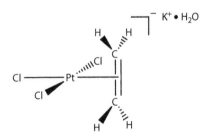

FIGURE 1.13 Coordination of a π-donor ligand (ethylene) to platinum metal.

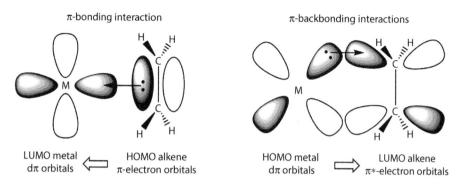

π-bonding interaction | π-backbonding interactions

LUMO metal dπ orbitals ⟸ HOMO alkene π-electron orbitals

HOMO metal dπ orbitals ⟹ LUMO alkene π*-electron orbitals

FIGURE 1.14   A π-donor ligand (ethylene) forms a π-complex with metal; the complex is stabilized by π-donation to the metal and dπ-back donation to the antibonding orbitals on the ligand.

(highest occupied molecular orbital) → metal LUMO (lowest unoccupied molecular orbital) interaction but the story does not end here. The question here is: Would an electron rich species donate its electrons for free and become electron deficient? Parallel to these sigma type interactions, there is also metal HOMO → ligand LUMO interactions, that is, π-back donation from filled dπ-orbitals into unoccupied C=C π* antibonding orbitals of olefin (Figure 1.14).

The two bonding interactions are shown separately just for clarity, in reality these two interactions take place at the same time. This is a synergistic interaction that is beneficial to both the metal and the π-actor ligand. The greater the extent of forward sigma donation to the metal is, the greater the reciprocal π-back bonding effect is. The same type of π-bonding also exists in other π-donor ligands such as allyl, alkynes, carbonyls, or any other ligands having π-clouds.

The π-bonding picture of allyl group ligands encountered in organometallic chemistry such as allylic radicals, anions, and cations is somewhat complicated. All these forms of allyl group derive their stability from resonance spread over three adjacent sp²-hybridized carbons (Figure 1.15).

These π-allyl anion can bind to the central metal as monohapto or trihapto ligands depending on the coordination needs of the metal. In monohapto mode, these ligands interact with sigma-type bonding with the available electron pair contributing one electron (η^1-allyl), whereas the trihapto mode utilizes the delocalized π-clouds over three contiguous carbons contributing three electrons (η^3-allyl). In the transition state of a complex, the monohapto allyl ligand can function as trihapto ligand when the other two electron donors withdraw its coordination.

1.4.3.3.2   Cyclic π-Systems

Similar to the allyl group, the cyclic π-systems also have the option to function as monohapto or polyhapto depending on the need of the metal's valency or electronic

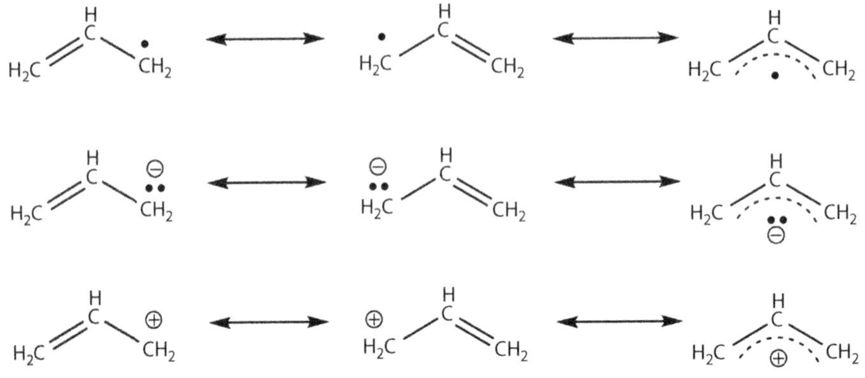

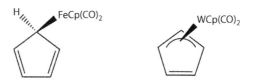

FIGURE 1.15   Forms of allylic ligands in organometallic chemistry.

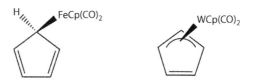

FIGURE 1.16   Cyclic π-ligands in organometallic chemistry.

configuration. Most common examples of cyclic π-ligands encountered in organo-metallic chemistry are cyclopentadienyl and benzene. The number of electrons con-tributed by these ligands is measured by the bonding modes present (η^1, η^3, η^5, etc. bonding modes contribute 1, 3, 5, etc. electrons). The most common bonding mode for cyclopentadienyl ligand is pentahapto coordination via all five contiguous carbons η^5-Cp, whereas one of the carbons of this anion Cp ligand contributes one electron, the two double bonds contribute four electrons to the metal, for example, titanocene dichloride [$(\eta^5$-Cp)$_2$TiCl$_2$], Ferrocene [$(\eta^5$-Cp)$_2$Fe]. Also Monohapto coordina-tion mode like in [$(\eta^1$-Cp)FeCp(CO)$_2$], where only one carbon binds as a anion, the π-cloud of four carbons is not involved in bonding, and trihapto coordination modes like in [$(\eta^3$-Cp)WCp(CO)$_2$], where only three carbons are coordinated to metal as an allyl anion have been observed for cyclopentadienyl ligands (Figure 1.16).

1.5   OXIDATION STATES

To express it in simple terms, oxidation state of a metal represents the charge on the metal ion in a complex. It is simply a numerical representation used to express the number of electrons that the metal usually loses or in some cases gains in order to form stable compounds. Since metals are known to be electropositive they pre-fer loosing electron over gaining. It is common for most metals to exhibit multiple potential oxidation states.

For the neutral complex, the oxidation state of the central metal ion is equal to the number of X⁻-type (anionic) ligands since each X-type ligand carries a charge of −1. For example, *cis*- or *trans*-platin [(NH₃)₂PtCl₂] has two X-type ligands (Cl⁻) and two L-type ligands (NH₃) and can be symbolized as MX_2L_2. Therefore counting the number of X ligands gives the oxidation state of Pt, which in this case is +2. By convention, Roman numerals placed in bracket after the symbol of element are used to represent the oxidation numbers of metals. Hence, Pt in *cis*-platin is said to be in +2 oxidation state which is represented as Pt(II). For charged complexes, the counter ion is also taken into account. For example, $[Co(NH_3)_5Cl]Cl_2$ has overall three X-type ligands in the form of Cl⁻, two outside and one inside the coordination sphere. In order to maintain neutrality the complex carries a net charge of +2, that is, $[Co(NH_3)_5Cl]^{2+}$ 2Cl⁻. Since NH₃ is a L type ligand it is neutral, and the one X-type ligand in the coordination sphere contributes −1 charge to the electron count. So for the complex ion $[Co(NH_3)_5Cl]^{2+}$ to have a net charge of +2 (x − 1 = +2, x = +2 + 1 = +3), Co ion must be in the oxidation state of +3, represented as Co(III).

Therefore, oxidation state of a metal can be calculated using a simple formula as follows:

$$\text{Oxidation state of metal} = (\text{Charge on complex ion})$$

$$- (\text{Charge contribution from ligands})$$

This formula is applicable for the neutral as well as charged complexes. To check its validity we shall take example of a neutral complex, [Pt(NH₃)Cl₂]. Oxidation state of Pt = 0 − (−2) = +2. In the second case of charged complex, $[Co(NH_3)_5Cl]^{2+}$, oxidation state of Co = +2 − (−1) = +3. Let us check the oxidation state of tungsten in [W(CO)5]²⁻, oxidation state of W = −2 − 0 = −2.

Most of the transition elements exhibit variable oxidation states. This is because the valence electrons in transition metals are located in (n − 1)d and ns orbitals, and electrons in both these orbitals can be used in bonding. In contrast, the alkali metals and alkaline earth metals have their valence electrons in ns orbitals. Alkali metals have one electron and alkaline earth has two electrons in their outermost subshell and hence can exhibit oxidation states of +1 and +2, respectively. The transition metals exhibit oxidation states from +1 up to +7. The loss of electrons in the ns orbitals helps it exhibit +1 and +2 oxidation states when (n−1)d orbital does not participate in bonding. For transition metal to exhibit higher oxidation states, loss of electrons from both ns and (n−1)d orbitals is necessary. There is a little difference in energy between (n−1)d and ns orbitals.

Except the first and last member of the transition metal series, all the other members display at least two stable oxidation states. The element in the middle of this series exhibits maximum number of oxidation states and also the highest oxidation state in the whole period since it has five unpaired electrons. The oxidation states of the first, second, and third row elements are given in Table 1.4.

TABLE 1.4
Possible Oxidation States of 3d, 4d, and 5d Metals

3d Metals	Sc	Ti	V	Cr	Mn	Fe	Co	Ni	Cu	Zn
Oxidation states	+3	+2, +3, +4	+2, +3, +4, +5	+2, +3, +4, +5, +6	+2, +3, +4, +5, +6, +7	+2, +3, +4, +5, +6	+2, +3, +4	+2, +3, +4	+1, +2	+2

4d Metals	Y	Zr	Nb	Mo	Tc	Ru	Rh	Pd	Ag	Cd
Oxidation states	+3	+3, +4	+2, +3, +4, +5	+2, +3, +4, +5, +6	+2, +4, +5, +7	+2, +3, +4, +5, +6, +7, +8	+2, +3, +4, +6	+2, +3, +4	+1, +2, +3	+2

5d Metals	La	Hf	Ta	W	Re	Os	Ir	Pt	Au	Hg
Oxidation states	+3	+3, +4	+2, +3, +4, +5	+2, +3, +4, +5, +6	+1, +2, +4, +5, +7	+2, +3, +4, +6, +8	+2, +3, +4, +6	+2, +3, +4, +5, +6	+1, +3	+1, +2

1.6 THE 18-ELECTRON RULE

Before we start with 18-electron rule, let us first recall the Octet rule. The Octet rule states that atoms such as carbon with one 2s and three 2p orbitals require eight electrons to completely fill the valence shell and compounds, which can fulfill this requirement are left with no or little tendency to participate in chemical reactions, thus achieve stability. There is a similar tendency with metals to react with ligands forming organometallic or coordination complexes in such a way as to attain the electronic configuration of next inert gas in the series $[ns^2(n-1)d^{10}np^6]$. The 18-electron rule states that a metal complex with 18 electrons (filled valance shell or completes the insert gas configuration) in its outer shell will be stable as compared to the metal complexes either less than or more than 18 electrons. This is just a rule of thumb and there can be many exceptions to this rule; nevertheless, it offers guidelines to the chemistry of coordination and organometallic compounds. For example, if there is a competition of any given metal with two sets of ligands, one which yields the complex 18-electron configuration and the other less or more than 18-electron configuration, then the metal will prefer the former ligand over later due to the obvious reason of stability. The valence electron count of a complex includes the valence electrons of the central metal ion and the electrons shared or donated by the ligands, and for inert gas configuration it comes to 18. Counting the valence electrons in a complex allows us to predict the stability associated with complex and also to predict the mechanism and mode of reactivity to achieve 18-electron configuration.

1.6.1 HOW TO COUNT THE ELECTRONS IN METAL COMPLEXES?

Traditionally, there are two methods used for counting the valence electrons in a metal complex: One is neutral or covalent method and the other one is ionic or charged method. The neutral or covalent model considers the bonding between ligand and metal as essentially covalent, whereas the latter model considers this bonding as ionic. But both these methods lead to precisely the same net result. However, for the use of either of the electron count methods, it is necessary to know how many electrons the ligands contribute in bonding to the metal according to the covalent or ionic model. Table 1.5 displays the electron contribution of some of the common ligands for the two models; the symbol L and X for ligands have been already discussed in the earlier section.

In neutral atom method the oxidation state of the metal is not taken into account and the metal is assumed to be neutral. For example, a d^{10} metal such as Ni will always be counted for 10 valence electrons and Ti valence electron count will be 4. In the ionic model of electron count the oxidation state of the metal ion in the complex is taken into account. For example in Cp_2TiCl_2, titanium will be counted for zero valence electrons and cobalt in $CoCp_2$ will be counted for seven valence electrons. Also each metal–metal bond contributes one electron to the total electron count (Table 1.6).

Although the tendency to achieve 18 valence electrons in complexes is looked upon as a tendency to achieve stability, however, there are many exceptions and a generalization can be drawn on metals that are expected to achieve 18 valence electron counts. The middle member of the transition metal series generally follows the 18-electron rule, whereas the earlier and the later metals of the transition metal

TABLE 1.5

Counting the Electron Contribution of Ligands in Covalent or Ionic Models

Ligands	Type	Neutral Method	Ionic Method	Ligands	Type	Neutral Method	Ionic Method
Terminal ligands				Tholate (OS)	X	1	2
Carbonyl (−CO)	L	2	2	Alkylidene carbene (CR_2)	L	2	4
Thiocarbonyl (−CS)	L	2	2	Alkylidyne carbine (−CR)	LX	3	6
Phosphine (−PR_3)	L	2	2	η^1-allyl	X	1	2
Amine (−NR_3)	L	2	2	η^3-allyl	LX	3	4
Dinitrogen (−N≡N)	L	2	2	η^1-cyclo pentadienyl	X	1	2
Dihydrogen (H–H)	L	2	2	η^5-cyclo pentadienyl	L_2X	5	6
Alkene ($R_2C≡CR_2$)	L	2	2	η^6-benzene	L_3	6	6
Alkyne (RC≡CR)	L	2	2	η^8-cyclo octatetraenyl	L_4	8	10
Isocyanide (−C≡NR)	L	2	2				
Nitrosyl linear (N≡O)	LX	3	2	Bridging ligands			
Nitrosyl bent	X	1	2	μ-Carbonyl (−CO−)	L	2	2
Halide (X−)	X	1	2	μ-Halide (−X−)	LX	3	4
Hydride (H−)	X	1	2	μ-Alkyne (−CR−)	L_2	4	4
Alkyl (R−)	X	1	2	μ-Hydrogen (−H−)	X	1	2
Acyl (−C(O)−R)	X	1	2	μ-Alkyl (−CR_3−)	X	1	2
Aryl (Ph)	X	1	2	μ-Alkoxide (−OR−)	LX	3	4
Alkoxide (OR)	X	1	2				

series usually do not follow the 18-electron rule. The early transition metals usually have less than 18 valence electrons, for example, $[TiF_2]^{2-}$ (12 electrons), $[VCl_6]^{2-}$ (13 electrons), $[Cr(NCS)_6]^{3-}$ (15 electrons), and so on, whereas the late transition metals find stability with 18 or more valence electrons, for example, $[Ni(en)_3]^{2+}$ (20 e^-), $[Cu(NH_3)_6]^{2\pm}$ (21 e^-), $[Zn(NH_3)_6]^{2\pm}$ (22 e^-), and so on. Another important class of metals is d^8-metals with square planar geometry from group 8–11 which consistently follow 16-electron rule for stability than 18 electron. Some examples

TABLE 1.6

Counting the Electrons in a Metal Complex by Covalent or Ionic Method

Covalent Method		Complex	Ionic Method	
Ti	4 e⁻		Ti⁴⁺	0 e⁻
2Cl•	2 e⁻		2Cl⁻	4 e⁻
2Cp•	10 e⁻		2Cp⁻	12 e⁻
Total	16 e⁻		Total	16 e⁻
Rh	9 e⁻	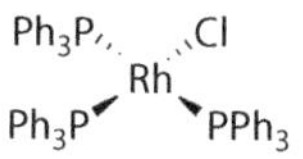	Rh¹⁺	8 e⁻
Cl•	1 e⁻		Cl⁻	2 e⁻
3PPh₃	6 e⁻		3PPh₃	6 e⁻
Total	16 e⁻		Total	16 e⁻
W	6 e⁻	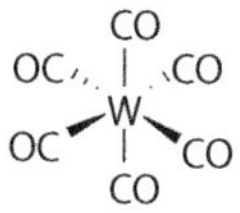	W⁰	6 e⁻
6CO	12 e⁻		6CO	12 e⁻
Total	18 e⁻		Total	18 e⁻
Fe	8 e⁻	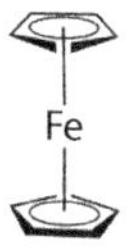	Fe²⁺	6 e⁻
2Cp	10 e⁻		2Cp⁻	12 e⁻
Total	18 e⁻		Total	18 e⁻

are $IrX(CO)L_2$, $RhClL_3$, PdX_2L_2, and $[PtX4]^{2-}$, where L and X are phosphine and halogen ligands, respectively.

1.7 ELEMENTARY STEPS IN METAL-CATALYZED REACTIONS

In Section 1.6.1, we discussed how the complex is formed by the organization of ligands with metal ion by covalent or coordinate bonding. Once a complex is formed it can further undergo reactions where the existing ligands undergo substitution by new ligands or new ligands may simply add up. With neutral ligand, a complex may simply add up new ligands if its coordination number is not satisfied or it may substitute the existing neutral ligands with newer ones. This is relatively simple and no changes in the metal oxidation number or electron count take place. However the coordination number may increase or remain unchanged. But if the anionic ligands react or add up it may change the coordination number, the oxidation state, and the valence electron count of the metal. An anionic ligand does not exist by itself. The anionic ligand is generated during its course of reaction with the complex. A molecule A–B may add up to the metal as A⁻ and B⁻ similar to that in H_2 or Me-I. This addition is accompanied by an increase in the electron count, the coordination number, and the oxidation state of the metal by two units.

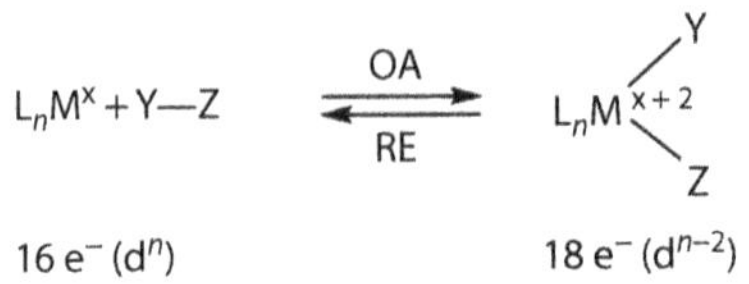

$$L_nM^x + Y-Z \quad \underset{RE}{\overset{OA}{\rightleftharpoons}} \quad L_nM^{x+2}\overset{Y}{\underset{Z}{\diagdown}}$$

$$16 \text{ e}^- (d^n) \qquad\qquad\qquad 18 \text{ e}^- (d^{n-2})$$

FIGURE 1.17 Representative oxidative addition (OA) and reductive elimination (RE) steps in an organometallic transformation.

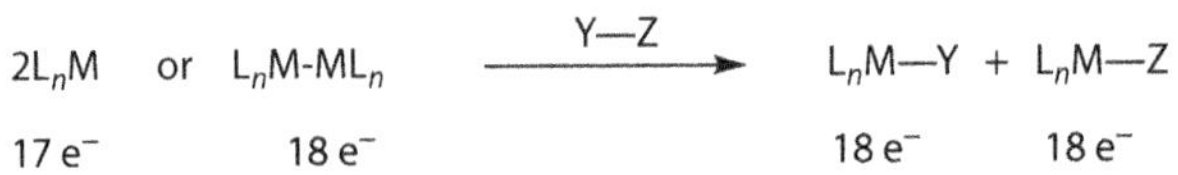

$$2L_nM \quad \text{or} \quad L_nM\text{-}ML_n \quad \xrightarrow{\quad Y-Z \quad} \quad L_nM-Y \; + \; L_nM-Z$$

$$17 \text{ e}^- \qquad\quad 18 \text{ e}^- \qquad\qquad\qquad 18 \text{ e}^- \qquad\quad 18 \text{ e}^-$$

FIGURE 1.18 Oxidative addition across metal–metal bond depicting an increase in the OS by one unit.

1.7.1 OXIDATIVE ADDITION

The oxidative addition (OA) may be described as the process of addition of a substrate molecule to the transition metal complex accompanied by an increase in the oxidation state of the metal ion by +2 units (Figure 1.17). The exact opposite of oxidative addition is reductive elimination (RE), where the two ligands (Y and Z) are eliminated from the metal center as Y–Z.

The oxidative addition of Y–Z to the complex, L_nM increases the oxidation state of the metal from d^n to d^{n+2} and electron count from 16 valence electrons to 18. Such oxidative addition reaction is favored in metals that have stable higher oxidation states.

In some special cases, 16 electron complexes, 17 electron complexes, and binuclear 18 electron complexes undergo oxidative addition to each of the metal changing the electron count, oxidation state, and coordination number of each metal by one unit instead of two. In oxidative addition of 16 electron complexes, two new ligands are added at the metal center, but in bimolecular complexes only one ligand at each metal center is added by oxidative addition. Thus, oxidative addition can be redefined as a change in the oxidation state of a metal by one or two units as depicted in Figure 1.18.

A practical example of oxidative addition could be hydrogenation of olefins with molecular hydrogen using the Wilkinson catalyst, chloridotris(triphenylphosphane) rhodium(I) as depicted in Figure 1.19. The oxidation state and coordination number of the Rh increase by +2 units and also the 16 electron species is now transformed into 18 electron species.

Another example of oxidative addition is seen in the palladium-catalyzed cross-coupling reaction (Suzuki coupling) of aryl boronic acid and aryl halides.

Suzuki coupling is well known to organic chemists for quite some time, but it is relatively new for polymer chemists. The same C–C coupling presented in Figure 1.20 can be extended to polymerization. Chapter 6 will present metal-catalyzed polycondensation methodologies that are being increasingly used, and oxidative addition is the first step in these reactions. Not only Suzuki coupling but also oxidative addition is the pivotal step in various C–C coupling reactions

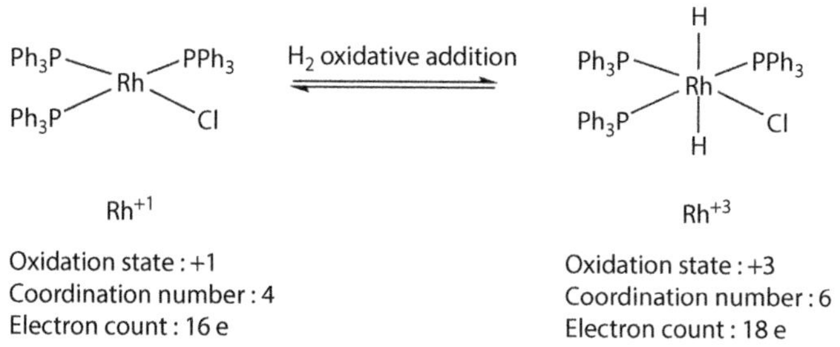

FIGURE 1.19 Oxidative addition of hydrogen in Wilkinsons hydrogenation catalyst.

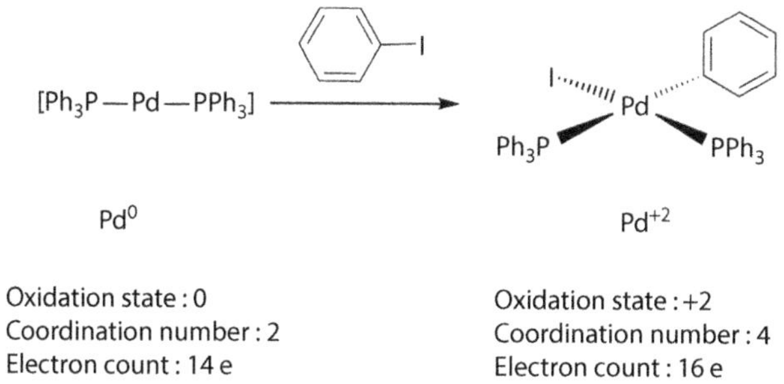

FIGURE 1.20 Oxidative addition on a palladium (0) catalyst in Suzuki type reaction.

such as Suzuki–Miyaura, Stille, Heck, Negishi, Sonogashira, and Kumada coupling that are now amenable to polymer chemists for metal-catalyzed polymerization. The oxidative addition and reductive elimination thus are highly important steps in metal-catalyzed polycondensation reactions.

1.7.2 Reductive Elimination

The reversal of oxidative addition is reductive elimination. In other words, a reaction in which the oxidation state and the coordination number of a metal are reduced by two (sometimes one) units is called as reductive elimination. Reductive elimination involves the elimination or expulsion of two anionic ligands coordinated to the metal in a complex accompanied by reduction of the formal oxidation state of the metal and the coordination number by two units (Figure 1.21). Reductive elimination leads to coordinative unsaturation at the metal center.

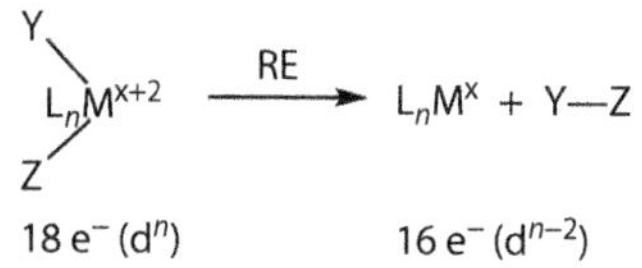

FIGURE 1.21 Representative reductive elimination (RE) steps in an organometallic transformation.

$$L_nYM^{x+1}\!\!-\!\!M^{x+1}L_nZ \xrightarrow{\text{RE}} L_nM^x\!\!-\!\!M^xL_n + Y\!\!-\!\!Z$$

FIGURE 1.22 Reductive elimination (RE) from a binuclear complex reducing the oxidation state of both metals by one unit.

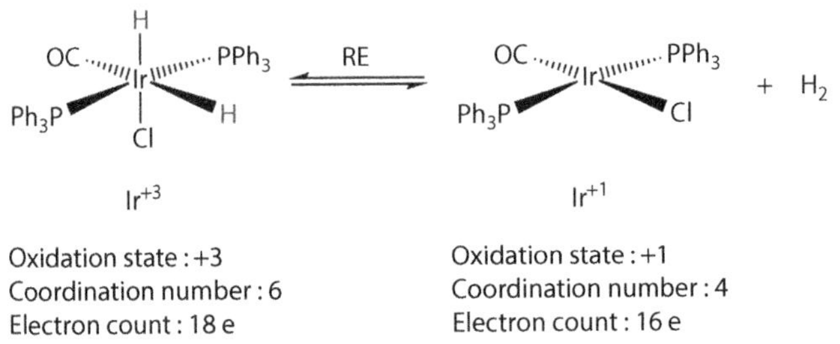

Oxidation state : +3
Coordination number : 6
Electron count : 18 e

Oxidation state : +1
Coordination number : 4
Electron count : 16 e

FIGURE 1.23 Reductive elimination step in dehydrogenation reaction using iridium complex.

Similar to what has been observed in the case of binuclear oxidative addition, the binuclear reductive elimination is also observed in some instances. As expected, the oxidation state and the coordination number decrease by one unit in the binuclear reductive elimination pathway (Figure 1.22). Binuclear reductive elimination is also feasible in some binuclear complex where the coordination number and oxidation state are decreased by one unit.

An example of reductive elimination from an iridium complex to eliminate H_2 molecule is presented in Figure 1.23. A dihydride iridium complex, chloroc arbonylhydrogenbis(triphenylphosphine)iridium(III) as presented later (Vaska's complex) reductively eliminates hydrogen molecule and the oxidation state of the iridium is reduced from +III to +I (Figure 1.23). The two expelled ligands are always anionic (X type) ligands in reductive elimination. Reductive elimination is part and parcel of C–C coupling reactions and is usually accompanied by oxidative addition. In Chapter 6 you will encounter this reaction (RE) that plays a pivotal role in metal-catalyzed condensation polymerization reactions such as Stille, Suzuki–Miyaura, Negishi, Heck, Sonogashira, and Kumada coupling. By tuning the electron density on the metal, the rate of reductive elimination can be controlled to obtain the desired product.

1.7.3 INSERTION AND ELIMINATION REACTIONS

Apart from oxidative addition and reductive elimination, a very frequently encountered reaction in metal-catalyzed polymerization (especially Ziegler–Natta type polymerization) is insertion of monomer in a M-E (E = CR_3, H, etc.) bond to create a long chain molecule called polymer. An insertion reaction in organometallic chemistry represents the interposition of a neutral unsaturated (π) 2e donor ligand L (such as ethylene) into the precoordinated metal E-type ligand bond, leading to the creation of new covalent σ bond between the two ligands (L and E) and between the inserting

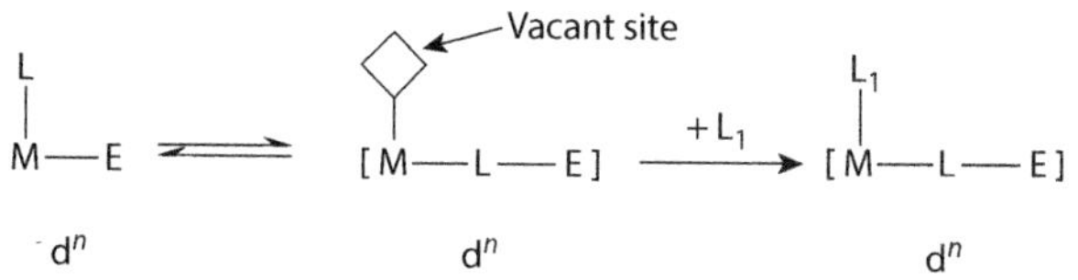

FIGURE 1.24   Insertion of L-type ligand (ethylene) in M-E bond.

ligand and metal (M-L) to produce a species M-L-E. The inserted ligand L now functions as X-type ligand. If the inserting ligand is from the same complex then the
insertion is referred as migratory insertion. The insertion reaction causes no change in
the oxidation state of the metal as the neural unsaturated ligand L reinserts as X-type
ligand in between the precoordinated X-type ligand (E) and metal M. However, the
loss of L-type coordination causes a decrease in the valence electron count by two
units and creation of a vacant site on the metal is enabled (Figure 1.24).

The reversal of these reactions is known as the elimination reaction. Similar
to oxidative addition and reductive elimination, the insertion and elimination
reactions play a vital part in organometallic synthesis and catalysis, especially in
insertion/coordination polymerization reaction, which will be discussed in details
in Chapter 2.

There are two main types of insertion reactions commonly encountered in organometallic chemistry and catalysis: 1,1-migratory insertion and 1,2-migratory insertion.
In 1,1-insertion the same atom (1,1) of the inserting group (M-Y^{1,1}(E)-Z atom of say
Y = Z) binds the metal and displaces X-type ligand, whereas in 1,2-insertion the
metal and the displaced X-type ligands E are bound to the adjacent atoms (1,2) of the
inserting ligands (M-Y^1-Z^2-E). The nature of the inserting 2e$^-$ unsaturated π-ligand
and topology of coordination has an influence on the type of insertion that takes
place. The best examples are of C=O and C=C insertions, carbon monoxide being
a monohaptic ligand, η^1-C=O, η^1 (C) always inserts in 1,1 fashion (Figure 1.25a)
while the dihaptic alkene such as ethylene, η^2->C=C<, η^2 (C, C) prefers 1,2 insertion

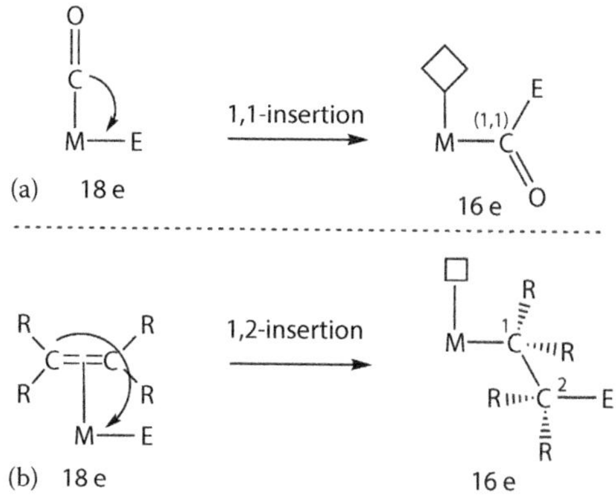

FIGURE 1.25   Illustration of 1,1 insertion of C=O into the M-E bond (a) and 1,2 insertion
of >C=C< into the M-E bond (b).

(Figure 1.25b). The only common ligand that can undergo both 1,1 and 1,2 insertion is SO_2, since it can be a monohaptic, η^1, (S) or dihaptic, η^2, (S, O) donor.

The 16e⁻ species left behind after 1,1 or 1,2-insertion reaction takes up a 2e⁻ neutral donor ligand (such as ethylene) to fill its vacant coordination site as well as its 18 valence electron count. In addition/insertion/coordination polymerization, this process can be repeated several times until a long chain molecule is built.

1.7.3.1 Migration and Insertion

Migration is when the anionic ligand (X type) makes the first move leaving its original coordinated site and performs an intramolecular nucleophilic-like attack on the unsaturated 2e⁻ neutral donor ligand (Figure 1.26a). A vacant site on metal is created where the anionic ligand (Me) originally resided. In contrast, in insertion the unsaturated 2e⁻ donor ligand (i.e., CO) shifts itself first and interpolates itself between the metal and anionic ligand generating a new 1e donor anionic ligand (COMe) (Figure 1.26b). A vacant coordination site on metal is created where originally the neutral ligand resided. The type of ligand decides whether the reaction is migration or insertion. If the anionic ligand moves over to the site where neutral ligand is coordinated it is migration and conversely it is insertion.

Since the net effect of both the reactions is same, the exact pathway of the reaction does not mean much and hence the term *migratory insertion* is commonly used to indicate that either of the pathways is possible and the exact mechanism is difficult to establish.

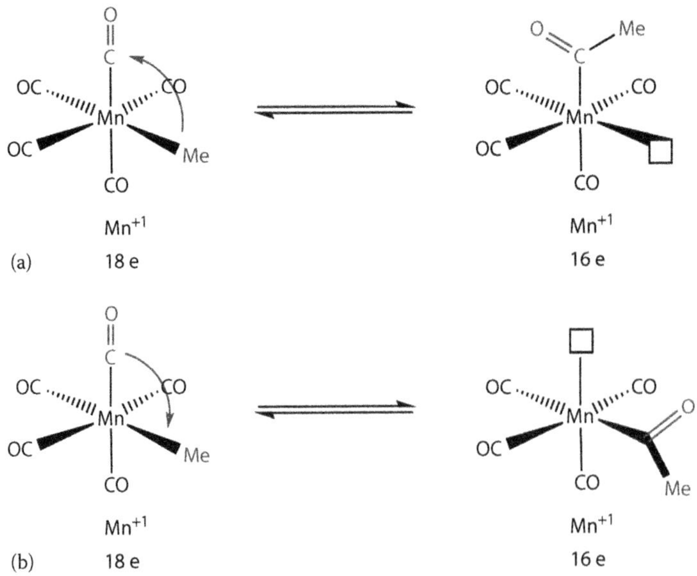

FIGURE 1.26 Illustration of migration (a) versus insertion (b).

1.7.3.2 Migratory Insertion

If the two steps of migration and insertion occur simultaneously, the net result is called *migratory insertion*. An insertion reaction is said to be migratory insertion when intramolecular migration of a cisoidal anionic ligand to a π-type ligand (neutral 2e⁻ ligand) takes place. In this process the neutral ligand rearranges to form a new coordinated anionic ligand by forming a bond with former anionic ligand, which appears as if the neutral ligand has inserted between the metal and the anionic ligand. The cisoidal arrangement between the migrating group and inserting group is a prerequisite for migratory insertion.

1.7.3.3 Migratory Insertion of Alkenes

The alkene insertion reaction into the metal-alkyl or metal-hydride bond is the fundamental reaction step in several important catalytic reactions such as carbonylation and olefin polymerization. Among these, the olefin polymerization with transition metal is industrially practiced. The insertion reactions of olefins such as ethylene and propylene into group 4 metal alkyl complexes are at the center of coordination insertion polymerization. The repeated sequence of coordination of alkene in the metal vacant site on the catalyst and migratory insertion into the metal alkyl bond lead to the growth of polymer chain. As an example, ethylene insertion into metal-alkyl bond of a zirconocene complex is depicted in Figure 1.27. The insertion of alkene species into the M-H bond is also one of the important reactions that are frequently encountered in coordination insertion polymerization reaction with transition metals. Also it is a key step in hydrogenation and hydroformylation reactions.

1.7.3.4 Elimination Reactions

The reverse reaction of an insertion reaction is called elimination reaction. There are at least two types of elimination reactions that we will encounter in organometallic chemistry: (a) α-elimination and (b) β-hydride elimination. Among these, α-elimination is very rarely observed and the elimination products are highly unstable.

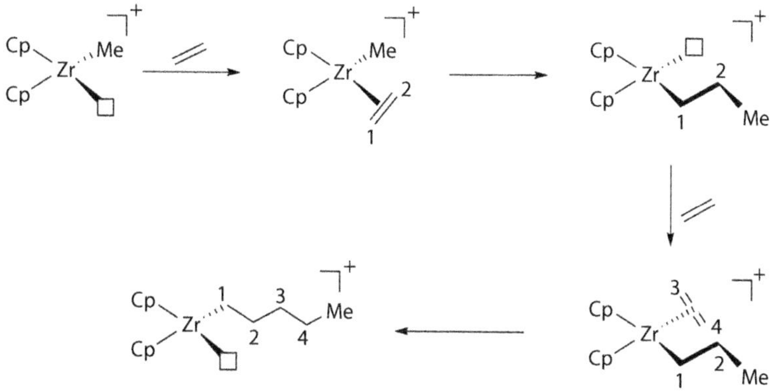

FIGURE 1.27 Coordination and insertion of ethylene into zirconium-methyl bond and subsequent chain growth.

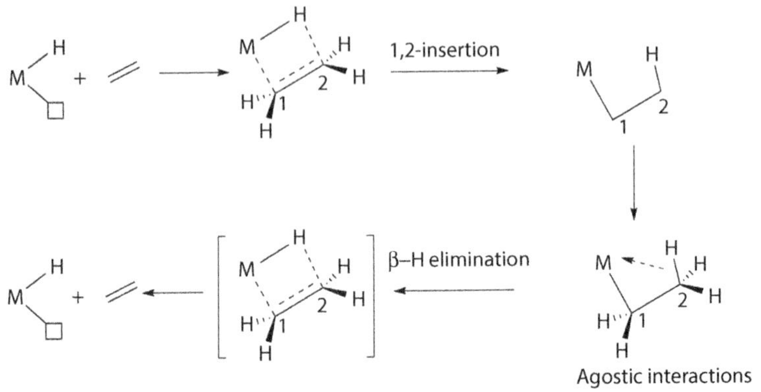

FIGURE 1.28   Illustration of β-hydride elimination reaction.

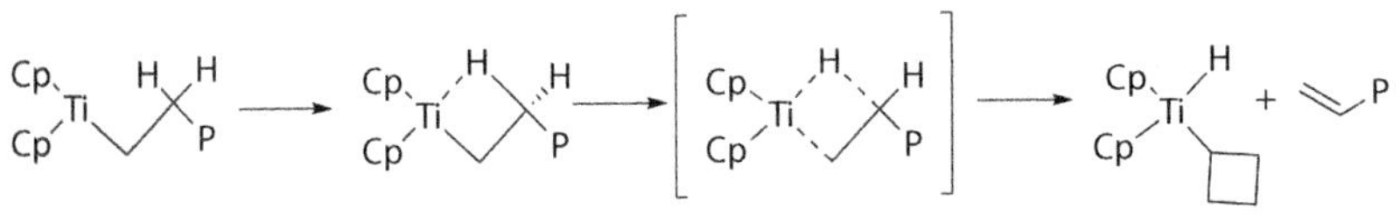

[Transition state]

FIGURE 1.29   Representative β-hydride elimination in titanocene-catalyzed olefin polymerization.

A more relevant reaction in metal-catalyzed polymerization is β-hydride elimination. The β-hydride elimination is actually reversal of 1,2-insertion (Figure 1.28) and is believed to proceed through agnostic interaction (see Section 1.7.3.5). The β-elimination is feasible if the complex meets the following conditions: (a) the metal should have 2e$^-$ vacant site and (b) the hydride and the metal should be in the same plane (coplanar M-C-C-H arrangement). Such β-hydride elimination is one of the major chain-transfer or chain-termination pathway in olefin polymerization and leads to lower molecular weight polymers. An example of β-hydride elimination in metallocene-catalyzed polymerization of ethylene is presented in Figure 1.29. A detailed discussion on β-hydride elimination in olefin polymerization is presented in Chapter 2.

1.7.3.5   Agostic Interactions

Agostic interactions play a central role in organometallic chemistry. The word agostic is derived from the Greek word *to hold on to oneself* first coined by Malcolm Green and later used by Maurice Brookhart and others. Agostic interaction in organometallic chemistry is used to represent the σ-like interaction of the electron density associated with the C-H bond with the coordinative unsaturated transition metal, resulting in the three center, two electron bond (Figure 1.28). The agostic interactions are believed to facilitate many important catalytic transformations such as oxidative addition, reductive elimination, chain transfer, or chain termination reactions in polymerization. Agostic interactions help in temporarily stabilizing the transition state in a catalytic reaction, which possesses less than 16 valence electrons before they take up ligands to saturate their valency.

REFERENCES

1. Odian, G. *Principles of Polymerization* (4th Eds.), Wiley Interscience, John Wiley and Sons, Hoboken, NJ, 2004.
2. The search was conducted in October 2016, therefore this number is not for the full year but still it shows more number of publications than the previous year (2015), indicating that more number of publications are expected in this area.
3. Elschenbroich, C. *Organometallics* (3rd Eds.), Wiley-VCH Verlag GmbH & Co. KGaA, Germany, 2005.
4. Lee, J. D. *Concise Inorganic Chemistry* (5th Eds.), Blackwell Science Ltd., Oxford, UK, 1996.
5. Detailed guidelines for the nomenclature of inorganic compounds can be found in the IUPAC red book. Available at: http://old.iupac.org/publications/books/rbook/Red_Book_2005.pdf.
6. Crabtree, R. H. *The Organometallic Chemistry of the Transition Metals* (5th Eds.), John Wiley and Sons, Hoboken, NJ, 2009.
7. Tolman, C. A. 1977. Steric effects of phosphorus ligands in organometallic chemistry and homogeneous catalysis. *Chem. Rev.* 77:313–348.
8. Tolman, C. A. 1970. Electron donor-acceptor properties of phosphorus ligands. Substituent additivity. *J. Am. Chem. Soc.* 92: 2953–2956.
9. Casey C. P., Whiteker, G. T. 1990. The natural bite angle of chelating diphosphines. *Isr. J. Chem.* 30: 299–304.
10. van Leeuwen, P. W. N. M., Kamer, P. C. J., Reek, J. N. H., Dierkes, P. 2000. Ligand bite angle effects in metal catalyzed C-C bond formation. *Chem. Rev.* 100: 2741–2770.

2 Insertion or Ziegler–Natta Polymerization of Olefins
Science and Technology

Samir H. Chikkali, Ketan Patel, and Sandeep Netalkar

CONTENTS

2.1   INTRODUCTION

One of the most disruptive discoveries of twentieth century has been metal-catalyzed polymerization of ethylene to polyethylene by Prof. Karl Ziegler from the Max Plank Institut Fuer Kohlenforschung, Mulheim, Germany. This discovery marked the beginning of an era, which revolutionized the field of chemistry, in particular, the field of polymer chemistry in many aspects and saw an avalanche of patents and publications. Unlike typical academic discoveries, this invention did not stop at the academic labs, but led to the development of a large-scale process that produces about 180 million tons of polyolefins (annually) today.[1] But the very obvious question that comes to our mind is *what prompted Prof. Ziegler to react ethylene with the two components (that we will discuss later in this chapter)*? It was not an overnight realization, but instead, it was rational analysis of a serendipitous observation and systematic experimentation by Prof. Ziegler.

Before we get in to the Ziegler–Natta (ZN) polymerization and to put the topic in the right context, let us quickly look at the history or the series of events that occurred prior to the discovery of polyethylene by Prof. Ziegler. One might argue about the exact seeding of this discovery but it is very clear that Ziegler was exploring the possibility of manufacturing long-chain alkyls for various reasons during World War II. In the following years he continued this passion and employed alkyl-lithium reagents. However, those usually yielded β-hydride elimination products with insoluble lithium-hydride (LiH).[2] In order to enhance the performance of this reaction, the Ziegler school used ether-soluble $LiAlH_4$ as an initiator for ethylene polymerization. After detailed investigation, Ziegler was convinced that *in situ*-produced triethyl aluminum ($AlEt_3$) is a more efficient initiator than the originally thought ethyl-lithium (LiEt). It was not the efficiency alone, but all the alkyl aluminum compounds and the alkyl aluminum hydride compounds were soluble in organic solvents, which allowed Ziegler to study the organometallic reaction in solution. The focus of his research now completely shifted to $AlEt_3$ and he started the synthesis of alkyl aluminum compounds in a step-by-step manner, which was called as *Aufbau reaction* by Ziegler. A representative *Aufbau reaction* is depicted in Figure 2.1. He observed that alkyl aluminum compounds such as triethyl aluminum react with ethylene at elevated temperature and high ethylene pressure to produce long-chain alkyl-aluminum compounds. The long-chain aluminum compounds can be hydrolyzed to obtain the linear alkyl chains and sometimes the molecular weight of each chain was found to be about 3000 Da. Although mechanism of this reaction was not the target, it was assumed that

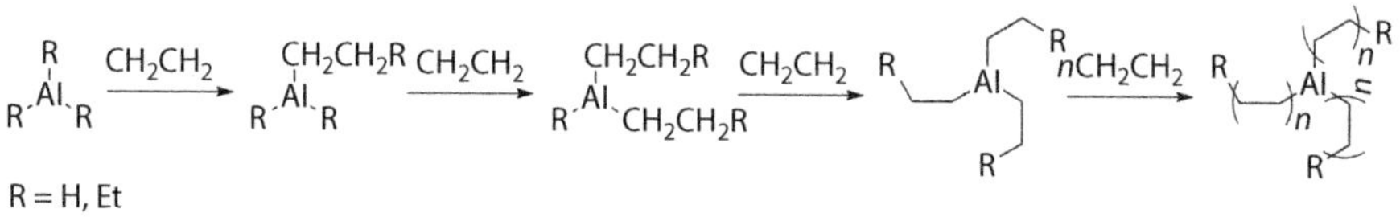

$$R \overset{R}{\underset{R}{\overset{|}{Al}}} \xrightarrow{CH_2CH_2} R \overset{CH_2CH_2R}{\underset{R}{\overset{|}{Al}}} \xrightarrow{CH_2CH_2} R \overset{CH_2CH_2R}{\underset{CH_2CH_2R}{\overset{|}{Al}}} \xrightarrow{CH_2CH_2} R \overset{R}{\underset{R}{\overset{|}{Al}}} \xrightarrow{nCH_2CH_2} R \overset{R}{\underset{R}{\overset{|}{Al}}}$$

R = H, Et

FIGURE 2.1 Ziegler's Aufbau reaction between alkyl/hydro-aluminum compound and ethylene to produce step-by-step addition products (*n* = a small number).

the double bond (C=C) containing monomer adds to the Al-C bond in a step-by-step fashion at elevated temperature and pressure. The molecular weight presumably could not build, as the termination or transfer reactions are favored at higher temperature and pressure (see Section 2.2 for termination and transfer reactions). This was the common feature of the AlEt$_3$-catalyzed reactions and waxy materials were obtained all the time. However, on one fine day Ziegler and coworker noticed an anomalous behavior and recovered an unexpected product, 1-butene. Fortunately, this was not ignored by the student and Ziegler. After a lot of speculations and concomitant investigations, it turned out that a nickel impurity was responsible for this change in the reactivity. The source of nickel was speculated to be the leached nickel from the metal reactor (which was washed with acid before the polymerization) or reminiscent colloidal nickel from a previous hydrogenation experiment. Ziegler school termed this as *nickel effect* (Figure 2.2).[3] To prove the nickel effect, Ziegler designed various control experiments and purposely added different Ni(0) or Ni(II) precursors to the AlEt$_3$. It was found that the combination of alkyl aluminum and nickel dimerizes the ethylene to butenes. The Ziegler group had an additional handle now to modify the Aufbau reaction. Having understood the nickel effect and the dimerization process, Ziegler investigated a series of transition metal salts in the presence of AlEt$_3$ to find out a suitable catalyst for polymerization. It is said that since the late transition metal (nickel) was dimerizing ethylene, Ziegler moved to early transition metals and started screening group III and IV metals. According to the Ziegler school, the discovery experiment was performed by Breil, who studied the reaction of ethylene with triethyl aluminum and zirconium acetylacetonate. This combination of early transition metal with triethyl aluminum produced a white powder, which was nothing but high molecular weight polyethylene, the *white gold*.[4] Zirconium along with other group IV to VI metals was active in ethylene polymerization, but the most active catalyst was

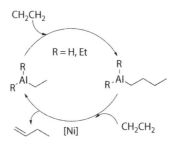

FIGURE 2.2 The *nickel effect* as anticipated by Ziegler school.

found to be a combination of $TiCl_4$ with $AlEt_3$. This combination was then scaled up for a large-scale industrial production of high-density polyethylene, which was named as *Muelheim atmospheric polyethylene process* by Ziegler.[5] Thus, careful observations and rational analysis at every step, be it the simple *Aufbau reaction* or the lucky *nickel effect*, enabled Ziegler to reach to the humongous Muelheim atmospheric polyethylene process.

Any fundamental scientific discovery that would like to meet the societal needs has to go through an industrial process. Therefore, Prof. Ziegler first patented his invention, which indicates that he was probably aware of the above-mentioned fact. He disclosed his patents and catalyst system to Montecatini company, Italy before he made it public to the scientific community, another visionary decision by Ziegler.[6]

During the same time, Prof. G. Natta, a crystallographer by training, was a consultant to Montecatini and had access to Ziegler catalyst. The Natta school was already active in kinetics of ethylene addition and was in an ideal position to investigate the Ziegler catalyst. Natta, being a crystallographer, was keen to investigate the stereoselectivity offered by Ziegler catalyst. In the early 1950s, Natta reported stereocontrolled polymerization of α-olefins including propylene using titanium tetrachloride and triethyl aluminum (Muelheim catalyst) catalyst combination.[7] The Natta school identified that titanium trichloride is a better catalyst for stereoregular polymerization of propylene. They went on to evaluate various metal halides and concluded that metal halides in lower oxidation state provide better control over the selectivity. The above-mentioned catalyst combination produced mixture of amorphous and crystalline PP in the same reaction. Natta was able to separate these two polymers by simple solvent extraction. The amorphous PP was soluble, whereas the crystalline PP was insoluble and thus the two were separated by Natta. The Natta school used various spectroscopic and analytical methods such as X-ray diffraction and differential scanning calorimetry (DSC) to further demonstrate that the insoluble fraction is actually the crystalline PP. Natta designated these polymers as *isotactic* polypropylene. Subsequently, Natta initiated a group that specifically investigated the stereoregularity in polymers and stereoregular polymerization of various α-olefins was established. The prestigious Natta school demonstrated the basic principles of stereocontrolled polymerization, such as in small molecules, and opened up a new era in polymer science.[8] This was followed by a flood of publications and patents on the synthesis of stereoregular polymers and their applications.[9,10] Apart from developing stereoregular polyolefins, the Natta school made fundamental contributions to the field of polymer physics, added new tools to characterize polyolefins, and laid a solid foundation for the future generations.[11] The founding fathers, Prof. Ziegler and Prof. Natta, were awarded with Nobel Prize in 1963 for their humongous contributions.

In the due course, a significant progress has been achieved and various generations of ZN catalysts have been developed. The development of ZN polymerization can be roughly arranged into seven generations[12] and Table 2.1. summarizes the important developmental stages in ZN polymerization.

The first-generation ZN catalyst consists of titanium tetrachloride and alkyl aluminum halide as the two fundamental components. The activities were low and therefore development of a new catalytic system was essential. The second-generation catalysts were mainly based on morphological $TiCl_3$ precursor and displayed improved

TABLE 2.1
Progress of Ziegler–Natta Polymerization

Generation	Year	Catalyst	Cocatalyst	Support	Activity (Kg. PP/g Cat.)
First generation	1957	$TiCl_4$, $AlCl_3$, and $TiCl_3$	$AlEt_2Cl$	NA	0.8–1.2
Second generation	1970	$TiCl_3$	$AlEt_2Cl$		3–5
Third generation	1979	$TiCl_4$ and $TiCl_3$ and di-ester donors	Trialkyl aluminum	$MgCl_2$	5–15
Fourth generation	1980	$TiCl_4$; internal and external donors	AlEt3	$MgCl_2$	20–100
Fifth generation	1991	Cp_2ZrCl_2 or Cp_2TiCl_2	MAO	NA	5000–9000
Sixth generation	1992	Cp_2ZrCl_2 or Cp_2TiCl_2	MAO	Silica	5000–9000
Seventh generation	1996–2000	Postmetallocenes	MAO	Silica/$MgCl_2$	NA

NA: Not applicable.

activities. Between 1978–1980, the third-generation catalysts were introduced. There is a significant difference between the second- and the third-generation catalysts. The third-generation catalysts were supported on a solid magnesium dichloride bed and were activated using trialkyl aluminum cocatalysts. It was during this time when donors were introduced to ZN system. Subsequently, donors became the focus of next-generation catalysts and the fourth-generation ZN system uses internal as well as external donors. Discovery of methylaluminoxane (MAO) by Prof. Sinn and Kamnisky opened up new avenues and the fifth-generation catalysts came to light in early 1990s. These are mainly cyclopentadienyl metallocenes, in combination with MAO as a cocatalyst. There was a revolutionary increase in the activity from 60–90 Kg of PP to 5000–9000 Kg of PP/g of metal. The metallocenes have reached the commercial levels and are being marketed worldwide. The history, chemistry, and market opportunities for metallocene grade polyethylene will be separately discussed in Section 2.4.

The last generation of ZN system consists of various classes of catalysts, which are commonly known as postmetallocenes. This includes, but is not limited to, ansa-metallocenes, bridged metallocenes, constrained geometry catalysts (CGCs), half-metallocenes, and FI catalysts among others. Some of these are finding niche applications and have entered the market in a small way. Postmetallocenes will be discussed in detail in Section 2.7 followed by ultrahigh molecular weight polyethylene (Sections 2.8 and 2.9) and functional polyolefins (Section 2.10).

2.2 MECHANISM OF ZIEGLER–NATTA POLYMERIZATION

Although the initial developments were dominated by the product characteristics, later investigations were focused on understanding the mechanism. Over the years, various proposals were presented to pin down the mode of action of ZN catalyst in olefin polymerization. Majority of them can be classified into two classes: (a) bimetallic mechanism and (b) monometallic mechanism. It should be kept in mind that with increasing advent of advanced tools the mechanistic proposal evolved into a reasonable and acceptable mechanism over the years. Therefore, the initial reports may not be valid today, but deserve their own credit in reaching the accepted mechanism. The founding fathers, Prof. Ziegler and Prof. Natta, initially believed that polyolefins are produced via a bimetallic mechanism.[13,14] But later developments, especially by Cossee and Arlaman, established that the ZN polymerization follows a monometallic pathway and the polymerization takes place at transition metal-carbon bond.[15] After these findings, the monometallic Cossee–Arlman mechanism is now generally accepted and is widely used in olefin polymerization studies.[16-18] Thus, in this section we will present only the *monometallic mechanism*, the bimetallic mechanism is discussed elsewhere.[19]

As stated in Chapter 1, a classical free radical polymerization consists of three steps, namely, initiation, propagation, and termination or chain transfer. Metal-catalyzed insertion polymerization or coordination polymerization is slightly different, in that, the first step is formation of active metal complex or the active species. The monometallic Cossee–Arlman mechanism assumes that the first step is *trans*-metallation between $TiCl_4$ and AlR_3 and an alkyl group is transferred to titanium center (The assumption is based on various investigations that probed reactivity of transition metal halide with trialkyl aluminum).[20] Thus, the classical Cossee–Arlman mechanism proposes an octahedral titanium complex 1, along with a vacant site, as an active species (Figure 2.3). The octahedral complex consists of three chlorides and one R or vacant site at its four corners in the square base with the titanium at the center. This square plane is anchored on the fourth chloride (located opposite to the R group in species 1) that is embedded in the $TiCl_4$ crystal.

The metal center is electron deficient and as soon as the olefin approaches, it forms a π-complex to generate species 2. In addition to the electron deficiency of the metal, a precondition for π-complex formation is the parallel approach of the olefinic double bond to the metal-alkyl bond. The said π-complex is stabilized through two types of interactions: first, the olefins donate its π-electron density to the vacant metal $d_{x^2-y^2}$ orbital and second, there is back-donation from the filled metal d (d_{yz}) orbital to the empty π^* orbital of the olefin. In the process, the olefin coordinates to the

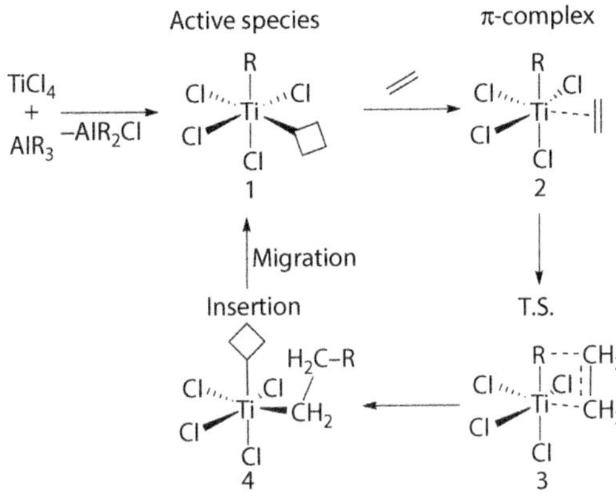

FIGURE 2.3 The classical Cossee–Arlman mechanism ($\square$ represents vacant site and R stands for alkyl group).

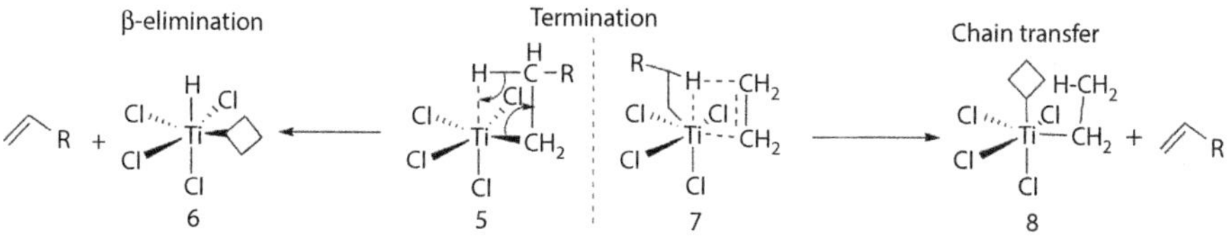

FIGURE 2.4 Termination of a growing polymer chain via β-hydride elimination or chain transfer (where R is either an alkyl group or a long alkyl chain).

metal and a relatively stable π-complex is obtained. Subsequently, the metal-R bond is polarized and the nucleophilic attack of R on one of the olefinic carbons leads to a four-membered transition state 3. Finally, the olefins insert into the metal-R bond and one olefin is added to the metal-R species. After insertion, a new vacant site is created in different places than the starting vacant site. In some cases, the inserted alkyl chain migrates to create the vacant site as it was in the parent complex 1 or in other cases the next olefin coordinates without migration. Insertion or migratory insertion closes the catalytic cycle and one olefin is inserted. Repetitive insertion as shown in Figure 2.3 results into a long alkyl chain, which is called propagation in classical polymer chemistry terms. It should be noted that the migration step is very crucial and differentiates the stereoregularity in polymers, which will be discussed in more details in Section 2.6. The final step is termination or chain transfer to release the polymer chain from the metal center. It is now well understood that both these mechanisms are operative, but the final outcome depends on which one is dominant. The growing polymer species 5 can undergo β-hydride elimination (if it satisfies all the conditions of β-H elimination process) to generate a hydride complex 6, along with vinyl-terminated polymer chain (Figure 2.4). The second competitive termination involves chain transfer to monomer or to cocatalyst or the active hydrogen compound. The growing species 7 can transfer β-hydrogen to the incoming monomer to produce species 8 and a vinyl-terminated polymer chain.

Thus, the first step of formation of active species can be termed as *initiation*, the next steps can be compared with *propagation*, and the final step of termination completes the polymerization process.

2.3 ZN POLYMERIZATION: INDUSTRIAL DEVELOPMENT AND APPLICATION

Tremendous progress has been made in ZN polymerization over the years and polyethylene, polypropylene, and various copolymers are being manufactured worldwide for various applications. Although various trade names are commonly used in the market, polyethylene can be roughly classified into: (1) very low-density polyethylene (VLDPE), (2) low-density polyethylene (LDPE), (3) linear low-density polyethylene (LLDPE), (4) medium-density polyethylene (MDPE), (5) high-density polyethylene (HDPE) (Figure 2.5), and (6) ultrahigh molecular weight polyethylene

(UHMWPE). However, we will restrict our discussion to only three polyethylenes: LLDPE, LDPE, and HDPE; the UHMWPE will be discussed in Sections 2.8 and 2.9. The linear low-density polyethylene (LLDPE, density = 0.915–0.925 g/cm^3) can be produced using ZN catalytic system by incorporating small amount of α-olefins into the polyethylene (PE) backbone. LLDPE is a substantially linear polymer with a significant number of short-chain branches. This class of polymers is regularly produced by copolymerization of ethylene with short-chain alpha-olefins (e.g., 1-butene, 1-hexene, and 1-octene). LLDPE has higher tensile strength than LDPE; it exhibits higher impact and puncture resistance than LDPE. LLDPE is routinely used to manufacture food-packaging films, in agricultural packaging, cast films, blown films, and various film and packaging applications. The global demand for LLDPE is around 22 million metric tons per year. The second class is LDPE, which is significantly branched (long-chain branching) and part of it is produced by a high-pressure/high-temperature process. Some of the common applications of LDPE include, but are not limited to, caps, closures, food-packaging containers, houseware articles, toys, trays, and so on. The last class is high-density polyethylene (HDPE, density = 0.940–0.965 g/cm^3), which is a linear and more crystalline polymer and it is produced by heterogeneous (ZN) or homogeneous catalyst. Due to the high density and increased crystallinity, HDPE is used to manufacture LAN cables, telecom wires, thermoplastic jacketing, industrial pails, vegetable crates, boxes, heavy duty crates, and so on, to name a few. The three grades are depicted in Figure 2.5.

A scientific discovery cannot deliver a societal impact unless it is scaled up to produce sufficient quantities to serve the society. Although the chemists were busy in understanding the mechanism and structure–activity relationship, chemical engineers (also polymer engineers) were engaged in scaling up the ZN polymerization process for commercialization. In a very short time, various polymerization processes were developed. There are mainly three types of processes that are commonly operated in the polyolefin industry, namely, (a) solution process, (b) slurry process, and (c) gas-phase process in a fluidized bed reactor. Each one of these processes has its own advantages and limitations. Production of polyethylene on a commercial scale is a complex operation and has to integrate various elements of the polymerization method in a controlled manner. For instance, polymerization of ethylene is an exothermic (22–26 kcal/mol) reaction and the plant should be equipped with efficient heat removal mechanisms. Typical features of these three processes are summarized in Table 2.2.

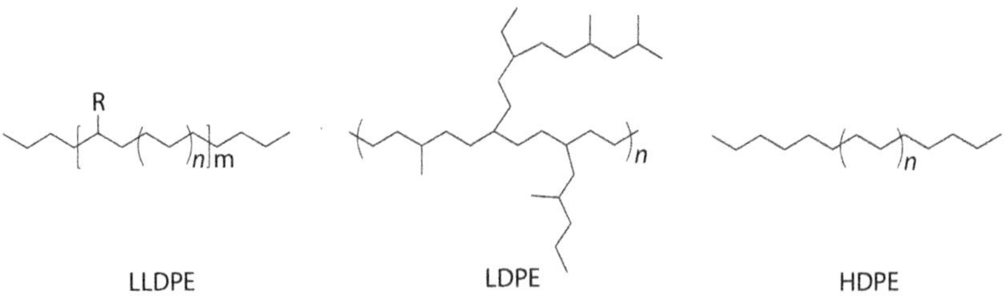

FIGURE 2.5   Representative structural difference between LLDPE, LDPE, and HDPE.

TABLE 2.2

Typical Features of the Polymerization Processes Using ZN Catalysts

Parameters	Solution Process	Slurry Process	Gas-Phase Process
Solvent	High boiling hydrocarbon	Low boiling hydrocarbon	No solvent
Operating temperature (°C)	150–220	70–110	70–110
Operating pressure (bar)	30–100	10–30	10–30
Catalyst residence time	Typically in minutes	One hour or more	2–5 hours
Catalyst stability	Should be stable at high temperature	Should be stable upto 100°C	Should be stable upto 100°C
Catalyst morphology	Not very important	Important	Very important
Polymerization	Takes place in solution	Polymerization on solid catalyst surface	Polymerization on solid catalyst surface
Comonomers	Wide range can be accommodated	Wide range can be accommodated	Limited range of comonomers
Recovery	Solvent and monomer	Solvent and monomer	Monomer recovered

1. *Solution Process*: The solution polymerization of ethylene uses continuous stirred tank reactors (CSTR). Typically, the catalyst, the resultant polymer, and the monomers are dissolved in a hydrocarbon solvent in the reactor. The reactor operates at temperatures that are higher than the melting temperature (T_m) of the resultant (co)polymer. Hence, high-boiling hydrocarbon solvents are preferred. A typical ethylene pressure in a solution polymerization reaction is between 30 and 100 bars at temperatures between 150°C and 220°C. The polymerization takes place in solution as a homogenous medium and is typically completed in minutes. As soon as the polymerization is completed, solvent is distilled back to the polymerization tank; the remaining solid is extruded and pelletized to obtain granular resin. In this process, a wide range of olefinic comonomers can be accommodated, and the monomers and solvent are generally recovered. This type of solution polymerization is practiced by Dow, DSM, Nova, and Borealis among others.

2. *Slurry Process*: As the name indicates, in this process the catalyst and the polymer formed remain suspended throughout the polymerization. Slurry-phase polymerization is carried out in CSTR or tubular reactors. Usually the diluents are saturated hydrocarbons such as propane, isobutane, and hexane, which do not react with the catalyst. The plant operates at lower temperature and pressure than the solution polymerization, typically at 70°C–110°C and at ethylene pressure of 10–30 bars. More than one reactor can be used in series or parallel to obtain desired polyolefin grades. The polymer residence time is in the range of few hours in contrast to few minutes in solution polymerization. The polymers can be easily separated from the rest and then extruded, pelletized, and dried to obtain granular resin.

Major multinational polyolefin manufacturers such as Basell, Chevron Phillips, and Mitsui operate slurry-phase polymerization reactors.

3. *Gas-Phase Process*: In gas-phase polymerization reaction, the gaseous monomer is contacted with the solid catalyst in a fluidized reactor to produce solid polymer. It is a heterogeneous reaction between a solid catalyst and gaseous monomer to yield solid polymeric product. The supported catalyst and the monomers are blown into the reactor and the polymerization takes place on a fluidized bed of the solid catalyst powder in the reactor. Better heat removal increases the productivity. The gas-phase polymerization reactor is typically operated at 10–30 bars pressure and at 70°C–110°C. An improved *condensed mode* gas-phase polymerization reactor technology was developed in 1990 by Univation in their trade mark *Unipol PE* reactor technology. The *condensed mode* technology enabled the gas-phase reactors to use higher alpha-olefin comonomers such as 1-hexene or 1-octene. Thus, this process greatly expanded the capacity and capability of the gas-phase reactors to produce various ethylene-alpha-olefins copolymers. The polymerization time varies between 2 and 5 hours and the monomer is recovered. The solid polymer is blown out from the reactor, which is then degassed and pelletized, and granules can be made. Although this is not an exclusive list, the gas-phase technology is mainly utilized by ExxonMobil, Total, Ineos, BP, Univation Technologies, Reliance Industry Limited, Basell, and Mitsui.

Among these three processes, the slurry-phase and gas-phase polymerization processes contribute around 42% each to the total polyethylene production. The remaining 15% is produced using the solution process. The polyolefin manufacturers have access to one of these or combination of these polymerization processes. Obviously, there are pros and cons to each of these reactor technologies. For example, if a desired grade of PE has to be an elastomer/plastomer, it cannot be produced using gas-phase reactor technology. Similarly, a solution-phase reactor cannot be operated at low temperatures such as a gas-phase reactor. Table 2.3 summarizes polyolefin manufacturers/licensors and the reactor technology they operate with.

TABLE 2.3
Global Major Polyethylene Manufacturers with Their Reactor Technologies

Solution Phase	Slurry Phase	Gas Phase
Dow	Basell	ExxonMobil
DSM	Chevron Phillips	Total
Nova	Mitsui	Ineos
Borealis		Univation Technologies
		British Petroleum
		Reliance Industries Limited
		Basell
		Mitsui

2.4 METALLOCENES: THE DISCOVERY EXPERIMENT AND STATE-OF-THE-ART DEVELOPMENT

The industrial development of ZN process was rapidly growing by 1980s, until the problems associated with these heterogeneous catalysts began to surface. The classical ZN catalysts are heterogeneous in nature; therefore the resulting polymer typically has broad molecular weight distribution. Due to the diversity of active sites, it is inherently difficult to control the polymer microstructure. In order to control the active sites, it was necessary to understand the formation of active sites and the subsequent events. However, it was difficult to deal with a heterogeneous system with available analytical tools in those days. Therefore, homogenous systems were being explored in academic labs to understand the active site formation and subsequent events.

Sir Geoffrey Wilkinson[21] along with Ernst Otto Fischer[22] (and the references therein) reported the syntheses of Cp_2TiBr_2 and Cp_2ZrBr_2 in 1953 for which they were awarded the Nobel Prize in 1973. Shortly after its synthesis, application of these metallocenes as polymerization catalysts was attempted. Titanocene complexes in conjunction with $AlClEt_2$ as cocatalyst were the first candidates employed for catalytic olefin polymerization reactions by Natta and Breslow. Unfortunately, the activities of such catalyst–cocatalyst combinations were not satisfying enough in ethylene polymerization and moreover they were not active in propylene polymerization (only dimerization of propylene to 2-methyl-1-pentene resulted).[23] Hence, these homogeneous catalysts were largely ignored and were limited to mechanistic studies only.[24] In one such mechanistic investigation, Prof. Sinn and coworkers at the University of Hamburg mixed toluene solution of titanium complex and trimethyl-aluminum (TMA) in a Schlenk flask equipped with an nuclear magnetic resonance (NMR) tube (see Figure 2.6) at −78°C and transferred it to the NMR tube. The NMR tube was then separated by sealing and NMR was recorded. However, a cleaver PhD

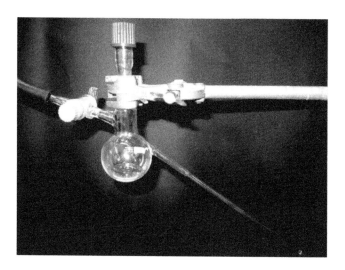

FIGURE 2.6 Schlenk flask equipped with NMR tube used in the discovery of MAO.

student wanted to simplify this tedious procedure and he recorded the NMR of the above-mentioned mixture in a regular NMR tube with a plastic cap. While comparing the two NMR spectra (one recorded in a sealed tube and another in a regular tube), he noticed that a new peak at 8.59 ppm appears in the regular NMR tube, which was absent in the sealed NMR tube. This was discussed in their group and it was decided to scale up the experiment to 1L reactor and to sample out the reaction solution after certain fixed time intervals. After mixing [Cp_2TiMe_2] complex and trimethylaluminum, the autoclave was pressured with ethylene and slow ethylene consumption was observed. The reactor was opened to take out the first sample for NMR measurements and was immediately pressurized with ethylene. Surprisingly, higher ethylene consumption was observed in the second pressure experiment as compared to the first pressure cycle. In this counterintuitive experiment, increasing ethylene consumption was observed in the subsequent sampling experiments. In their attempts to nail down the reasons for increased activity of this catalyst after exposures, the authors deliberately added controlled amount of water to the autoclave, which turned out to be the discovery experiment. Indeed, this led to increased activity with increasing amount of water until the molar ratio of trimethylaluminum to water reached 1:1.[25] In fact, similar observations were reported in separate investigations by Reichert[26] and Breslow,[27] though they missed the discovery experiment.

In as early as 1973–1975, Reichert and Breslow unexpectedly discovered that addition of small amount of water actually increases the activity of the trimethylaluminum cocatalyst, although these catalysts are extremely sensitive to hydrolysis. Thus, the observed increase in the activity is most likely due to the partial hydrolysis of trimethylaluminum. *Often accidents are the root cause for new discoveries!* If not, adding water to known pyrophoric, trimethylaluminum does not make any sense.

Having observed the 10,000-fold increase in activity after addition of water, the Sinn group sets out to investigate the reactivity of trimethylaluminum with water. By now it was almost certain that trimethylaluminum reacts with water in toluene and therefore identifying the products of this reaction would be very crucial to solve the puzzle of increased activity in ethylene polymerization. Trimethylaluminum reacts violently with neat water, therefore it was treated with inorganic salts containing water [such as $CuSO_4 \cdot 5H_2O$, $Al_2(SO_4)_3 \cdot 14H_2O$] and the resultant product was named as MAO.[28] Various tools were used to identify MAO, but even today, it is a puzzle, no one knows what exactly MAO is. Over the years, it has been found that MAO is a mixture of dimers, trimers, tetramers, and oligomers including some ring structures. It is not just the above-mentioned mixture, but this mixture is in equilibrium with each other and with unreacted TMA. This equilibrium mixture poses serious synthetic challenges that are yet to be addressed. In the past 30–40 years, only few species from the equilibrium mixture could be identified. The currently accepted MAO structures are presented in Figure 2.7. The aluminum atoms in these structures are coordinatively unsaturated; MAO therefore exists largely in the form of clusters and cages (Figure 2.7). MAO cannot be fully defined, but it can be stated that, MAO is a compound in which oxygen and aluminum atoms are arranged alternatively and the remaining valences are satisfied by methyl groups. Sinn[29] and Barren[30] stated that [$Al_4O_3Me_6$] is the repeat unit of MAO.

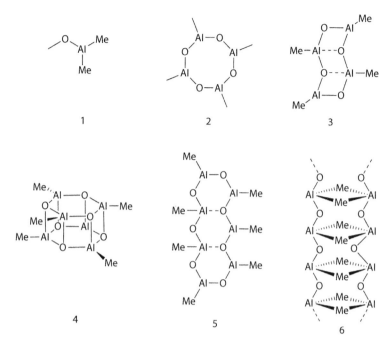

FIGURE 2.7 Proposed structures of MAO showing different possible structural arrangements.

After the discovery of MAO the situation changed quickly leading to metallocenes gain the forefront space in polyolefin chemistry. Since the discovery of MAO as efficient cocatalyst combination, metallocenes have, for more than 40 years, revolutionized the field of olefin polymerization. MAO has become the workhorse of metallocene-grade polyethylene manufacturing, still going strong, holding many more surprises in its hat and being young as ever. In addition, unlike a few active sites on the surface of heterogeneous ZN catalysts, every single molecule of metallocenes is an active site for polymerization and hence shows catalytic activity, which is much higher than that of the classic Zeigler–Natta systems. The application of separately produced MAO with metallocenes further increased the activity by about 100 folds (to 10,00,000 g. PE/g. Zr.h.bar).[31] A further important landmark development in the historical timeline of homogeneous metallocene catalysts was the stereoselective alkene polymerization achieved through the introduction of chiral bridged metallocenes by Brintzinger and coworkers in 1982.[32] This work was noticed by Prof. Kamnisky, who contacted Prof. Brintzinger and borrowed few milligrams of ansa ethylidenebis(4,5,6,7-tetrahydroindenyl)titanium dichloride and the zirconium counterpart reported in the above-mentioned publication. The complex was tested in propylene polymerization with MAO as cocatalyst and within few minutes Prof. Kamnisky and coworkers observed the formation of insoluble isotactic polypropylene in the reactor, whereas all their earlier studies produced clear solution indicating formation of atactic PP.[33] Around the same time Ewen, working for Exxon, employed a C_2 symmetric titanium catalyst and reported stereoregular polymerization of pro-pylene.[34] These historical and significant discoveries have elevated the status of homogeneous metallocenes and today metallocenes are commercially challenging the previous heterogeneous ZN catalysts. There are at least two scientific limita-tions that will decide the commercial fate of metallocenes, apart from many other nonscientific parameters: (a) high loading of MAO and (b) heterogenizing the MAO-metallocene catalyst on a support.

Although a combination of metallocenes and MAO catalyzes polymerization of ethylene roughly 10,000 times faster than classical ZN system, often a large excess of MAO over the metallocene precursor (Al/metal = 100–10,000) is required. This was one of the limitations of the metallocene system right from the beginning and detailed attempts were made to address this bottleneck. These concentrated efforts lead to the development of alternative boron-based cocatalysts, such as tris(pentafluorophenyl) borane and organic salts of the noncoordinating tetrakis(pentafluorophenyl)borate $[(C_6F_5)_4B]^-$, to name a few.[35–38] The activation mechanism is almost similar to the Cossee–Arlman mechanism. The cocatalyst undergoes rapid ligand exchange with metallocenes replacing chlorides with alkyl groups and subsequently abstracting an alkyl group on metal centers, generating the active cationic catalysts stabilized by counteranion (MAO).[39–41] The activated metallocene is cationic, highly Lewis acidic-14 electron species with a pseudo-tetrahedral structure with the two coordi-nation sites occupied by cyclopentadienyl group. The open coordination site along with the methyl group directly takes part in polymerization process and the next steps in the catalytic cycle follow Cossee–Arlman type mechanism, which will be discussed in the Section 2.4.1. Figure 2.8 illustrates a general and simplified activa-tion mechanism.

Ethylene polymerization studies conducted on metallocenes by Kamnisky and coworkers employing different group IV metals of the type $Cp_2M(CH_3)_2$ (M = Ti, Zr, and Hf) with MAO under similar condition concluded that zirconium metal is more active than titanium or Hafnium.[42] This trend was further attested by Gianetti et al. who also observed that zirconium catalyst was highly active and hafnium catalyst, the least (Figure 2.9).[43] Though less active than their zirconium analogs, in some cases, Hf metallocenes do produce higher molecular weight polyethylenes.

The variation of ligand electronic environment around the metal centres also has profound influence on activity and selectivity of catalysts. Based on their exten-sive investigations on the ring-substituted metallocenes, Mohring and Coville con-cluded that the catalytic activity increases with the increasing electron-donating

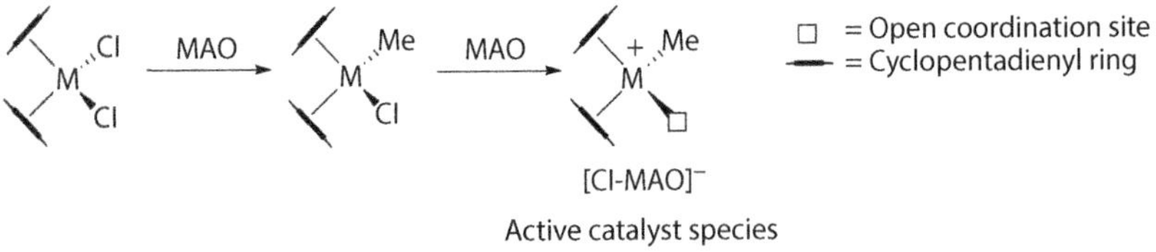

[Cl-MAO]⁻

Active catalyst species

□ = Open coordination site
— = Cyclopentadienyl ring

FIGURE 2.8   The role of MAO in activation of homogeneous metallocene catalysts.

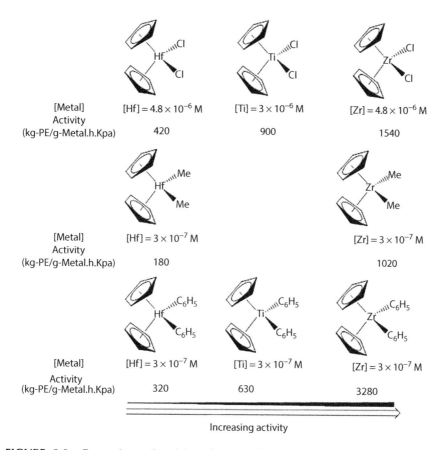

| [Metal] | [Hf] = 4.8 × 10⁻⁶ M | [Ti] = 3 × 10⁻⁶ M | [Zr] = 4.8 × 10⁻⁶ M |

[Metal]
Activity
(kg-PE/g-Metal.h.Kpa)

[Hf] = 4.8×10^{-6} M   420

[Ti] = 3×10^{-6} M   900

[Zr] = 4.8×10^{-6} M   1540

[Metal]
Activity
(kg-PE/g-Metal.h.Kpa)

[Hf] = 3×10^{-7} M   180

[Zr] = 3×10^{-7} M   1020

[Metal]
Activity
(kg-PE/g-Metal.h.Kpa)

[Hf] = 3×10^{-7} M   320

[Ti] = 3×10^{-7} M   630

[Zr] = 3×10^{-7} M   3280

Increasing activity

FIGURE 2.9 Comparison of activity of group IV metallocene catalysts in ethylene polymerization.

ability of the R group on the Cp ring.[44] However, if large α-olefins and small metal ions are employed, then not only electron-donating ability but also steric factors can influence the catalytic activity. Further studies by Tait on substituent effect on cyclopentadienyl ring $(C_5H_4R)_2ZrCl_2$ where R = H, Me, i-Pr, t-Bu) revealed that the rate of propagation for ethylene polymerization follows the trend t-Bu < H < i-Pr < n-Pr < Me (Figure 2.10).[45] The electron-donating ability of alkyl substituents on the cyclopentadienyl ligands leads to increased electron density at the active metal site, which is reflected in the increased rate of polymerization. The increased activity can be attributed to the fact that electron-donating ligands stabilize the transition state facilitating ethylene insertion leading to chain growth. Conversely, increasing the size of alkyl substituents may have adverse effects even though they increase the electron density at the metal center. The approach of the incoming monomer may be hindered by the bulky substituents leading to decreased polymerization rates. Hence, it is a combination of electronic and steric effects that decide the rate of polymerization.

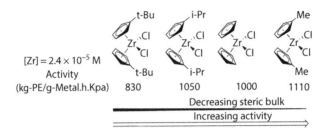

$[Zr] = 2.4 \times 10^{-5}$ M
Activity
(kg-PE/g-Metal.h.Kpa)   830   1050   1000   1110

Decreasing steric bulk

Increasing activity

FIGURE 2.10   Effect of steric-bulk of Cp-ring on ethylene polymerization activity.

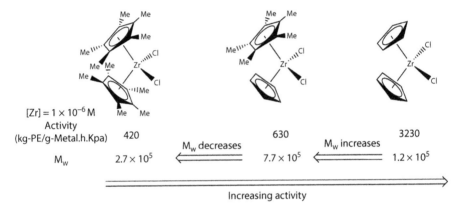

$[Zr] = 1 \times 10^{-6}$ M
Activity
(kg-PE/g-Metal.h.Kpa)   420   630   3230

M_w decreases M_w increases

M_w 2.7×10^5 $\Longleftarrow$ 7.7×10^5 $\Longleftarrow$ 1.2×10^5

Increasing activity

FIGURE 2.11   Catalyst structure, polymerization activity, and polymer molecular weight correlation.

In their efforts to correlate the catalyst structure, polymerization activity, and molecular weight, Kaminsky and coworkers observed that the molecular weight of the polyethylene produced by bis(pentamethylcyclopentadienyl)zirconium dichloride, $[(CpMe_5)_2ZrCl_2]$ ($M_w = 2.7 \times 10^5$, 420 kg-PE/g-Zr.h.Mpa), was more than twice to that obtained from Cp_2ZrCl_2 (1.2×10^5, 3230 kg-PE/g-Zr.h.Mpa) but the activity of the latter was nearly eight times more than the former (Figure 2.11).[42]

It has been observed that bridged metallocene complexes generally display better productivities than their unbridged analogs.[31,46–48] The systematic studies by Alt et al. revealed that often silyl-bridged ansa-metallocene complexes are found to increase the activity of a metallocene catalyst than their carbon-bridged derivatives.[49] This fact was further confirmed by Kaminsky who studied the effect of different bridged and unbridged metallocene complexes on the catalytic activity and molecular weights under the same experimental setup.[50] It was found that silicon-bridged metallocene with indenyl ligands was highly active and provided high molecular weight polyethylenes than those bridged by carbon (Figure 2.12). These comparative investigations allow to conclude that ligand electronic effects are particularly important for the insertion reaction.[51]

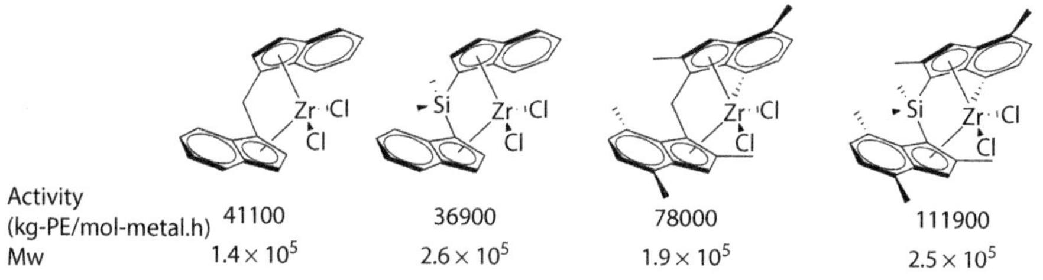

| Activity (kg-PE/mol-metal.h) | 41100 | 36900 | 78000 | 111900 |
| Mw | 1.4×10^5 | 2.6×10^5 | 1.9×10^5 | 2.5×10^5 |

FIGURE 2.12 Effect of bridging group on polymerization activity and polymer molecular weight.

Thus, the field of metallocene-catalyzed ethylene polymerization has been intensively investigated and the relationship between catalyst structure, polymerization activity, and molecular weight of the polymer is considerably established. There is a large amount of literature dealing with structure–property correlation in metallocene-catalyzed ethylene polymerization, which is beyond the scope of this chapter and cannot be fully accommodated. In the next Sections 2.4.1 through 2.5, we will discuss the mechanism of ethylene polymerization, followed by supported metallocenes and finally brief industrial development will be discussed before we switch over to the metallocene-catalyzed propylene polymerization.

2.4.1 MECHANISM OF ETHYLENE POLYMERIZATION

Metallocene-catalyzed ethylene polymerization follows the Cossee–Arlman mechanism discussed in Section 2.2. The ethylene monomer initially forms a π-ethylene-coordinated complex with the cationic metallocene, which then undergoes insertion into the M-C bond. Initially, π-coordinated ethylene that undergoes insertion via the elusive four-membered transition state creates a new vacant coordination position. Such multiple ethylene insertions lead to polymer chain propagation (Figure 2.13).

In an alternative proposal, Brookhart and Green in 1983 proposed that the ethylene insertion and formation of new carbon–carbon bond are assisted by the agostic interaction of α-hydrogen of the growing alkyl chain with the metal enabling the alkyl chain on the α-carbon to move away from incoming ethylene (Figure 2.14).[52] This mechanism was supported experimentally by Brintzinger,[53] Bercaw,[54] and density functional theory (DFT) calculation by Ziegler.[55] The partial migration of α-hydrogen atom to the metal helps in stabilizing the forming metal C-H-M bond. This α-agostic interaction also assists the olefin insertion into the M-C bond and

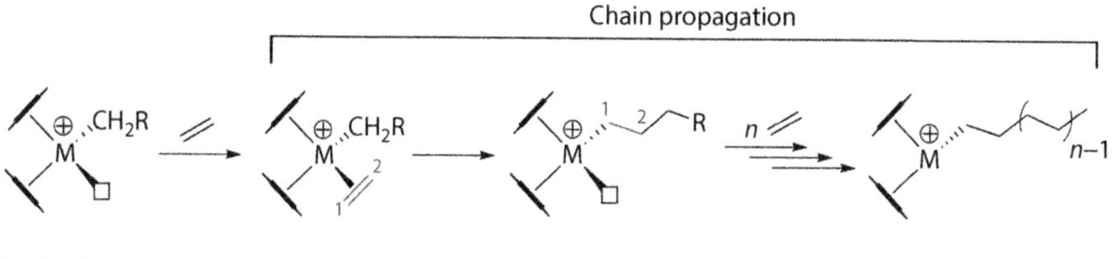

FIGURE 2.13 General mechanism of ethylene polymerization by activated metallocenes.

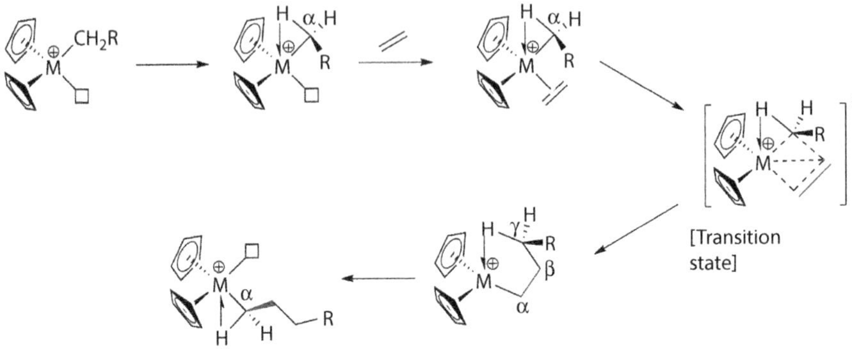

FIGURE 2.14 Brookhart–Green mechanism of alkyl chain propagation.

C-C bond formation leading to the formation of γ-C-H-M bond. This is followed by a rearrangement step, wherein the α-hydrogen of the newly inserted ethylene agostically stabilizes the metal center.

The chain termination or transfer can occur by many ways. The most common termination modes that are very specific to metallocene catalysts are discussed as follows:

1. The β-hydrogen transfer from growing polymer chain to an incoming ethylene is one of the common mechanisms of chain transfer that is illustrated later (Figure 2.15a)[56].
2. Chain termination (transfer) can also occur by β-hydride elimination from the growing polymer chain to give rise to the odd chain polyethylene with a terminal C=C group. The metal hydride formed in this process can further

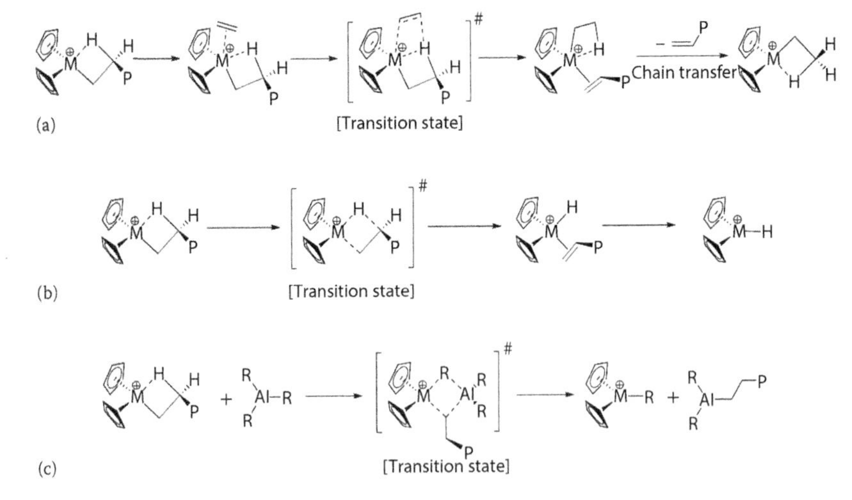

FIGURE 2.15 Termination of growing polymer chain *via* (a) chain transfer to monomer, (b) β-hydride elimination, and (c) chain transfer to cocatalyst.

undergo ethylene insertion to give rise to even polymer chain. Figure 2.15 illustrates the accepted mechanism of termination by β-hydride elimination.

3. In addition, the growing chain can be transferred to the residual AlMe$_3$ present in MAO, which usually happens at lower olefin concentration.[57]

The above-mentioned mechanism illustrates the general concept of group IV metallocene-catalyzed ethylene coordination and insertion leading to polyethylene. But there are a number of factors that decide how well the ethylene approaches the metal centre, coordinates, inserts, and grows into polyethylenes. This depends on type of cocatalyst employed, catalyst/cocatalyst ratio, temperature, pressure, and also the nature of group IV metal and ligands present.[58] These factors have been studied by various research groups and knowledge of these parameters is very crucial to design a better metallocene catalyst, which will guide the future generations.

2.4.2 SUPPORTING THE METALLOCENE CATALYST

As discussed in Section 2.4, metallocene catalysts are soluble in hydrocarbon solvents and are therefore considered as homogeneous catalysts. But most commercial plants around the world use slurry- or gas-phase reactors for olefin polymerization, except 15% manufacturers operate on the solution process. Inherently, the use of slurry- or gas-phase processes requires an insoluble catalyst system before they can be used in these processes. The general benefits of supported metallocene catalysts over their homogeneous analogs are: (1) the cocatalyst/catalyst ratio needed to attain the maximum activity is lower for the supported metallocene catalysts; (2) the average polymer molecular weight varies in supported systems; and (3) some metallocene catalysts can be activated using boron-based cocatalyst instead of MAO, which are easy to support. Silica is the most commonly used support for metallocene catalysts because (a) it is inexpensive, (b) it is chemically inert, (c) it is stable at high temperatures, and (d) it can be synthesized in a variety of pore sizes, pore volumes, and surface areas. There are many other examples of supports such as MgCl$_2$, clay minerals such as montmorillonite and kaolin, zeolites, alumina, and polymers.

In 1997, Quijada et al. studied the catalytic activity of metallocene catalysts supported on porous and nonporous silica, and their efficacy in polymerizing ethylene.[59] It was found that the catalytic activity depended on the amount of zirconium for metallocenes supported on (aerosol) nonporous silica; whereas the zirconium content had no such predominant effect on the catalytic activity of metallocenes supported on porous silica (Table 2.4). This is probably because pores may hinder the active sites, making the activity remain constant at higher percentage of zirconium. The effect of calcination temperature on the activity was investigated and it was found that the activity increased with calcination temperature up to 800°C for both porous and nonporous supports. At higher temperature, the activity of porous silica decreased whereas that of nonporous still increased (Table 2.5). At temperatures more than 800°C, the porous support probably lost a significant amount of surface −OH groups and hence only a lower amount of zirconium was present. However, for a nonporous support, the amount of zirconium was still higher, because the −OH groups were located on the surface and were more accessible.

TABLE 2.4
Effect of the Support and the Amount of Zr Incorporated in the Catalyst on the Activity for Ethylene Polymerization*

Type of Catalyst	Zr% in the Catalyst	Productivity (kg-PE/(mol-Zr-h))	Activity (kg-PE/(mol-Zr-h pressure))
ES-70EtInd$_2$ZrCl$_2$	1.4	64	106
	2.4	65	108
	6.0	72	119
EP-10EtInd$_2$ZrCl$_2$	1.2	65	108
	6.1	64	107
AER EtInd$_2$ZrCl$_2$	3.7	67	111
	6.0	122	204
ES-70Ind$_2$ZrCl$_2$	1.5	49	82
EP-10EtInd$_2$ZrCl$_2$	1.3	58	96
AER EtInd$_2$ZrCl$_2$	6.0	118	197

*ES-70 and EP-10 are porous silica. AER is nonporous silica. Calcination temperature: 500°C; temperature of polymerization: 60°C; time of polymerization reaction: 30 min; pressure of ethylene: 1.6 bar; Al/Zr: 200; and zirconium in the reactor: 1.20×10^{-4} mol.

TABLE 2.5
The Effect of Calcination Temperature of Silica Support on the Catalytic Activity for Ethylene Polymerization[a]

Catalyst	Zr% in the Catalyst			Activity (kg-PE/(mol-Zr-h pressure))		
	500°C[b]	800°C	1000°C	500°C[b]	800°C	1000°C
ES-70EtInd$_2$ZrCl$_2$	6.0	6.4	4.7	119	180	143
AER EtInd$_2$ZrCl$_2$	6.0	6.6	6.2	197	254	319

[a] ES-70 is porous silica and AER is nonporous silica. Time of polymerization: 30 min; temperature of polymerization: 60°C; Al/Zr: 200; and pressure: 1.6 bar.
[b] Calcination temperature of the supports.

In 2004, Alonso et al. reported that when silica was modified with $(Me_3Si)_2O$, catalytic activity increased due to higher zirconium content in the support than in the ordinary silica.[60] Along the same line, in 2007, Ketloy et al. studied the catalytic properties of [t-BuNSiMe$_2$Flu]TiMe$_2$/dMMAO on various supports toward ethylene/1-octene copolymerization.[61] They employed silica, titania, and mixed silica–titania (4:1 by weight) as support for the catalysts. The results revealed that titania contains the highest amount of $[Al]_{dMMAO}$ and the silica–titania combination possessed more $[Al]_{dMMAO}$ in comparison to bare silica. The polymerization activities of the homogeneous system and various supports using toluene as a solvent are shown in Table 2.6. The polymerization activities are in the order of:

TABLE 2.6

Supporting Metallocenes on Silica, Titania, or the Mixture of Two and Their Polymerization Activity[a]

System	Solvent	Yield (g)	Activity (kg of Polymer mol^{-1} Ti h)
Homogeneous	Toluene	3.25	3897
SiO$_2$ support		2.49	2984
SiO$_2$–TiO$_2$ support		2.61	3131
TiO$_2$ support		2.33	2795
Homogeneous	CB	3.23	3871
SiO$_2$ support		1.68[b]	10095
SiO$_2$–TiO$_2$ support		2.69[c]	12127
TiO$_2$ support		2.53	3032

[a] Polymerization condition: Ti = 10 μmol, Al/Ti = 400, temperature = 343 k, time = 5 min, 50 psi of ethylene.
[b] Polymerization time = 1 min.
[c] Polymerization time = 1.5 min.
CB—chlorobenzene.

homogeneous > SiO$_2$-TiO$_2$ > SiO$_2$ > TiO$_2$. According to Severn, the cocatalyst is connected to support by a O$_{support}$-Al$_{cocatalyst}$ linkage.[62] The stronger the interaction between the cocatalyst and support, the harder it was for the cocatalyst to react with the Ti-complex during the activation process, which resulted in lower catalytic activity. This is why TiO$_2$ exhibited the lowest catalytic activity. However, in SiO$_2$–TiO$_2$, even though there was a stronger interaction than silica, it exhibited better catalytic activity, because there was a higher concentration of [Al]$_{dMMAO}$ and TiO$_2$ acted as a spacer group between the silica and the cocatalyst.

When toluene was replaced with chlorobenzene, it was found that there was a dramatic increase in the activity of supported catalysts, especially for the SiO$_2$ and SiO$_2$–TiO$_2$-supported catalysts. Alonso et al. proposed that this increase in the catalytic activity was due to: (a) the change in the interaction between the support and the cocatalyst and (b) the change in the form of the active species. It was also found that the molecular weight of the polymer produced was higher when chlorobenzene was employed. This was due to the solvent separated ion-pair active species, which allowed the insertion of more monomers into the growing chain. The earlier observation suggests that overly strong interactions between the support and the cocatalyst result in lower activity. However, the spacer group between cocatalyst and support can increase the activity or even the solvent can play a crucial role in increasing the activity.

In 2010, Brambilla et al. reported that Cp$_2$ZrCl$_2$ fixed on silica-magnesia mixed-support system showed better catalytic activity in comparison with Cp$_2$ZrCl$_2$ fixed on bare silica.[63] They prepared three mixed-support catalysts by varying the Mg/Si weight percentages. The results are represented in Figure 2.16.

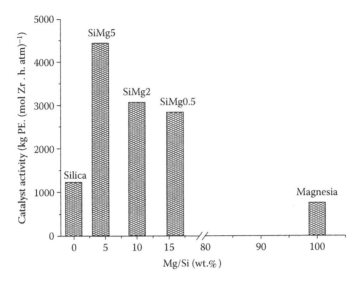

FIGURE 2.16   Catalyst activity in ethylene polymerization for different supported catalysts.

The highest catalytic activity was exhibited by the $Cp_2ZrCl_2/SiMg_5$ (Mg/Si equal to 4.3%) system. This was mainly due to a higher surface area and a greater pore volume, both of which led to better accessibility of monomers to the active site. They also reported that the silica-magnesia support partially played the role of cocatalyst in the activation of the system. This silica-magnesia-supported catalyst showed the highest catalytic activity at an Al/Zr ratio of 500:1. In contrast, the catalyst supported on silica showed the highest catalytic activity at an Al/Zr ratio of 1500:1. The role of MAO in the activation of catalysts was to exchange the chloride with the methyl group and to stabilize the cationic Zr species (Figure 2.17). They proposed that for $Cp_2ZrCl_2/silica$, the support was not acidic enough to reduce the quantity of MAO,

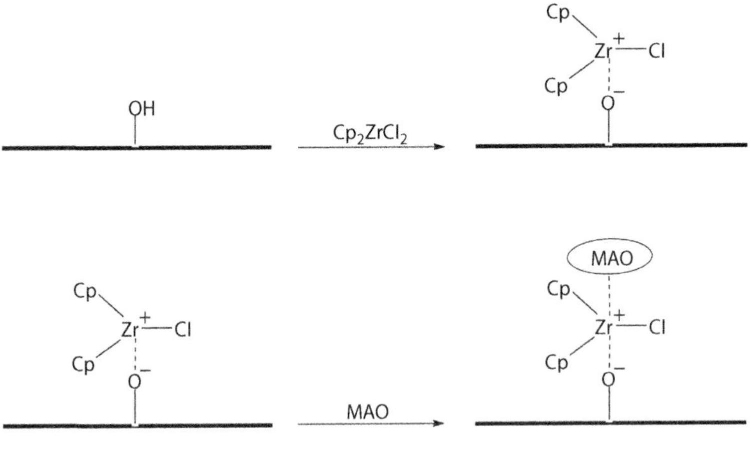

FIGURE 2.17   Activation of supported metallocene catalysts.

but for the $Cp_2ZrCl_2/SiMg_5$ system, the support had sufficient acidity necessary to reduce the amount of MAO.

The polyethylenes obtained from silica-magnesia-supported catalysts were characterized in terms of melting temperatures. They were seen to have melting temperatures in the range of 132°C–134°C, which is a characteristic of HDPE. The molecular weight of the polymer obtained from supported catalysts is higher than that obtained from homogeneous catalysts. This is most likely due to blocking of one of the sides of the active site by the support, thereby suppressing the deactivation step.

The molecular weight distribution and molecular weight are two important properties of a polymer, since they control the mechanical and rheological behavior. Metallocenes usually produce polymers with a narrow molecular weight distribution that have improved physical properties such as clarity and impact resistance. On the other hand, polymers with a broad molecular weight distribution show greater flow-ability and such polymers are good for blowing and extrusion applications. There are several methods to control the molecular weight distribution. One of the effective ways is to use a combination of two or more transition metal catalysts.

In 2003, Santos and coworkers prepared a series of supported catalysts by combining $(nBuCp)_2ZrCl_2$ and Cp_2ZrCl_2 that were sequentially grafted on silica in different ratios and immobilization orders.[64] To begin with, they determined the activity of soluble hybrid catalytic systems. Among three combinations of mixtures (Cp:nBu = 1:1, 1:3 and 3:1), the 1:1 combination of Cp and nBu exhibited the highest catalytic activity. It has already been reported that $(nBuCp)_2ZrCl_2$ showed better catalytic activity than Cp_2ZrCl_2. Therefore, a 3:1 combination of Cp and nBu is less active due to a lesser amount of the more active catalyst. A 1:3 combination of Cp and nBu is less active due to either bimolecular deactivation reactions or due to high initial activity. This initial polyethylene production leads to the hindering of further diffusion of the monomers to the catalytic center. In supported systems, the best activity was exhibited by 3:1 nBu:Cp and 1:3 Cp:nBu (Figure 2.18). Their activities are very close, which indicates that there is no influence of the grafting order on catalytic activity. Heterogeneous catalysts produce polymer with high molecular weight, which possess better mechanical properties. The polydispersities (Mw/Mn) are around 2.0, indicating that this combination of catalysts fails to achieve bimodality. The most likely reason could be that both the catalyst centers produce similar polymer chains in similar productivity.

The same group extended this study and investigated the effect of different supports on catalytic activity.[65] Different hybrid catalysts were prepared by sequentially grafting Cp_2ZrCl_2 and $(nBuCp)_2ZrCl_2$ (1:3 ratio) onto a set of aluminosilicates, alumina, silica, and chrysotiles. The highest catalytic activity was exhibited by the catalyst with a silica support (Figure 2.19). This was due to the highest pore diameter (270 Å), which allowed enough space and catalyst to be more accessible to the cocatalyst and the monomer. In aluminosilicates, SBA-15 (SBA: Santa Barbara Amophous) is more active, which is also due to the higher pore size. However, the activity cannot be explained only in terms of the pore diameter. For example, even though the pore diameter of alumina is 52 Å, it showed slower catalytic activity which may be due to its amphoteric character. For a given group of supports, the activity increases with a decrease in the particle size (Figure 2.20). For example,

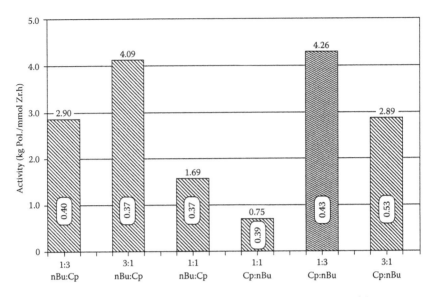

FIGURE 2.18   The catalytic activity of supported metallocene for the hybrid systems.

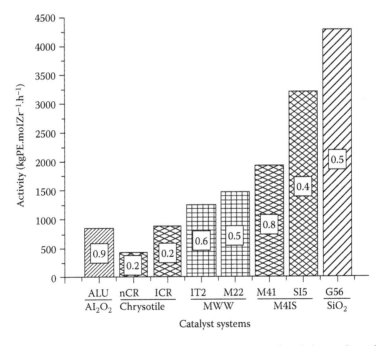

FIGURE 2.19   Catalyst activity of the supported metallocenes in ethylene polymerization. Grafted metal content (wt% Zr/SiO$_2$; for ALU wt% Zr/Al$_2$O$_3$) was also included.

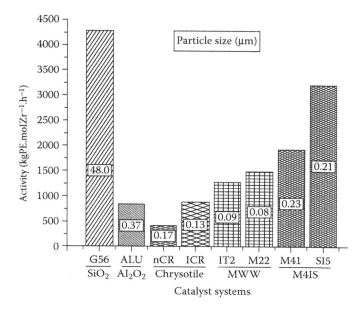

FIGURE 2.20 The particle size influence on catalyst activity.

SBA-15 (particle size: 0.21 µm) showed higher activity in comparison to MCM-41 (particle size: 0.23 µm, MCM: Mobil Composition of Matter). It is also observed that a small pore size leads to a high molecular weight polymer.

2.5 APPLICATION OF METALLOCENE-GRADE POLYETHYLENE

Driven by their superior activity and their ability to tailor the properties of polyethylenes, metallocenes have started challenging the commercial dominance of classical ZN catalysts for olefin polymerization. The ability of metallocene catalysts to incorporate a wide range of α-olefins in ethylene polymerization makes them stand out from ZN system. In addition to their superior chemistry, an attractive feature (for commercial production) of metallocenes is their ability to serve as a *drop-in substitute* to classical ZN polymerization plants. Thus, the polyolefin manufacturers do not have to build separate plants or design separate processes, but rather use their existing infrastructure to manufacture metallocene-grade polyethylene. Henceforth metallocene-grade polyethylene will be referred as mPE. mPE is a LDPE, which can be seen as a substitute to LLDPE produced by classical ZN method. The major application area of mPE is in films and packaging industry. The mPE offers the following superior properties as compared to the polyethylene manufactured by ZN catalyst: (1) mPE displays superior film toughness and strength, (2) exceptional film-sealing performance, (3) better optical properties, (4) down-gauging potential, and (5) improved puncture resistance.

Metallocene-grade polyethylene has attracted a significant attention in the last 20 years and leading manufacturers of polyolefins have started producing metallocene-grade polyethylene all over the world. Roughly 8% (about 6 million tons)

TABLE 2.7
mPE Manufacturers and Corresponding Trade Names

Entry	Manufacturer	Brand/Trade Name/s
1	ExxonMobil	Exceed™ and Enable™
2	Total	Lumicene™ and Supertough™
3	ChevronPhillips	mPACT™
4	Mitsui	Evolue™
5	Borealis	Queo™
6	Ineos	Eltex™
7	Dow	ENGAGE™, AFFINITY™, and ELITE™

of the total polyethylene is now being produced using metallocene catalyst technology (mPE) and this share is projected to increase significantly in the next few years. Almost every major polyolefin manufacturer has started switching to metallocene-grade polyethylene. Having said that, metallocene technology cannot fully replace the ZN technology for various technical, as well as, commercial reasons. Table 2.7 summarizes the manufacturers of mPE with the respective grades (trade names) offered by these polyolefin producers.

To conclude, a discovery that started with a serendipitous observation but rational analysis of the observation has led to a successful process and the above-mentioned manufacturers produce the mPE on huge scale, which is then marketed worldwide.

2.6 METALLOCENE-CATALYZED POLYMERIZATION OF PROPYLENE

After the successful implementation of metallocene in ethylene polymerization, the very obvious question was: Can it produce stereoregular polypropylene? As briefly discussed in the Section 2.4, Prof. Kaminsky and Prof. Brintzinger joined forces to demonstrate the stereoregular polymerization of propylene catalyzed by metallocene catalysts. Rather than summarizing the available literature, we have selected few metallocene systems to demonstrate how catalyst structure can change stereoregularity in the polymer. Therefore, following Sections (2.6.1 to 2.6.3) will only focus on those systems that have a profound impact on the stereoselection process.

The polymerization of propylene and higher 1-olefins, unlike the symmetric ethylene, bring in the complexity of regioselectivity because of their prochiral nature. Depending on the arrangement of the pendant methyl group on the backbone chain, the polypropylenes can be classified into three types[*]: (1) isotactic polypropylene, (2) syndiotactic polypropylene, and (3) atatic polypropylene, (Figure 2.21). These can be defined as follows:

---

[*] There are two more types, namely, (1) hemiisotactic polypropylene in which every alternate pendant methyl group is randomly oriented and (2) isotactic-atactic stereoblock polypropylene which has isotactic block at one end and atactic block at the other. However, these are not very commonly encountered.

Isotactic
polypropylene

Syndiotactic
polypropylene

Atactic
polypropylene

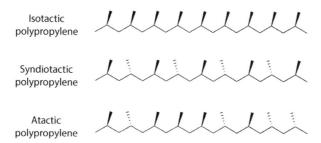

FIGURE 2.21 Polymer architectures commonly known for polypropylene.

1. Isotactic polypropylene: The pendant groups are arranged on the same side of the polymer backbone.
2. Syndiotactic polypropylene: The pendant group is alternatively above and below the plane of the polymer backbone.
3. Atactic polypropylene: In this case, there is no ordered arrangement of pendant groups and they are randomly placed on the polymer backbone.

The properties of the final polymer are largely dictated by the arrangement of the pendant groups on the polymer backbone. Therefore, precise control over the placement of pedant groups is highly desired. Thus, controlling the stereoregularity of the polymer by rational catalyst design has been the subject of large scientific discussion and section 2.6.1 will highlight the origin of stereocontrol in the metal-catalyzed propylene polymerization.

2.6.1 ORIGIN OF STEREOCONTROL IN PROPYLENE POLYMERIZATION

The propylene insertion occurs predominantly through 1,2-insertion (primary) mode, although a few metallocenes do insert propylene with 2,1-regioselectivity (secondary or regio-irregular) and via 3,1-regioselectivity (Figure 2.22). Insertion of propylene in 1,2-fashion leads to primary alkyl group coordinated to metal, whereas 2,1-insertion causes secondary alkyl group coordination. This slight change in electronic parameter also has an influence on the next propylene insertion. The 2,1-inserted propylene has less activating effect on incoming propylene and is more likely to undergo chain transfer by β-hydride elimination and reinsertion, ultimately leading to 3,1-inserted propylene. Also secondary α-olefin insertion results in steric crowding around the metal and hence is an infrequent incident usually accounting to fewer than 1% of the enchainment[46] and so is the 3,1-insertion, which has to occur *via* isomerization from 2,1-inserted propylene.

Each prochiral propylene monomer that inserts into the growing polymer chain, regardless of its orientation, generates a new stereocenter (Figure 2.23). Hence, the tacticity of the resulting polymer is defined by the stereochemical relationship(s) that the stereogenic carbons share among themselves in the resultant polymer chain. In simple terms, the configuration of the new stereogenic center generated after each propylene insertion is defined by which propylene enantioface coordinated with

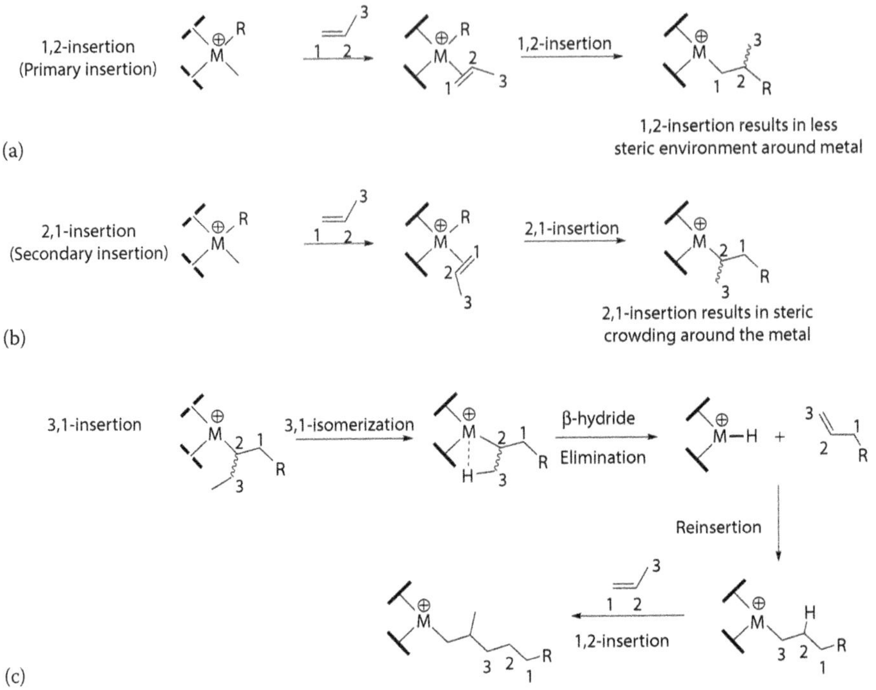

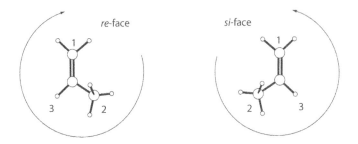

FIGURE 2.22 (a) 1,2-propylene insertion, (b) 2,1-propylene insertion into metallocenes, and (c) 3,1-propylene insertion via isomerization from 2,1-inserted propylene.

FIGURE 2.23 Two enantiofaces *re* and *si* of prochiral propylene.

metallocene precedes insertion.[66] The two metallocene-coordinated polypropylene enantiofaces, *re* and *si* are depicted in Figure 2.24.

An activated metallocene-type catalyst offers two possible coordination sites on metals (a vacant site and a methyl position, which directly participates in chain growth) for olefin coordination and insertion. Since propylene can coordinate to metallocene with its two enantiofaces, (*re* and *si*), the pendant methyl groups can come on the same side of polymer chain when propylene inserts with same enantioface successively (*si* after *si* or *re* after *re*) resulting in isotatic propylene. Conversely,

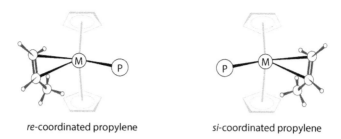

re-coordinated propylene   si-coordinated propylene

FIGURE 2.24 *si* and *re* coordinated enantiofaces of propylene (P = Polymer chain).

syndiotactic polypropylene results when the metallocene catalyst prefers propylene monomer insertion with opposite enantiofaces consecutively (*si* after *re* or *re* after *si*) bringing the pendant methyl groups on opposite side of the polymer chain, whereas random enantioface insertion results in polymer with no stereoregularity or atactic polypropylene.

The symmetry, the steric surrounding of the metal active site as well as that of the growing polymer chain, influences the stereochemistry of α-olefin insertion. The stereochemical effects arising due to stereogenicity in metallocene catalyst (chiral catalyst site) are referred as *enantiomorphic site control*. In the absence of stereogenic centre in catalyst, the stereochemical regulation may come due to the chiral induction from the growing polymer chain. This type of stereocontrol originating from growing polymer chain is referred as *chain-end control*. Regardless of the fact that whichever insertion selectivity mechanism operates, it is the *steric pressure* between the growing polymer chain and the arriving α-olefin that dominates the enantioface insertion selectivity of α-olefin. As has been discussed earlier, each α-olefin insertion creates a new chiral center on the growing polymer chain and this new stereogenic center formed by last inserted monomer influences the enantiofacial selectivity of the incoming α-olefin entity. If this effect regulates the enantioface preference countering the influence of stereogenicity of the metal active site, the process is said to be *chain-end controlled*. The chiral induction to the polymer chain can also come from the *chiral pocket* at the active metal site, imposed by the presence of asymmetric auxiliary cyclopentadienyl ligands which when substantial enough to overrule the stereochemical influence due to polymer chain end, the process is said to be *enantiomorphic site control*. Figure 2.25 depicts the two stereocontrols.

As a consequence of the two selectivity centers, the growing polymer chain can shuttle between the two controlling centers. The symbols *m* and *r* are the abbreviations for *meso* and *raceme*, respectively, which describes the stereochemical relationship between the neighboring pendant groups. *meso* (*m*) indicates the pendant groups are on the same side of the polymer chain, whereas *raceme* (*r*) indicates the opposite relationship between pendant groups. The operation of enantiomorphic site-controlled insertion mechanism (Figure 2.26a) does not lead to the propagation of stereoerrors, since site-controlled insertion mechanism is influenced by *chiral pocket* at the catalyst-active center. Therefore, the next insertion is not affected by the isolated cases of previous misinsertion and the resultant isotatic and syndiotactic

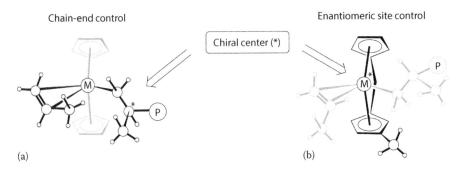

FIGURE 2.25 (a) Chain-end control mechanism and (b) enantiomeric site control mechanism in polypropylene polymerization.

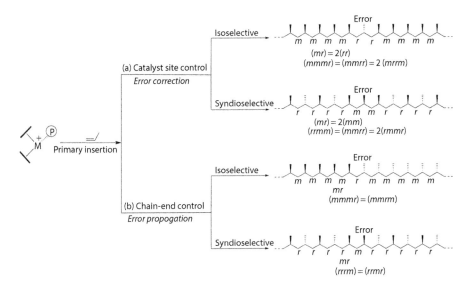

FIGURE 2.26 Mechanism of (a) enantiomorphic site-controlled and (b) chain-end-controlled primary propylene insertion.

polymers have *rr* and *mm* triads, respectively. Conversely, chain-end controlled insertion mechanism (Figure 2.26b) propagates stereoerrors in view of the fact that chiral center formed by previous monomer misinsertion directs the next insertion and hence the resultant isotactic and syndiotactic polymers have remote *r* and *m* diads, respectively.

The understanding of the insertion mechanism makes it possible to predict the microstructure of the poly-propylene (poly-1-olefin) made with metallocene catalyst of known symmetry in majority of the cases. As has been stated in the latter part of introduction to metallocene catalyst, first Ewen[67,68] and later Kaminsky developed the relationship between the symmetry of the metallocene catalysts employed and

TABLE 2.8

Correlation between the Symmetry of the Metallocene and the Polypropylene Produced

Symmetry	Metallocene	Coordination Sites[a]	Polymer[b]
C_{2v} Achiral	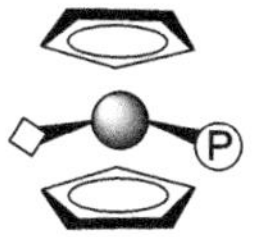	N, N Homotopic	Atactic CEC
C_2 Chiral	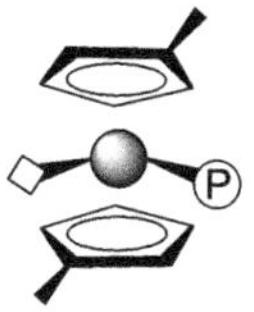	E, E Homotopic	Isotactic ESC
C_s Achiral	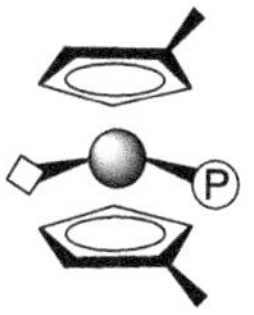	N, N Diastereotopic	Atactic CEC
C_s Prochiral	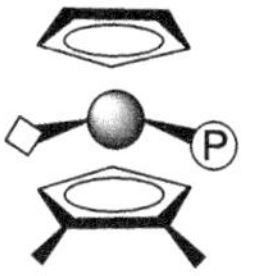	E, E Enantiotopic	Syndiotactic ESC
C_1 Chiral	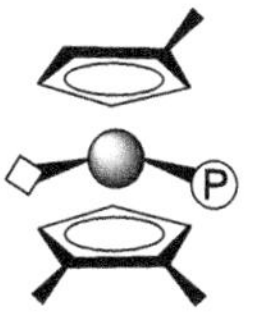	E, N Diastereotopic	Variable ESC

[a] E = Enantioselective site; N = Nonselective site.
[b] CEC = Chain-end control; ESC = Enantiomeric site control.

polymer microstructure obtained. This correlation between the metallocene structure and polypropylene properties is also known as *Ewen's symmetry rules*, which are summarized in Table 2.8.

Due to lack of chirality in catalyst, metallocenes with C_{2v} or *meso-C_s* symmetry characteristically generate aspecific polymers (atactic or moderately stereoregular), wherein the chain-end controlled mechanism operates. While isotactic polymers *via* enantiomorphic site control are expected with chiral metallocenes possessing C_2 symmetry. The prochiral C_s symmetric metallocenes commonly produce syndiotactic polymers, whereas for the metallocenes with least symmetry, that is, C_1 metallocenes, hemiisotactic polymer microstructure is predicted. However, the microstructure is largely influenced by the ring substituents giving hemiisotactic to isotactic to syndiotactic polymers with increasing steric bulk on the substituents.

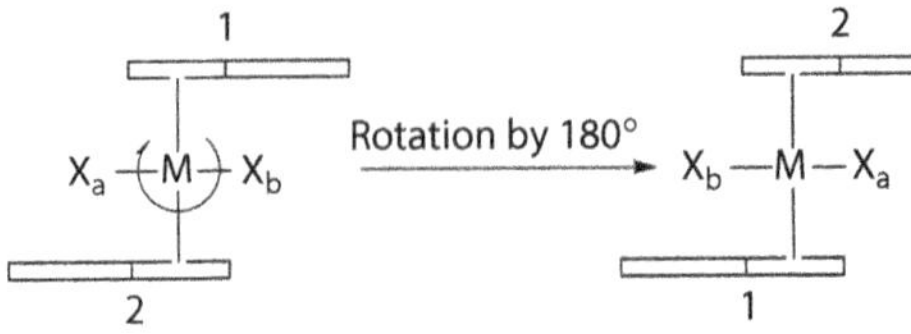

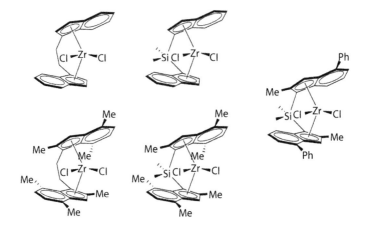

FIGURE 2.27   Representation of C_2 symmetric metallocene.

FIGURE 2.28   Representative indenyl-derived zirconocenes.

2.6.1.1 C_2 Symmetric Metallocenes and Isotactic Polypropylene

Metallocenes possessing C_2 rotational axis fall into C_2 symmetry category (Figure 2.27). Because of this C_2 axis the two coordination sites on metal can interchange position by C_2 rotational axis that makes them equivalent (homotopic) and hence prefer coordinating to monomer with same enantioface on either side of metal.

Examples of metallocene catalyst that can effectively catalyze the polymerization of propylene and can also control the tacticity of the resulting polymer chain producing isotactic polypropylene are C_2 symmetric indenyl zirconocenes presented in Figure 2.28.[50]

Due to steric effects of the indenyl ligands and chiral pocket at the metal center, the propylene coordinates only *via re*-face, on whichever side of metal it is bound. Multiple *re*-face insertion of propylene results in isotactic polypropylene. The isotactic growth of propylene chain in C_2 symmetric metallocene is schematically represented in Figure 2.29.

Similarly, the propylene monomer coordinates only *via re*-face to the other enantiomer of these C_2 indenyl metallocenes illustrated earlier producing isotactic polypropylene.[50]

Apart from the ligand scaffold, metal center can also influence the catalytic activity. It was observed by Akhihiro Yano that, the same ligands with varying metals of the type $Me_2Si(3\text{-}MeCp)_2TiCl_2/MAO$ under similar conditions, the catalytic activity follows the trend Zr > Hf > Ti (Figure 2.30).[69] The observed low activity of Ti

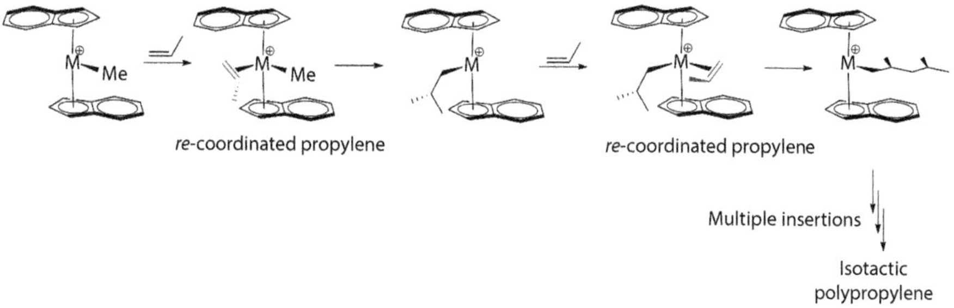

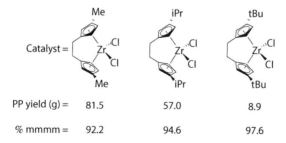

FIGURE 2.29   Indenyl zirconocene-catalyzed synthesis of isotactic polypropylene.

Catalyst:	Ti	Zr	Hf
Activity (kg.PP/mmol.M.h):	0.01	11.9	0.7
Mol. Wt.:	30000	22000	206000

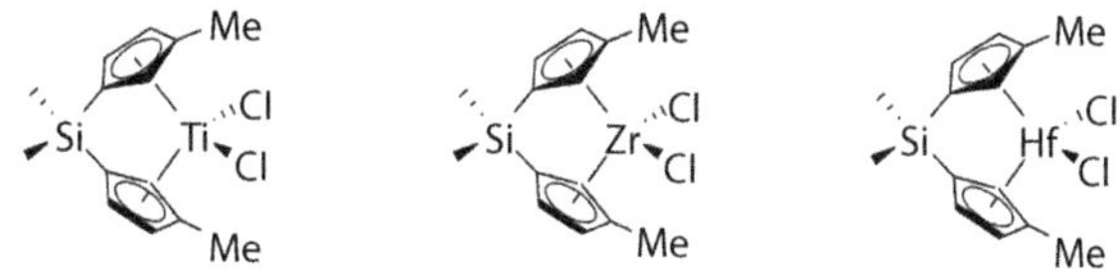

FIGURE 2.30   Effect of metal center on catalytic activity and molecular weight.

metallocenes can be attributed to catalyst deactivation due to the tendency of titanium to undergo reduction to Ti(III) at elevated temperature. Although the silyl bridged zirconocene exhibited the highest activity, the Hf analogs produced poly(propylene) with the highest molecular weight. Also the amount of 3,1-inserted propylene units in polypropylene produced from rac-Me$_2$Si(3-MeCp)$_2$TiCl$_2$/MAO is significantly more than the analogous Zr or Hf metallocenes, which show 2,1-inserted irregularities.

Studies by Collins et al. on the effect of C(3)-substituted biscyclopentadienyl ligands on the *i*PP isotacticity and molecular weight revealed that the methyl and isopropyl groups at C(3) position have positive impact on the activity as well as molecular weight.[70] However, the isotacticity was slightly better for 3-t-Bu substituent (Figure 2.31).

Catalyst =	Me / Me	iPr / iPr	tBu / tBu
PP yield (g) =	81.5	57.0	8.9
% mmmm =	92.2	94.6	97.6

FIGURE 2.31   Effect of ligand substitution on PP yield and isotacticity.

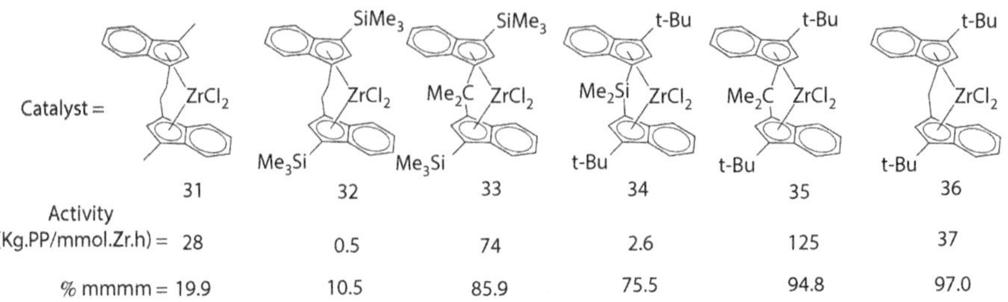

Catalyst =					
31	**32**	**33**	**34**	**35**	**36**
Activity (Kg.PP/mmol.Zr.h) = 28	0.5	74	2.6	125	37
% mmmm = 19.9	10.5	85.9	75.5	94.8	97.0

FIGURE 2.32 Influence of indene substitution on the polymerization of propylene and isotacticity.

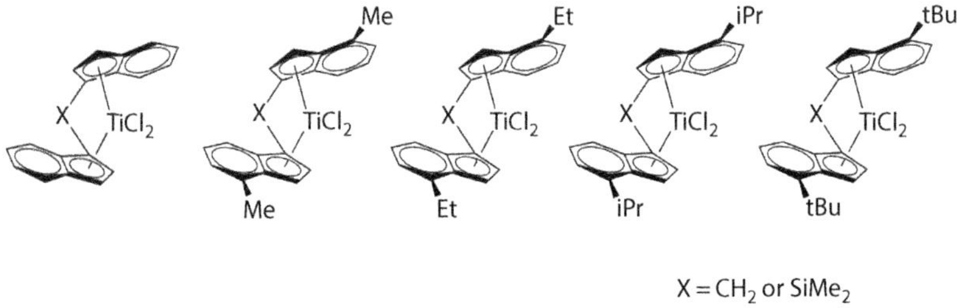

X = CH$_2$ or SiMe$_2$

FIGURE 2.33 Influence of indenyl substitution by computation.

Along the same line, Resconi[71,72] and Miyake[73] investigated the effect of 3-substituted indenyl zirconocenes with varying bridging groups. The aspecific nature and low molecular weight polypropylene generated by both [rac-C$_2$H$_4$(3-Me-1-Ind)$_2$ZrCl$_2$] and [rac-Me$_2$Si(3-Me-1-Ind)$_2$ZrCl$_2$] were reported by Britzinger[74] and Ewen.[75] The activity, molecular weight of polypropylene, and isoselectivity reach maximum when the *tert*-butyl group is present at 3-position in the CMe$_2$-bridged indenyl zirconocenes (Figure 2.32, **35**). The same is not true for the SiMe$_2$-bridged 3-*tert*-butyl-substituted indenyl zirconocene, which is almost 50 times less active than CMe$_2$-bridged counterpart. This observation indicates that the single C-bridge plays its part too, imparting a high rigidity to the molecule. Secondly, comparatively large bite angle in the C-bridged zirconocenes might be responsible for higher activity, which is also the case with the CH$_2$-bridged 3-*tert*-butyl-substituted indenyl zirconocene **36**. It must be noted that the situation is different for unsubstituted indenyl metallocenes where the molecular weight and isospecificity for the bridging group follow the trend CH$_2$ < CH$_2$CH$_2$ < (CH$_3$)$_3$C < Si(CH$_3$)$_2$. Computational studies by Rappé and Morokuma on the effect of increasing the steric bulk of 4- and 4'-substituent from hydrogen to methyl, ethyl, isopropyl, and *tert*-butyl of ethyl and silyl-bridged indenyl zirconocenes, [C$_2$H$_4$(1-Ind)$_2$[76] and H$_2$Si(1-Ind)$_2$][77] predicted that steric of methyl and isopropyl group would be most ideal to increase isotacticity of polypropylene (Figure 2.33). However, increasing steric size beyond this has a negative impact on isotacticity and may also lead to more syndiotactic defects as compared to 4- and 4'-methyl-substituent catalyst.

2.6.1.2 C_s Symmetric Metallocenes and Syndiotactic Polypropylene

Metallocenes that possess only the plane of symmetry are classified as C_s symmetric metallocenes. Because the two ligands on either side of metal are not same, there exists only a plane of symmetry passing through the center of the molecule making the two sites on metal enantiotopic, and each site would prefer to coordinate monomer with opposite enantioface. This type of coordination and subsequent insertion results in the formation of syndiotactic polypropylene (Figure 2.34).

Following are the examples of C_s symmetric metallocenes that produce highly syndiotactic polypropylene (Figure 2.35). Ewen and Razavi were the first to demonstrate that [Me$_2$C(Cp)(fluorenyl)ZrCl$_2$] polymerizes propylene to syndiotactic polypropylene in 1988.[78] The steric bulk of the fluorenyl ligand forces the approaching propylene monomer as well as the growing polymer chain away from it. Secondly, in order to diminish the repulsive steric interaction with the propagating polymer chain the propylene inserts preferably with its methyl pendant anti to the polymer chain resulting in insertion of opposite propylene enantioface alternately. A consequence of this type of insertion with alternate enantiofaces is generation of syndiotactic polypropylene (Figure 2.36).

The effect of the metal center on the polymerization activity in case of C_s symmetric metallocenes follows similar trends as that of C_2-symmetric metallocenes. C_s symmetric zirconocenes are the most active and hafnocenes produce high molecular weight polypropylene. However, the major difference is that the titanocenes are the least active and least enantiospecific, producing completely stereoirregular polypropylenes than their Zr and Hf analogs (Figure 2.37). Structure–property correlation studies by Ewen[79] on [1,1′, 2,2′-(Me$_2$Si)$_2$(4-i-PrCp)(3,5-i-Pr$_2$Cp)MCl$_2$]/MAO revealed

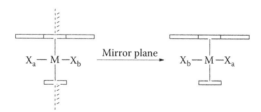

FIGURE 2.34 C_s symmetry around the metal.

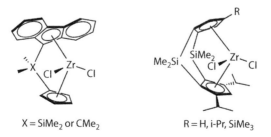

FIGURE 2.35 C_s symmetric metallocenes for syndiotactic polypropylene

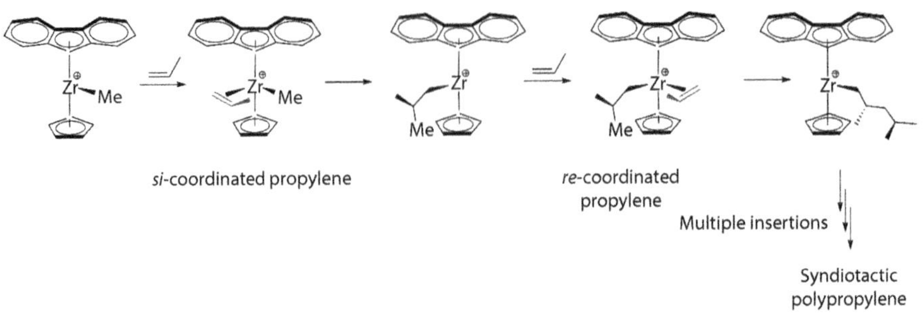

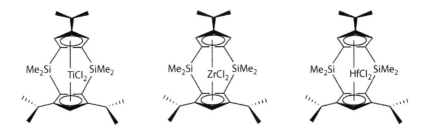

FIGURE 2.36 Fluorenyl-based zirconocene as a catalyst for syndiotactic polypropylene.

FIGURE 2.37 C_s symmetric group 4 metallocene complexes in propylene polymerization.

that the C_s titanocene is completely nonsyndioselective. Two possible reasons for the nonselective polymerization by Ti-complex could be: (a) the pendant methyl group on the propylene might experience nonbonded interactions with neighboring titanium centers or (b) the rapid interconversion of the catalytic sites.

Apart from the metal center, the type of bridging group present in C_s symmetric metallocenes has profound influence on the tacticity of the produced polypropylene. The effect can be realized by the fact that for unsubstituted *bridged*-(Cp)(9-Flu) zirconocenes, the molecular weight is dependent on the bridging group and follows the trend $Ph_2C > PhP > CH_2CH_2 > Me_2C \_ Me_2Si > Ph_2Si$, whereas the syndiotacticity increases in the order $Me_2Si < Ph_2Si < CH_2CH_2 < PhP < Ph_2C < Me_2C$. Also the C_s symmetric metallocenes produce highly syndiotactic polypropylene at lower temperature and high propylene concentration because of epimerization of growing chain by back skip to the other coordination site on the metal.

2.6.1.3 C_{2v} Symmetric Metallocenes and Atactic Polypropylene

Metallocene possessing a C_2 rotational axis and a plane of symmetry is referred as C_{2v} symmetric metallocenes (Figure 2.38). The presence of C_2 axis and a mirror plane makes the two coordination sites on metal homotopic but nonselective.

Atactic polypropylenes are produced by achiral C_{2v} symmetric unsubstituted (e.g., Cp_2ZrCl_2) or substituted (XCp_2ZrCl_2, $X(MeCp)_2ZrCl_2$; $X = Me_2Si$ or C_2H_4) unbridged metallocene dichlorides as well as achiral bridged C_{2v} symmetric metallocenes [(*meso*-Xt(1-Ind)$_2$ZrCl$_2$, *meso*-X(4,5,6,7-H4-1-Ind)$_2$ZrCl$_2$][70,80] and [XFlu$_2$ZrCl$_2$] (Figure 2.39).[81]

1
Rotation by 180°
2
X_a M X_b
X_b — M — X_a
2
1
1
Rotation by 180°
X_a M X_b
X_b — M — X_a
2

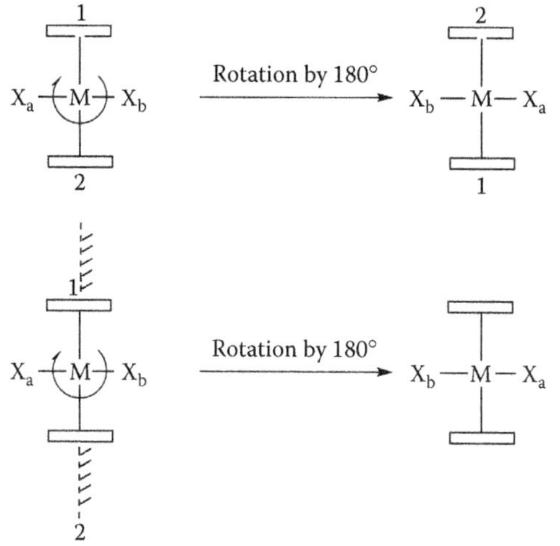

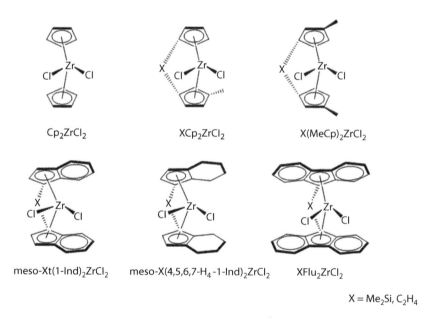

FIGURE 2.38 Demonstration of C_{2v} symmetry in metallocenes.

Cp_2ZrCl_2
XCp_2ZrCl_2
$X(MeCp)_2ZrCl_2$
meso-Xt(1-Ind)$_2$ZrCl$_2$
meso-X(4,5,6,7-H$_4$-1-Ind)$_2$ZrCl$_2$
$XFlu_2ZrCl_2$
X = Me_2Si, C_2H_4

FIGURE 2.39 C_{2v} symmetric metallocenes employed in the synthesis of atactic polypropylene.

In C_{2v} metallocenes the metal cannot influence the chiral induction from site control mechanism because of achiral nature. Moreover, the chain-end control mechanism is also not effective enough and hence neither the metal active site nor the chiral site generated by previous insertion of growing polymer chain can differentiate between *re*- and *si*-faces of the propylene. Such multiple insertions lead to atactic polypropylene as represented in Figure 2.40.

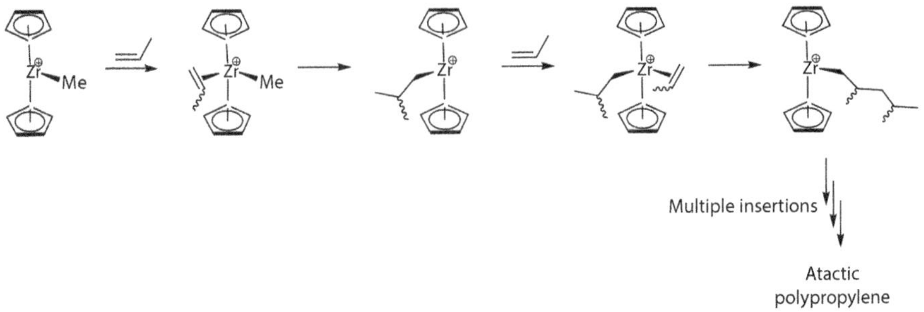

FIGURE 2.40 C_{2v} symmetric zirconocenes as a catalyst for atactic polypropylene.

However, there is also a tendency of these metallocene to form isotactically or syndiotactically enriched atactic polypropylene depending on temperature and propylene concentration.

2.6.1.4 C_1 Symmetric Metallocenes and Hemiisotactic Polypropylene

Metallocenes that lack symmetry elements fall into C_1 symmetry group (Figure 2.41). The two coordination positions on metal in C_1 symmetric metallocenes are nonequivalent and diastereotopic. This has a considerable influence on the stereoselectivity, polymer molecular weight, and activity depending on the substituents on the Cp ring and bridging group. These parameters decide the microstructure of polypropylene produced by C_1 metallocenes, which can vary from syndiotactic to hemiisotactic to isotactic. The asymmetric nature of the metallocene by appropriate substitution [e.g., $Me_2C(3\text{-}R\text{-}Cp)(Flu)MCl_2$] differentiates the two coordination position on metal, one closer to the substituent behaving as isospecific and other as aspecific. The alternation between the two different sites by migratory insertion causes hemiisotactic polymer.[10] However, the hemiisotacticity decreases as the size of the 3-substituent on Cp ring increases from methyl to *tert*-butyl group and also with increasing temperature of the reaction and decreasing monomer concentration.

The isoselectivity of the 3-Cp substituted [$Me_2C(3\text{-}R\text{-}Cp)(Flu)MCl_2$] increases with the increasing size of the substitutent at 3-position (Figure 2.41). Going from methyl to ethyl, substituent has minimal effect, the outcome is visible with 3-*i*Pr substituent

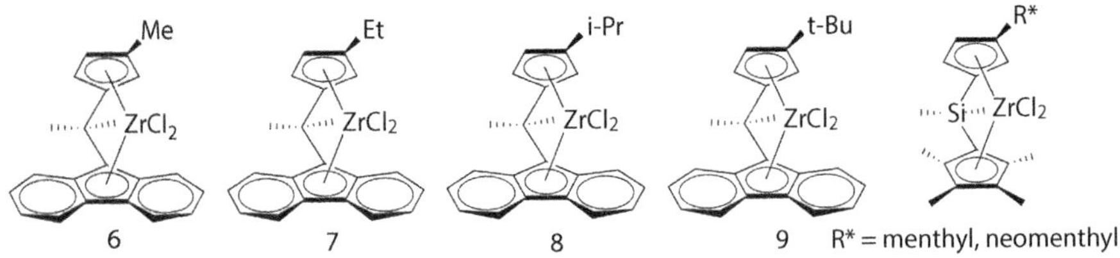

FIGURE 2.41 C_1 symmetric zirconocenes as a catalyst for hemiisotactic polypropylene synthesis.

($mmmm$) = 0.64 at 60°C and reaches highest with 3-$tert$-Bu, ($mmmm$) = 0.88 at 50°C and 0.95 at 30°C.[10] The isospecificity further increases when the CMe_2 bridge is replaced by $SiMe_2$. Also C_1 metallocenes of the type [$Me_2Si(Me_4Cp)$- (3-R*Cp) $ZrMe_2$/[Ph_3C][$B(C_6F_5)_4$], introduced by Marks and coworkers show highly isospecific propylene polymerization with ($mmmm$) = 0.95.[82,83]

2.6.2 ATACTIC-ISOTACTIC BLOCK POLYPROPYLENE

In 1995, Coates and Waymouth reported an exciting catalyst (Figure 2.42) that isomerizes between chiral, C_2 symmetric, and achiral $meso$ C_s symmetric geometries.[84–87] This switching between the two stable rotamers takes place by rotation through the C_2-axis facilitated due to the absence of bridge connecting the two indenyl ligands. The chiral C_2 symmetric rotamer produces isospecific polypropylene, whereas the $meso$-isomer produces aspecific polypropylene. The slow oscillation between the two conformations produces atactic-isotactic stereoblock polymers.

This kind of behavior would be quite expected as C_2 symmetric metallocenes and meso-C_s symmetric metallocenes individually tend to be isoselective and non-selective, respectively. Combining both these symmetries in a single metallocene by switch over between the two stable conformers in reaction medium would be challenging. However, if designed, such a catalyst would most likely result into alternating isotactic and atactic block polymers. Waymouth and Coates successfully designed such a catalyst that switches between two symmetries in a polymerization medium. Influence of metal is limited to a slight change in isotacticity, hafnium yields considerably less isotactic polymers than their zirconium analogs, and the molecular weight and activities are comparable for the two. It was also found that increasing the monomer concentration and decreasing the temperature have a positive impact on the molecular weight.

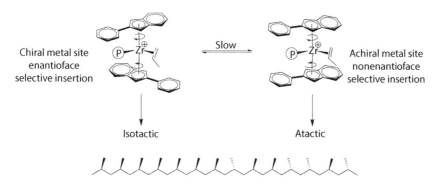

Chiral metal site enantioface selective insertion   Slow   Achiral metal site nonenantioface selective insertion

Isotactic   Atactic

FIGURE 2.42   The oscillating catalyst of Waymouth and Coates.

2.6.3 Industrial Application of Metallocene Polypropylene

As detailed in the above-mentioned section, in-depth understanding of the catalyst structure and resultant polymer properties provide an additional handle to synthesize desired grades of polypropylene. The academic world has provided various tools to tune the structure of the catalysts to selectively synthesize, only isotactic or only syndiotactic or fully atactic PP. This laboratory development has been taken forward by the polyolefin industry and commercial scale production of metallocene-grade PP (mPP) has been realized. mPP provides certain unique properties and these properties define the final application of mPP. Some of the most characteristic features of metallocene PP are listed as follows:

1. It provides excellent impact resistance.
2. It provides a better control over the molecular weight distribution.
3. It has a precisely controlled incorporation of an array of comonomers.
4. It has ultrahigh melt flow rate.
5. It can be sterilized in steam.
6. There are very low extractables.

The metallocene-derived polypropylene is commercially manufactured on large scale and sold under various trade names; two such trade names and their manufacturers are listed as follows:

1. Achieve™: ExxonMobil
2. Metocene™: Lyondellbasell

The market share of mPP is growing day by day and it finds applications in various sectors as indicated below:

1. Medical packaging
2. Lab wares
3. Healthcare applications such as medical gowns and covers, and so on
4. Hygiene products such as diapers, wet wipes and disposals
5. Housewares such as disposable items, and so on
6. Melt-blown fibers

2.7 POSTMETALLOCENES, ADVANTAGES, AND APPLICATIONS

Metal-catalyzed polymerization has been a very fertile area of research both academically and industrially and the search for new catalyst continues past metallocenes. This era of olefin polymerization catalysts after the metallocenes is generally described as *postmetallocenes*. The thrust to produce polyolefins with various structural features, the desire to serve a new market, and an access to various combinations of olefinic monomers are driving the research in this area, and new grades are being added. Over the years, various classes of postmetallocenes have been introduced. The following paragraphs will present

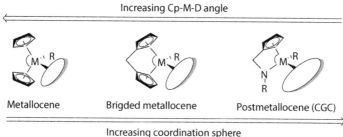

Increasing Cp-M-D angle

Metallocene   Brigded metallocene   Postmetallocene (CGC)

Increasing coordination sphere

FIGURE 2.43   Comparison of bite angle in different Cp-based catalytic systems.

an elaborate discussion on the different classes of postmetallocenes and their performance in polymerization.

Postmetallocenes are family of transition metal complexes consisting of one η^5-coordinated cyclopentadienyl ring, which is covalently bonded to an amido-group through a bridging spacer. The bridging spacer could be an $-SiMe_2$ or $-CMe_2$ group leading to a strained coordination complex as depicted in Figure 2.43. The small Cp-M-D bite angle ($\sim107°$) leads to a large open space around the metal center, which is believed to be responsible for the high activities displayed by this class of postmetal-locenes. Another approach that has been successfully employed in a new ligand design is based on the isolobal analogy introduced by Hoffman.[88] The orbital symmetry theories suggest that when two systems are electronically isolated, their shapes, symmetry, and energies of the frontier orbitals are similar. Hence, the correct combination of these parameters might allow developing powerful ligands for olefin polymerization. On the basis of these findings, the following steric and electronic parameters have been considered as an important factor for postmetallocene catalyst design.

(1) d^0, 14-electron metal system, (2) highly unsaturated metal centre, (3) available vacant orbital sites *cis* to alkyl ligands, and (4) electronically deficient metal centre.

Thus, ligands and subsequent catalyst designed based on the above-mentioned criteria may offer an excellent opportunity to improve and control properties of the resultant polymer. The number of complexes prepared using the symmetry rules of Hoffmann is vast and beyond the scope of this book. Therefore, only scientifically challenging and industrially relevant postmetallocene systems that are employed in olefin polymerization are discussed in depth. The known postmetallocene catalysts for olefin polymerization can be roughly classified into: (a) monocyclopentadiene (CGC), (b) bis(imino), (c) bis(imino)pyridyl, (d) bis(phenoxy-imino), (e) phosphine-sulfonate, (f) phosphinimide, (g) imino-amido, and (h) diamido complexes. The transition metal complexes derived from these ligand families have been developed in the last four decades, which hold great potential for practical applications. The early 1990s witnessed flood of literature in the development of a new type of homo-geneous single-site catalyst for olefin polymerization, including the development of postmetallocene catalyst. During this period, a significant attention was paid to com-plexes of group IV metals with one cyclopentadienyl ligand and other ligand con-taining heteroatom donor. The metal complexes derived from one cyclopentadienyl ring and one heteroatom are often referred to as *constrained geometry complexes*

(*CGC*) (The CGC catalyst development has been summarized in Braunschweig and Breitling[89]). Section 2.7.1 will discuss the significance of CGC in olefin polymerization. A path-breaking achievement was development of late transition metal complexes containing diimine type ligands by Brookhart in 1995. These postmetal-locene nickel and palladium complexes were active in olefin polymerization and revealed comparable activities to the metallocene catalysts.[90–92] The discovery that even late transition metal catalysts can polymerize ethylene accelerated the research on postmetallocene catalysts. In 1998, Brookhart[93] and Gibson[94] reported that iron or cobalt complexes possessing diimine-pyridine ligands exhibited very high ethylene polymerization activities. In the same year, Grubbs and coworker published a phenoxy imine ligand framework in conjunction with late transition metal and demonstrated that it can polymerize ethylene,[95] whereas Fujita and coworkers prepared early transition metal complexes using phenoxy imine ligand and tested them in ethylene polymerization.[96] The above-mentioned catalyst families display better performance compared to the group IV parent metallocene catalysts in olefin polymerization. This apart, few postmetallocene catalysts produce branched polyethylenes, interestingly, without using comonomers,[97] and some promote copolymerization of ethylene or α-olefins with polar vinyl monomers to produce functionalized polyolefins. Similar to metallocenes, postmetallocenes do require activation step and MAO is the most commonly used cocatalysts for the activation postmetallocenes. These cocatalysts are Lewis acids, which produce cationic-active metal centers with a high olefin affinity.

The following Sections (2.7.1 through 2.7.4) will present a chronological discussion on the above-mentioned (a–e) ligand classes, corresponding metal complexes, and their performance in olefin polymerization.

2.7.1   CONSTRAINED GEOMETRY COMPLEXES

Ligand-centered catalysts design for olefin polymerization resulted in the development of constrained geometry complexes (catalysts). In constrained geometry complex (CGC) systems, one of the cyclopentadienyl (Cp) rings is replaced by donor atom that electronically stabilizes the metal center and offers acute bite angle to leave considerable open space on the metal. Thus, short bridging groups (spacers) pull the donor atom (D) away from its ideal position, thereby reducing the Cp-M-D bite angle by a factor of 20°–30° than Cp-M-Cp in metallocenes (Figure 2.43).

Thus, the formed strained geometry leads to a sterically open metal center and provides an opportunity for incoming olefin to approach the metal center that is presumed to be the reason for their high comonomer incorporation and activities.[89] The term *constrained geometry complex* was originally coined by Stevens et al.[98] and the basic framework of a CGC complex is depicted in Figure 2.44.

The ligand framework consists of three main parts: (a) a Cp-anionic ligand part; (b) a briding group or spacer, which separates the η^5-Cp from the σ-donor heteroatom; and (c) a donor atom, which can form a sigma bond or which can donate a lone pair of electrons to the metal center. Most common cyclopentadienyl fragments used are cyclopentadiene or tetramethylcyclopentadiene and donors are typically neutral or monoanionic and are usually alkyl- or aryl-amido fragments. However,

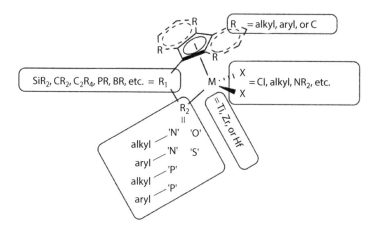

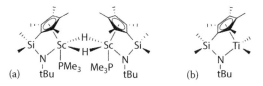

FIGURE 2.44 Schematic representation of CGCs catalyst.

FIGURE 2.45 First reported constrained geometry catalysts based on scandium (a) and titanium (b).

electronic and steric variation in Cp ring or/and donor in CGCs (Figure 2.44) can have different effects on the polymerization activity and olefin incorporation rates of these complexes. These stabilizing effects render high thermal strength to CGC's, which enable them to polymerize olefins at elevated temperatures than conventional metallocenes.[99] The journey of CGCs for olefin polymerization began as early as 1990s. The first report on constrained geometric catalysts was based on scandium-constrained geometry complex by Shapiro and Bercaw[100] followed by titanium-CGC by Okuda.[101] (Figure 2.45)

The features that make CGC as the ideal candidates for olefin (co)polymerization than the parent metallocenes (both bridged and unbridged) can be ascribed to the following characteristics:

1. Sterically less hindered open coordination pocket.
2. The acute bite angle offered by Cp-amido ligand.
3. Higher thermal stability than metallocenes.
4. Reduced chain-transfer rate due to sterically less demanding coordination sphere.

Over the years, CGC catalysts have been tailored to suite the stringent industrial requirements to deliver polymers with desired properties. Various copolymers of ethylene-propylene such as ethylene-propylene-diene rubber (EPDM), rubber, and so on have been prepared using CGC systems. Due to their high thermal stability, CGC catalysts have been successfully applied in solution (co)polymerization of ethylene with various higher α-olefins. The detailed synthesis of CGC catalysts, various modifications to the ligand backbone, supporting the CGC systems on solid supports, and their performance in olefin (co)polymerization have been summarized in recent reviews by Braunschweig,[89] Mecking,[102] and Figueroa.[103]

The commercial potential of the CGC system was recognized early on and Dow chemical company (with their INSITE™ technology) and Exxon chemical were the first to develop commercial processes based on CGC-catalyzed olefin polymerization. Since then the number of reports in open literature and also the number of patents on CGCs have increased with many folds. The specific properties such as narrow molecular weight distribution and homogeneous comonomer incorporation lead to copolymers with enhanced mechanical and processing features, which are otherwise inaccessible through the traditional ZN polymerization methods. Thus, the (co)polymers obtained from CGC systems find niche applications with value addition that is encouraging the polyolefin manufacturers to get into this market.

2.7.2 α-DIIMINE LIGANDS AND THEIR METAL COMPLEXES

An important event in the search for new catalysts in the area olefin polymerization occurred when Brookhart and co-workers in 1995 synthesized a new class of Ni(II) and Pd(II) complexes stabilized by bulky α-diimine ligands (Schiff bases) (Figure 2.46).[97] This finding represents one of the most important breakthroughs in the development of late transition metal catalysts for efficient olefin polymerization. In a ligand-centered catalyst design, Brookhart and coworkers discovered that α-diimine ligands stabilize Ni(II)/Pd(II) complexes. The ligands display a neutral coordination with the two imine-nitrogens donating their electron lone-pairs to the metal center and assuming *cis*-coordination. This neutral donor ligand was proven useful to stabilize the metal center by forming a five-membered chelate ring. In addition, the α-diimine ligands set also offers an opportunity to modify electronic and steric properties by introduction of substituents onto the aryl rings.[97] In a typical Brookhart diimine complex, the ligand backbone consists of a bidentate diimine ligand with the possibility of variation at R_1 or R_2 positions, or the imine nitrogen's donate their lone pair to the electrophilic metal center. The metal center displays square-planar geometry, with the *cis*-coordinating diimine ligand flanked by aryl substituents. The net charge on the metal complex is zero, whereas the oxidation state of the metal is +2, which is satisfied by the two anionic chlorides.

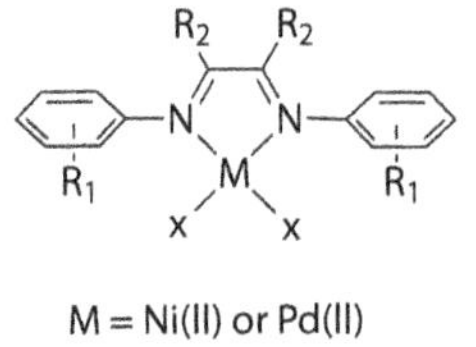

M = Ni(II) or Pd(II)
R_1 = Alkyl
R_2 = H, CH_3, 1,8-$C_{10}H_8$

FIGURE 2.46 Brookhart's α-diimine nickel and palladium complexes.

The above-mentioned bis(imino) nickel and palladium catalysts are active in the polymerization of ethylene, propylene,[104] styrene, norbornene, and cyclic olefins[105] and in copolymerization of ethylene with higher olefins and polar comonomers.* Under defined reaction conditions, the catalyst is able to produce high molecular weight polyolefins as well. A unique feature of the α-diimine catalysts is that they produce branched polyethylene from ethylene monomers without the use of any additional α-olefin comonomer (as is required for early metal catalysts). The special polymerization mechanism they adopt and their ability to produce branches lead to the production of amorphous PE or even elastomeric homopolymers. The significant reactivity and unique variation in polymer topology in such catalytic system are most likely due to the chain isomerization or *chain-walking* event during polymerization. In a chain-walking mechanism, the active metal center undergoes β-hydride elimination, which is followed by reinsertion, instead of new monomer addition. This sequence is repeated several times to produce methyl, ethyl, propyl, and alkyl branches after every β-hydride elimination, reinsertion step. The overall chain-walking mechanism is presented in Figure 2.47. Due to the β-hydride elimination reinsertion steps, the catalyst moves along the polymer chain (Figure 2.47) attracting the name *chain walking*. This chain-walking step can be repeated several times before a monomer is added to the chain or the chain is terminated. Control of *ortho* substituents on aromatic rings of (α-diimine)nickel(II) catalysts is important to achieve a desired activity. The *ortho* substituents have been reported to be located at axial site of the metal, and they retard chain transfer reactions, promote the chain walking, and accelerate the rate of migratory insertion. Therefore as steric bulkiness of the *ortho* substituents increases, molecular weights of polyethylenes, branching densities of polyethylenes, and polymerization activities increase. The structural variations of the α-diimine ligand-coupled with the conditions of polymerization (temperature and ethylene pressure) can be used to control branching and molecular weight of a polymer in a *predictable* manner. The unique behavior of the metal-diimine catalysts provided access to ethylene polymers, which could not be produced using ZN systems or even the metallocene catalysts.

The uniqueness of the resultant polyethylenes and the predictable control over the polymer topology attracted industrial interest. The pioneer of this method, Prof. Brookhart, established collaboration with DuPont's Central Research to evaluate the commercial potential of this process. In the course, a technology was further developed at the University of North Carolina at Chapel Hill and was licensed to Dupont. Later, Dupont commercialized this technology and the polyethylenes produced are sold by Dupont under the trade name *Versipol*. Under the licensing agreement in 2001, between Akzo Nobel, Albemarle Corporation, Grace Davison, and DuPont, the former companies also manufacture and sale Versipol catalysts for the olefin and polyethylene industries. The licenses include the manufacturing and sale of Versipol catalysts to licensed users for the production of polyethylenes with a density range of 0.86 to 0.97 grams per cubic centimeter, as well as for the oligomerization of ethylene into a variety of olefin products.

---

* Insertion copolymerization of ethylene with functional olefins will be separately discussed in Section 2.10. The diimine system will be referred again. As a representative example, please see Reference 87.

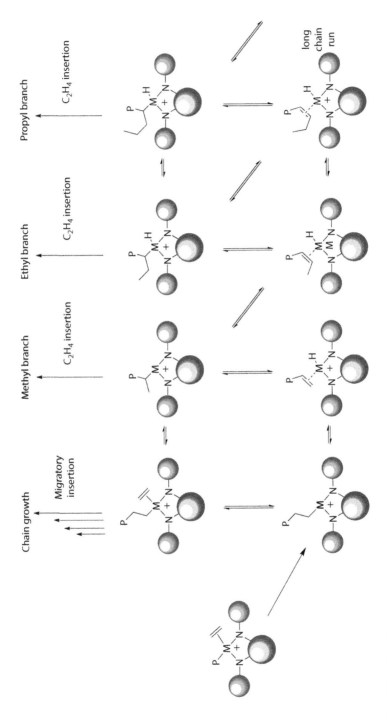

FIGURE 2.47   The chain-walking mechanism.

2.7.3 Bis(imino)pyridine (NNN Chelate Complex) Ligand Class and Their Metal Complexes

Bis(imino)pyridine ligand with iron and cobalt complexes were independently discovered in 1998 by Brookhart[106] and Gibson[107] (Figure 2.48). This bis(imino) pyridine ligand is tridentate U-shaped framework with two nitrogen donors connected by an odd number of sp^2-hybridized nitrogen atoms, which is in conjugation with the two terminal N-R donors. In the modern terminology, these can also be referred as *pincer ligands*.[108] They are on the borderline of hard and soft Lewis bases with well-established redox activity. The bis(imino)pyridine ligands are known to promote reversible transfer of 1–3 electrons between the chelate and to facilitate stabilization of five-membered chelate ring. Treatment of neutral bis(imino)pyridine ligand with metal precursors leads to pentacoordinate metal complex with pseudo-square-pyramidal geometry. The most notable feature of this ligand framework is the perpendicular arrangement of the aryl rings relative to the square plane as well as the *syn* conformation of the bulky sterically hindered *tert*-butyl groups in the complex. This catalyst family displays high ethylene polymerization activities and produces linear high-molecular weight polymers.[109] It is believed that the steric bulk provided by the substituents on the imino nitrogen plays a crucial role in stabilizing the metal complex. More importantly, the bulky substituents suppress β-hydride elimination, chain-transfer reaction, and also control the rates of propagation and chain termination. Thus, substituents appear to be very crucial and influence both activity and molecular weight of the polymer. After the initial ground work, the scope of the ligand design was expanded to various modifications in ligand framework and establishment of structure–activity relationship was attempted.

Although dedicated efforts were made to obtain high-temperature (which is applied in olefin polymerization on an industrial scale) stable cobalt and iron complexes, those were mostly futile. In addition, the metal complexes presented in the earlier discussion fall short of stringent industrial requirements. The commercialization efforts for these complexes were only partly successful and Sinopec in China operates a 500 ton pilot plant based on this catalyst family.[110]

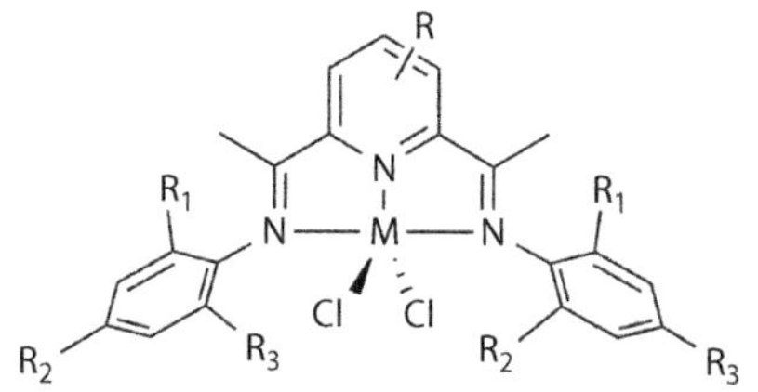

M = Fe, Co
R, R_1, R_2, R_3 = H, alkyl

FIGURE 2.48 Bis (imino) iron and cobalt complexes.

2.7.4 PHENOXY-IMINE [N,O] LIGANDS AND THEIR METAL COMPLEXES

Among the postmetallocene systems, phenoxy-imine ligand framework invoked a lot of interest and spurred intense research into the phenoxy-imine ligands and its associated chemistry. In the late 1990s, it was realized that salicylaldimine (phenoxy-imine) framework can serve as an excellent ligand for olefin insertion polymerization catalysts with early (Fujita at Mitsui Chemicals, 1997)[111] and late transition metals (Grubbs at Caltech, 1998). The design of these ligands was most likely inspired by a phosphine-based system discovered and commercialized by Keim[112] and Brookharts diimine system. Grubbs in 1998 reported neutral salicylaldiminato complexes of nickel for olefin polymerization (Figure 2.49).

The ligand design consists of a rigid aromatic backbone with a hydroxy substituent and bulky *ortho*-imine. The idea seems to revolve around providing enough rigidity to the ligand, at the same time some flexibility (the imine-substituent) should be built in. This bidentate ligand can form complexes with nickel that effectively prevent the chain-transfer reaction via β-H elimination, thus allowing the formation of high molecular weight polymers. The monoligated complex revealed square planar geometry around nickel center with the phenoxy group as an anionic donor and imine as neutral donor. The anionic and neutral ligands are usually situated *cis* to each other (O-Ni-N = 86.5°), and therefore the leaving group (PPh$_3$ in this case) and the phenyl substitutent are *cis* to each other. This arrangement provides an ideal *cis*-orientation for the olefin coordination and insertion. The notable characteristic feature in phenoxy-imine nickel complexes is that the Ni-C bond is preinstalled and there is no need for further alkylation by cocatalyst. These nickel complexes can be simply activated *via* dissociating the neutral ligand (i.e., PPh$_3$) from the nickel center, readily providing the active species. This catalyst in their activated form is neutral in nature, and therefore it displays even more enhanced tolerance toward polar functional groups than the α-diimine nickel or palladium complexes. Because this catalyst can tolerate polar functions, copolymerization of ethylene with functional olefin was successfully tested. Interestingly enough, the catalyst was found to be active in ethylene polymerization even in the presence of 1500 equivalents of water.[113] The use of this phenoxy-imine ligand was further extended to aqueous polymerization of ethylene.

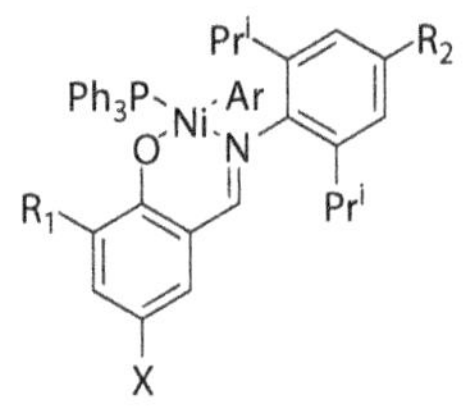

R$_1$ = H, Alk, Ar, CHO, NO$_2$
R$_2$ = H, i-Pr
X = H, OCH$_3$, Cl

FIGURE 2.49   Mono-phenoxy imine Ni complex.

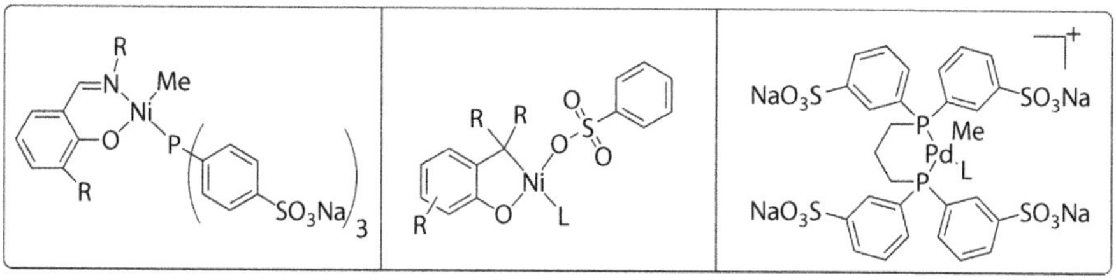

FIGURE 2.50 Representative cationic palladium complexes and neutral nickel complexes used in aqueous polymerization of ethylene.

Insertion polymerization of ethylene in water would primarily appear as contradiction for two reasons: (a) All those catalytic systems discussed in this chapter or for that matter the typical ZN catalysts are highly water sensitive and will not survive in the presence of water, and (b) polyethylene is a highly hydrophobic polymer and producing that in water would be a paramount challenge. These two arguments might leave an impression that aqueous-phase insertion polymerization of ethylene is unrealistic. However, in spite of the above-mentioned odds, consistent efforts by various research groups paved the way and a significant progress has been made in aqueous-phase/emulsion polymerization of ethylene using transition metal complexes. The discovery of late transition metal catalysts such as those presented in Figure 2.50 played a crucial role in realizing the aqueous-phase polymerization of ethylene. Mainly two classes of catalysts have been investigated: neutral nickel complexes and cation palladium complexes (Figure 2.50).

Among the above-mentioned general classes of catalyst, the phenoxy-imine-based nickel complexes have been very frequently investigated. The parent phenoxy-imine ligand developed by Grubbs was modified and water soluble versions were prepared [sulfonate (SO$_3$Na), halide containing groups were installed]. The resultant ligands displayed promising results in aqueous-phase polymerization of ethylene. Mecking and coworker in 2013 demonstrated the aqueous-phase polymerization of ethylene using phenoxy-imine-nickel complex and molecular weight as high as 420000 g/mol could be achieved. The thus-prepared PE revealed 90% crystallinity along with perfectly *ordered* nanocrystal of polyethylene.[114]

Polyethylene produced via aqueous route leads to polyethylene latexes, this could be yet another commercially important application of polyethylene to explore. Thus, the aqueous-phase insertion polymerization of ethylene has opened new avenues and future developments might lead to a totally different grade of polyethylene. Although there is a commercial potential, this topic is still in academic domain[115] and extra efforts are required to make it commercially attractive route. Apart from the commercial significance and academic development, a comprehensive picture of aqueous-phase ethylene polymerization is unclear. Future developments might shed light on the mechanism of this polymerization, role of ligand, kinetics of polymerization, and mechanism of crystallization.

Almost at the same time when Grubbs was developing the phenoxy-imine-based nickel complexes for olefin polymerization, Fujita at Mitsui chemicals started looking at early transition metal complexes with the same phenoxy-imine ligand systems

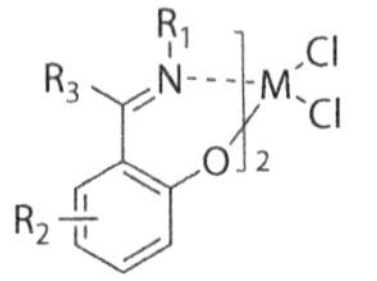

R$_1$ = H, Alk, Ar, cyclic
R$_2$ = H, Alk, Ar
R$_3$ = H, Alk, Ar
M = Ti, Zr, Hf

FIGURE 2.51   Bis(phenoxy) imine group 4 metal complexes (FI-catalyst).

(Figure 2.51). Fujita and coworkers investigated various phenoxy-imine ligands and established that early transition metals do form mononuclear metal complexes with phenoxy-imine ligands. These phenoxy-imine early transition-metal complexes were found to be active for olefin polymerization only in the presence of cocatalyst. A significant difference between these early transition metal complexes and the late transition metal complexes developed by Grubbs was that, two ligands complex with single metal center in the former and only one ligand complexes with the metal in the latter case. The molecular structure of one of the titanium complexes with phenoxy-imine ligand system revealed a distorted octahedral complex in which titanium is bound to two *cis*-coordinated phenoxy-imine ligands and the two chlorides assume *cis*-disposition.[116] It is most likely that during polymerization the two chlorides will be abstracted and will provide a *cis*-active site for olefin coordination and insertion. An in-depth investigation revealed that substituents on the *ortho*-position and electronic nature of the phenyl group on the imine nitrogen considerably affect the polymerization. The judicious exploration by Fujita and coworkers yielded a versatile class of olefin polymerization catalysts called FI catalysts, based on parent phenoxy-imine system.[117]

The following paragraph will present detailed analysis of ligand parameters and polymerization conditions that can affect the olefin polymerization process, either positively or otherwise. These group 4-bis(phenoxy-imine) complexes adopt an *Oh* (*Octahedral*) geometry around the metal center with two imine nitrogens, two phenolic oxygens, and two nonspectator ligands. Theoretically, five possible structural isomers arise from the coordination modes of bis(phenoxy-imine) ligands in an *Oh* configuration (Figure 2.52).

It is likely that all these five isomers are in equilibrium in solution state and some of them (Figure 2.52, D & E) might be inactive for olefin polymerization. Therefore, it is very important to analyze which of these five dominate during the complexation process, but it is equally tricky to distinguish a specific isomer responsible for ethylene polymerization. Out of these five isomers, three isomers (A-C) are considered as an active precatalyst in olefin polymerizations as they all have chloride ions in a *cis* position (Figure 2.52). Isomer A displays *cis*-nitrogens, whereas in isomer B the two oxygen atoms are *cis* to each other. In the third conformer, both the nitrogen and oxygen atoms occupy the *cis* orientation (C). The existence of all these isomers is experimentally verified by various techniques, including molecular structure

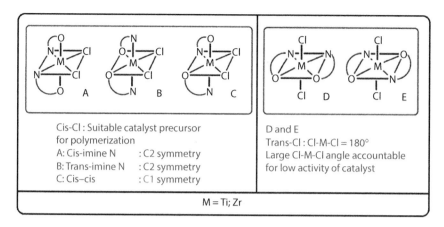

Cis-Cl : Suitable catalyst precursor
for polymerization
A: Cis-imine N   : C2 symmetry
B: Trans-imine N : C2 symmetry
C: Cis–cis   : C1 symmetry

D and E
Trans-Cl : Cl-M-Cl = 180°
Large Cl-M-Cl angle accountable
for low activity of catalyst

M = Ti; Zr

FIGURE 2.52 Five possible isomers of *Oh* complexes.

determination. Thus, the fluxionality of phenoxy-imine catalysts will be an important parameter in understanding their polymerization behavior. It is assumed that
the fluxional isomerization probably takes place via M-N bond dissociation due to
the labile nature of the imine-N donors as a function of temperature. Whenever the
conformation of the catalyst changes, one of the M-N coordination bonds is released,
after which rotation of the ligand can take place followed by the imine-donor recoor-
dination. This offers the option that the *cis* or *trans* isomers can be converted directly
into the *cis–cis* isomer, which is also the required intermediate isomer if ever the
exchange between *cis* and *trans* isomers has to take place (Figure 2.53). Thus, fluxional behavior plays a crucial role in olefin polymerization, and exchange between
the five isomers to desired *cis–cis* isomer provides access to a highly active olefin
polymerization catalyst.

Establishing catalyst structure, polymerization activity, and polymer properties is
necessary to improve the catalyst performance in the desired direction. Tremendous
work has been carried out on bis(phenoxy)imine group IV metal complexes to
develop the structure–activity–performance correlation. These investigations clearly

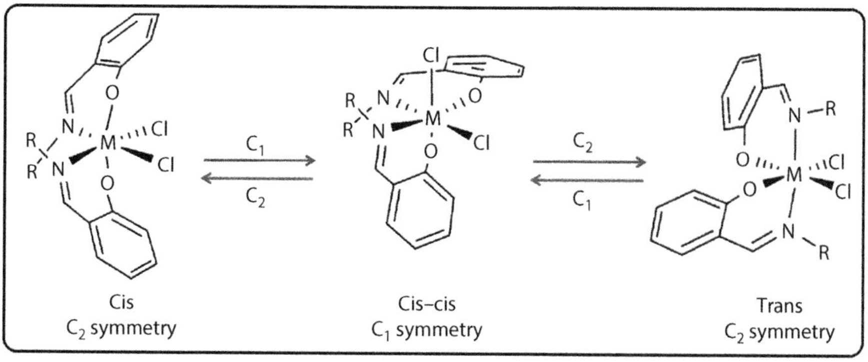

Cis
C_2 symmetry

Cis–cis
C_1 symmetry

Trans
C_2 symmetry

FIGURE 2.53 Fluxional behavior of FI-catalyst in solution.

delineate the effect of ligand modification (ligand environment, nature of metal, catalyst–cocatalyst ratio) on polymerization activity as well as polymer properties. Similarly, a significant progress has been made to understand the effect of polymerization reaction parameters and how these (time, temperature, pressure, etc.) can influence the polymer properties.[118]

To begin the structure–activity–property correlation, let us first understand the effect of metal center. Due to their higher Lewis acidity, group IV transition metals display higher propensity for enchaining ethylene molecules as compared to other (such as late transition metals) transition metals. The size of metal cations contributes significantly in building the molecular weight of the polymer. For example, due to increased ionic radii in Zr^{4+} (0.86 Å) as compared to Ti^{4+} (0.68 Å), Zr congener experiences less shielding and more steric openness as compared to Ti congener, though the two complexes possess the same configurations. Thus, it readily undergoes β-hydride transfer, which is considered as one of the most prominent causes in decreasing molecular weight (Table 2.9) of formed polymer. To summarize the effect of metal center, it is found that titanium complexes produce polymers with highest molecular weight, whereas zirconium and hafnium are at par, this follows the molecular weight order of Ti > Zr ~ Hf (Table 2.9).

In organometallic chemistry, ligands play a prominent role and tailor the reactivity, as well as selectivity of the metal catalyst. Appropriate ligand framework with correct balance of electronic and steric environment is very important in olefin polymerization as it greatly affects the rate of propagation (k_p) and rate of transfer reaction (k_{tr}) (k_p/k_{tr}) ratio. The phenoxy-imine-based catalyst systems exhibit increasing molecular weight with an increase in the steric bulk on the ligand (Table 2.10, Entries = 4–7, 8–11).[119] The increase in the molecular weight is probably due to the decrease in the β-hydride elimination rate as a result of the increased steric crowding around the metal center. Introduction of bulky group around the metal center seems to sterically block access to transition state that encourages β-hydride elimination.

TABLE 2.9
Effect of Size of Metal Cation on Molecular Weight and Molecular Weight Distribution

Catalyst Structure	Entry	Metal	Mn (10^3)	M_w/M_n
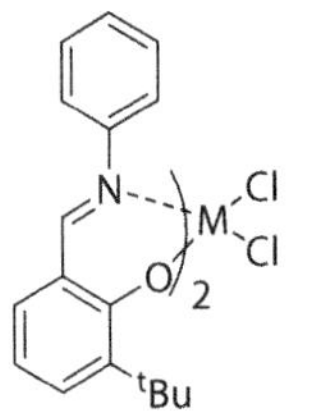	1	Ti	170	1.61
	2	Zr	4	1.81
	3	Hf	6	3.01

Conditions: Temperature: 25°C; pressure: 0.1 MPa; solvent: toluene (250 ml); catalyst: 20 μmol; cocatalyst [MAO]: 1.25 mmol (Al); and time: 5 min.

TABLE 2.10

Effect of Steric Bulk around Metal Center on the Molecular Weight and Molecular Weight Distribution[a]

Catalyst Structure	Entry	Metal	Substitution Pattern		M_w ($\times 10^3$)	M_v ($\times 10^4$)	M_w/M_n
			R_1	R_2		–	
	4	Ti	H	H	66	–	3.13
	5	Ti	Me	H	402	–	1.54
	6	Ti	^{t}Bu	H	1281	–	2.55
	7	Ti	SiMe$_3$	H	1105	–	2.51
	[b]8	Zr	^{t}Bu	H	–	0.9	Not determined
	[b]9	Zr	^{t}Bu	Me	–	32	2.31
	[b]10	Zr	^{t}Bu	i-Pr	–	113	2.61
	[b]11	Zr	^{t}Bu	t-bu	–	274	Not determined

Source: Furuyama, R. et al. *J. Mol. Catal. A: Chem.* 200, 31–42, 2003.

[a] Polymerization conditions: Toluene: 250 ml; complex: 5 mol.; DMAO: 1.25 mmol; ethylene: atmospheric pressure; polymerization time: 30 min; and polymerization temperature: 25°C.

[b] Conditions: Temperature: 25°C; ethylene pressure: 0.1 MPa; MAO (Al): 1.25 mmol; toluene: 250 mL; polymerization time: 10 min; and catalyst concentration (Entry 8,9,10 = 0.5 μmole; 11 = 5 μmole).

It is presumed that an appropriate steric bulk around the metal would suppress β-hydride elimination and would enhance the polymer molecular weight.

Apart from steric parameters, electronic factors also influence the reactivity of a metal center. Electron-donating substituents on the ligand can donate electron density to the metal making it less electrophilic, which results into lower polymerization activities, whereas electron-withdrawing substituents maintain electrophilic character of metal center, which can enhance the activity of the metal catalyst. Specifically, it has been observed in the case of phenoxy-imine titanium complexes that an increase in the electrophilicity of metal center enhances the rate of propagation (k_p), (Table 2.11, entries 12–17). Interestingly, installation of electron withdrawing groups at either R_1 and R_5 or R_1–R_5 (Table 2.11) position significantly suppresses the chain termination pathway, and under defined reaction condition the catalyst displays a living behavior.[120,121]

A unique feature of bis-phenoxy-imine titanium dichloride complexes is their fluxional behavior, with up to five different isomers under polymerization conditions. Literature precedence indicated that many octahedral bis(phenoxy-imine) group 4 metal complexes with C_2 symmetry were extremely active toward ethylene polymerization giving high molecular weight polyethylene. With an increase in the polymerization temperature, molecular weight of the polymer increased (Table 2.12, entry 18–22), but at the penalty of broad molecular weight distribution. The increased molecular weight distribution indicates loss of well-defined single-site catalyst to a multisite catalyst in solution due to inherent fluxional behavior of this catalytic system (entry 20–22). The ligand arrangement with differing symmetry might be also responsible for multimodal molecular weight distribution (entry 21, 22).

TABLE 2.11
Effect of Electronic Environment on Molecular Weight and Molecular Weight Distribution[121]

Catalyst Structure	Entry	Metal	Substitution					M_v ($\times 10^4$)
			R_1	R_2	R_3	R_4	R_5	
	12	Ti	H	H	H	H	H	32.6
	13	Ti	H	H	F	H	H	41.9
	14	Ti	H	F	H	F	H	62.3
	15	Ti	H	F	F	F	F	37.8
	16	Ti	H	H	CF_3	H	H	54.2
	17	Ti	H	CF_3	H	CF_3	H	136.5

Polymerization conditions: Toluene: 250 ml; MAO: 1.25 mmol; ethylene: atmospheric pressure; polymerization time: 5 min; and temperature: 25°C.

TABLE 2.12
Effect of Fluxional behavior on Molecular Weight and Molecular Weight Distribution

Catalyst Structure	Entry	Temperature (°C)	M_w (10^3)	M_w/M_n	Area Ratio of GPC Peaks
	18	0	5.3	2.03	100/0/0
	19	20	7.5	2.34	98/2/0
	20	25	23	8.64	94/5/1
	21	40	133.2	24.80	68/21/11
	22	75	136.7	15.89	40/39/21

M = Zr

Standard conditions: Ethylene:100 L/h; toluene: 250 mL; precatalyst: 0.02 μmol; MAO: 62,500 equiv.; polymerization time: 5 min.

In addition to the catalyst that tailors the polymer properties, the polymerization parameters and reaction conditions such as temperature, pressure, type of cocatalyst, and additives also play a crucial role in determining molecular weight and molecular weight distribution of a polymer. Temperature is considered to be one of the most crucial operational parameters for polymerization. It is well established that solubility of ethylene in the solvent decreases with increasing temperature hence the

TABLE 2.13

Effect of Polymerization Temperature on Molecular Weight and Molecular Weight Distribution

Catalyst Structure	Entry	Substituent (R)	Temperature (°C)	M_w (10^6)	M_w/M_n
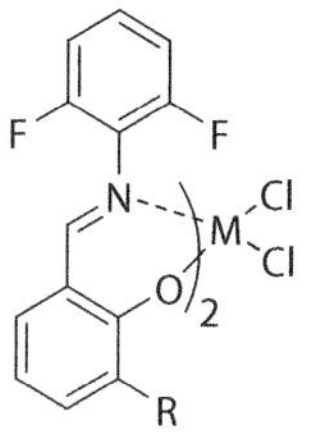	23	Cl	10	3.98	1.13
	24	Cl	20	4.20	1.11
	25	Cl	30	4.11	1.10
	26	Cl	40	3.61	1.17
	27	Cl	50	1.75	2.34
	28	Br	10	4.17	1.08
	29	Br	20	4.07	1.14
	30	Br	30	4.27	1.05
	32	Br	40	2.58	1.77
	33	Br	50	1.86	2.07
	34	I	10	3.88	1.15
	35	I	20	4.21	1.08
	36	I	30	3.59	1.16
	37	I	40	3.47	1.27
	38	I	50	3.45	1.26

Conditions: Catalyst concentration: 2 μmol; MAO: 1.25 mmol; ethylene: 1 atm; and toluene: 250mL.

monomer concentration decreases at higher temperatures hence reducing the propagation rate.[122] Increasing temperature enhances both the rate of propagation as well as the probability of chain-transfer reactions. Usually, chain-transfer reactions have higher activation barriers than insertion reactions, and a change in polymerization temperature strongly affects the rate of transfer reactions. This was directly evidenced by Praserthdam and coworkers and a significant effect of temperature on building up ethylene molecular weight was observed (Table 2.13).[123] Based on this study, the authors suggested that an initial increase in the temperature from 10°C to 30°C increases the molecular weight of polymer with narrow molecular weight distribution (Table 2.13, entry 23–25, 28–30, 34–35). The initial observation can be attributed to an increase in the rate of propagation. But, further increase in the temperature leads to enhanced chain-transfer reactions and at the same time the solubility of ethylene in the solvent decreases. A combined effect of the above-mentioned two parameters is reduced molecular weight (Table 2.13, entry 27, 33, 38) and broadening of the molecular weight distribution. In addition to this, another important aspect is that rate of catalysts decomposition increases with increasing temperature, leading to reduced activity.

Polymerization time is an indicative tool to map lifetime of employed catalyst under given reaction conditions. However, time becomes an important parameter in defining living nature of catalyst. Praserthdam and coworkers investigated the effect of time on molecular weight and molecular weight distribution (Table 2.14).[122,123] It was noted that the molar mass of the polymer increases with increasing polymerization

TABLE 2.14

Effect of Polymerization Time on Molecular Weights and Molecular Weight Distribution

Catalyst Structure	Entry	Time (min)	M_w (10^6)	M_w/M_n
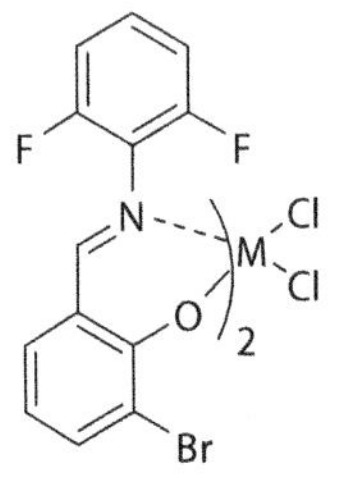	39	10	3.56	1.37
	40	20	3.60	1.23
	41	30	3.69	1.19
	42	40	3.80	1.15
	43	50	3.88	1.13
	44	60	4.03	1.10

(Catalyst structure: bis-(phenoxy-imine) complex with 2,6-difluorophenyl N-substituent and Br-substituted phenoxy; M = Ti)

Conditions: Complex: 2 µmol; MAO: 1.65 mmol; [Al]/[Ti] = 1825; ethylene: 1 atm; toluene: 500 mL; and temperature: 30°C.

time. Up to 60 min, a linear dependence between the polymerization time and the molar mass is observed, which indicates the living nature of the catalyst employed (entry 39–44).

Another influential parameter is monomer pressure. In the case of bis-(phenoxy-imine) titanium catalysts, no significant effect of increasing monomer pressure on molecular weight and molecular weight distribution (Table 2.15) was observed. However, it is well established that the increase in pressure generally increases activity because of high monomer concentration in close proximity of active metal center.

TABLE 2.15

Effect of Ethylene Pressure on Molecular Weights and Molecular Weight Distribution

Catalyst Structure	Entry	Ethylene (bar)	M_w (10^6)	M_w/M_n
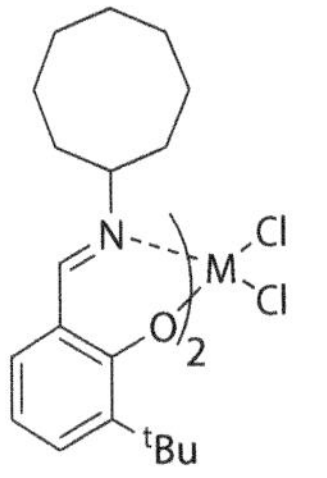	45	1	2.06	1.61
	46	2	1.94	1.65
	47	3	1.91	1.63
	48	4	2.13	1.65
	49	5	2.00	1.68

(Catalyst structure: bis-(phenoxy-imine) complex with cyclooctyl N-substituent and ^{t}Bu-substituted phenoxy; M = Ti)

Condition: Toluene: 750 ml; time: 20 min; temperature: 40°C; catalyst: 3 mol; and MAO: 4.35 mmol.

TABLE 2.16

Effect of Nature of Solvent on Activity and Molecular Weight

Catalyst Structure	Entry	Solvent	Temperature (°C)	Activity (kg PE mol^{-1} cat^{-1} h^{-1})	M_v (10^{-4})
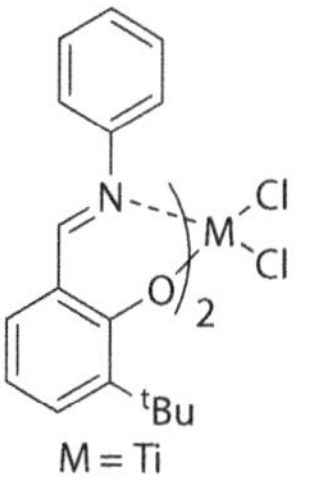	50	Toluene	50	4150	31.5
	51	Toluene	75	4010	32.7
	52	Heptane	50	3260	28.8
	53	Heptane	75	3150	29.1

M = Ti

Polymerization conditions: Solvent: 250 cm^3; catalyst concentration: 0.005 mmol; MAO: 1.25 mmol; and ethylene = 100 l.N h^{-1}.

Unusual solvent effects have been observed in organic transformations. Along the same line, the nature of solvent molecules influences polymerization reaction in many ways. Solubility of metal catalyst or monomer completely depends on the type of solvent. Most importantly, active metal catalysts are cationic in nature, hence polarity (or dielectric constant of the solvent) of the solvent will have considerable impact on solvation and ion separation of the catalyst. Fujita and coworkers investigated ethylene polymerization in different solvents and important results from this study are presented in Table 2.16. No significant change in molecular weight was observed upon changing the solvent. However, activity was found to increase when toluene was used as solvent. This is most likely due to higher polarity of toluene compared to heptane, which enhances effective separation of ion pair and an increase in the Lewis acidity of metal leading to an increased rate of propagation.

Nature of cocatalyst was considered as one of the critical operational factors as it has a decisive effect on the rate of propagation as well as the rate of transfer/termination. In principle, suitable cocatalysts are required for converting metal complexes into highly active cationic metal complex for ethylene polymerization. Such cationic species stabilized by highly noncoordinating anions can considerably influence the growth of polymer chain. Investigations by Zhu and coworkers reveal a significant effect of aluminum-based cocatalyst on the molecular weight of the resultant polymer (Table 2.17).[124]

The authors found increased molecular weight in case of MAO as compared to other alkyl aluminum (entry 54–57). The high molecular weights observed with MAO could be due to the formation of stable cationic species, which maintained electrophilicity of the metal cation and suppressed the growing PE chain transfer to the aluminum under given reaction conditions as compared to other alkyl aluminum. In search of aluminum-free cocatalyst, Fujita and coworkers observed that boron-based cocatalysts have higher propensity to produce high molecular weight polyethylene as compared to MAO (Table 2.18). The parameters that play a crucial role are: (a) stronger Lewis acidity of the cocatalyst, (b) larger steric bulk, and (c) weaker

TABLE 2.17
Effect of Cocatalyst on the Molecular Weight of Polymer

Catalyst Structure	Entry	Cocatalyst*	Al/M Ratio	M_w (10^4)
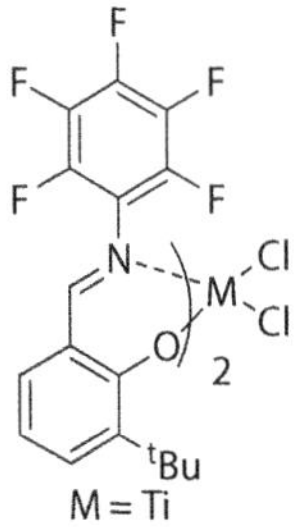	54	TEA	300	–
	55	DIBA	300	–
	56	DEAC	1000	1.93
	57	MAO	1000	7.28

Conditions: Catalyst concentration: 3×10^{-5} mol-Lit^{-1}; ethylene: 0.15 MPa; toluene: 250 mL; and time: 10 min.
*TEA: Triethylaluminum; DIBA: Di-iso-butyl-aluminum; DEAC: Di-ethyl-aluminum-chloride; and MAO: Methylaluminoxane.

TABLE 2.18
In Catalyst Structure M-C → M-Cl.

Catalyst Structure	Entry	Co-catalyst	M_v (10^4)
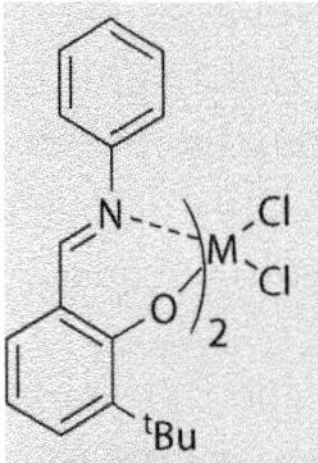	58	MAO	51
	59	$^iBu_3Al/Ph_3CB(C_6F_5)_4$	481

Polymerization conditions: Toluene: 250 mL; catalyst: 0.005 mmol; MAO: 1.25 mmol; $Ph_3CB(C_6F_5)_4$: 0.006 mmol; $^iBu_3Al$: 0.25 mmol; ethylene: 0.1 MPa (100 L/h^{-1}); temperature: 25°C; and time: 5 min.

reduction ability. Such weakly coordinating anions stabilize the resulting cationic species and thus reduce the chain transfer to aluminium, which is considered to be one of the most influential parameters that reduce the molecular weight. Thus, the boron-based cocatalyst was found to produce high molecular weight polyethylene compared to MAO.

Appropriate ratio of cocatalyst to catalyst has a remarkable effect in achieving high molecular weight polymers. In general, large excess of MAO is often required to achieve high activities in metallocenes and postmetallocene-based catalytic

TABLE 2.19
Effect of Al/M Ratio on Molecular Weight

Catalyst Structure	Entry	Catalyst Conc. (μmole)	Al: M	M_w (10^6)	M_w/M_n
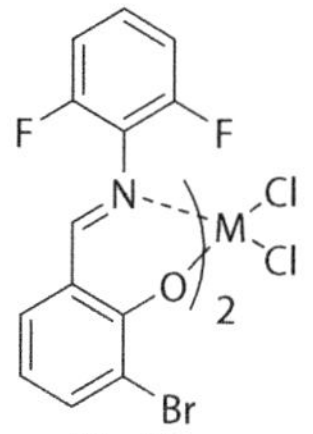	60	6	417	3.57	1.24
	61	5	500	3.68	1.18
	62	4	625	3.72	1.16
	63	3	833	3.90	1.13
	64	2	1250	3.95	1.11
	65	1	2500	4.01	1.10

M = Ti

Conditions: MAO: 2.50 mmol; ethylene: 1 atm; toluene: 500 mL; time: 20 min, and temperature: 30°C.

systems. This large excess of MAO also influences the polarity of polymerization medium, which might also affect the polymerization rate. MAO can be also considered as a potential chain-transfer agent. With increasing Al/M ratio, an increase in the polymer molecular weight was observed (Table 2.19). The increased Al/M ratio not only converts catalyst to the active cationic species but it also increases polarity of the medium, which assists in better ion separation and stabilizes the ion pair. However, in case of metallocene, further increase in this ratio had an adverse effect on molecular weight. This decrease in the molar mass upon increasing the Al/TM ratio can be explained by chain transfer to aluminum due to the increasing concentration of trimethylaluminum invariably present in MAO.

In 2001, using bis(phenoxy-imine) Ti complexes with pentafluoro substitution on *N*-aryl group, Fujita and coworkers polymerized ethylene at 25°C to produce linear PE with a high molecular weight and narrow molecular weight distribution (M_n = 412,000 g/mol, M_w/M_n = 1.13).[116] Furthermore, polymerizations at 25, 50, and 75°C (Table 2.20) exhibited a linear increase in M_n with reaction time although, at 75°C, molecular weight distributions were found to be broadened with longer reaction times (M_w/M_n = 2.05 at 15 min). Living behavior was further exemplified through the synthesis of PE and poly(E-co-P) containing di- and triblock copolymers.[125]

The authors claimed the existence of weak attractive [C–H–F–C] interactions between the fluorine atoms on the ligands and the β-hydrogen atom of the growing polymer chain. These secondary interactions in turn prevent termination and impart living character to the polymerization process. Thus, the catalyst family is able to produce polyethylene with extremely high molecular weight (UHMWPE), which will be discussed in Sections 2.8 and 2.9.

2.8 ULTRAHIGH MOLECULAR WEIGHT POLYETHYLENE

An almost linear polyethylene with a molar mass of more than one million g/mol is usually referred as UHMWPE.[126] Due to the ultrahigh molecular weight, it displays outstanding physical and mechanical properties. However, synthesis of such high

TABLE 2.20

Ethylene Polymerization with Ti-Complex/MAO Using Diluted Ethylene and Various Polymerization Temperatures

		Temperature (°C)					
		25		50		75	
Catalyst Structure	Time (min)	M_n (10^3)	M_w/M_n	M_n (10^3)	M_w/M_n	M_n (10^3)	M_w/M_n
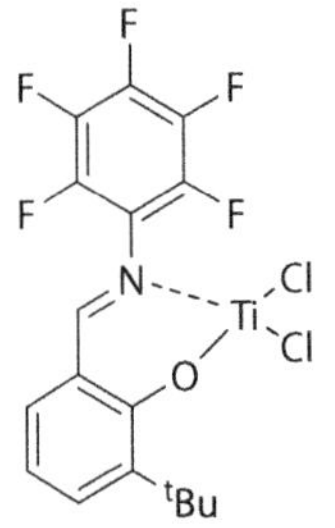	5	56	1.07	54	1.06	82	1.30
	10	103	1.09	95	1.14	135	1.65
	15	144	1.13	137	1.19	160	2.05

Conditions: Catalyst: 1.0 μmol; MAO: 1.25 mmol; ethylene: 1 atm; ethylene/N2 feed 2 L/50 L/h (25, 50°C), 5 L/50 L/h (75°C); toluene: 250 mL; M_n and M_w/M_n are determined by GPC using polyethylene standards.

molecular weight PE is highly challenging due to the following reasons: (a) To achieve molecular weight more than 1 million, the β-hydride elimination and chain transfer (either to monomer or to cocatalyst) should be suppressed, (b) the catalyst should be robust and avoid deactivation/decomposition, (c) the rate of insertion/propagation should be faster than the rate of termination and transfer, and (d) the metal center should be highly electron deficient. A catalyst design targeted for UHMWPE production should incorporate the above-mentioned features in the catalyst, apart from the conventional catalyst requirements for insertion polymerization of olefin. Suitably modified classical ZN catalyst system is capable of producing UHMWPE (Table 2.21, entry 1) with a molecular weight close to 10 million g/mol.[127] Similarly, a homogenous bulky zirconocene produces polyethylene with about 1.5 million molecular weight (Table 2.21, entry 2).[128] There is a huge scope to increase the molecular weight by tuning the catalyst structure or by supporting it on a suitable support. Very high molecular weight polyolefins could be obtained by designing

TABLE 2.21

Catalytic Systems for the Synthesis of UHMWPE

Entry	Catalyst	Cocatalyst	M_w (g/mol) × 10^{-6}
1	$MgCl_2$-supported $TiCl_3$	TiBAl	10.5
2	$Cp*ZrCl_2$	MAO	1.5
3	FI	MAO	11.3

a catalytic system that displays characteristics of *living polymerization* behavior. Fujita and coworkers reported a pentafluoro-substituted phenoxy-imine system that catalyzed living polymerization of ethylene with significantly high molecular weight.[117,120,129] Apart from these systems, several research groups around the globe have attempted the synthesis of UHMWPE.[130–133] The significant academic developments in the area of UHMWPE have been summarized in a review.[127] Thus, the ZN or metallocene or postmetallocene catalysts are capable of producing UHMWPE under suitable conditions. Cossee–Arlman type insertion mechanism is followed in all these systems leading to a very high molecular weight polymer.

The characteristic features of UHMWPE are its chemical inertness, lubricity, impact resistance, and abrasion resistance. UHMWPE is made up of extremely long chains of polyethylene, which all align in the same direction. It derives its strength largely from the length of each individual molecule (chain). Van der Waals bonds between the molecules are relatively weak for each atom to overlap between the molecules, but because the molecules are very long, large overlaps can exist, adding up to the ability to carry larger shear forces from molecule to molecule. Each chain is bonded to the others with many Van der Waals bonds due to which the intermolecular strength is high. In this way, large tensile loads are not limited as much by the comparative weakness of each Van der Waals bond. The simple structure of the molecule also gives rise to surface and chemical properties that are rare in high-performance polymers. For example, the polar groups in most polymers easily bond to water. Because olefins do not carry any such groups, UHMWPE does not absorb water readily nor gets wet easily, which makes bonding to other substrates difficult. For the same reasons, skin does not interact with it strongly, making the UHMWPE fiber surface feel slippery. In a similar manner, aromatic polymers are often susceptible to aromatic solvents due to π-stacking interactions, an effect for which aliphatic polymer such as UHMWPE is immune to. Since UHMWPE does not contain functional groups (such as esters, amides, hydroxyl) that are susceptible to attack by aggressive agents, it is very resistant to water, moisture, most chemicals, UV radiation, and microorganisms. Under tensile load, UHMWPE will deform continuously as long as the stress is present—an effect called *creep*. Due to its light weight, high abrasion resistance, and biocompatibility, it is used for demanding applications which are discussed later.

The above-mentioned special properties offered by UHMWPE attracted a significant industrial attention, and commercialization of UHMWPE was undertaken.[134] Dow (Figure 2.54) designed a catalyst based on Ti(IV) for the synthesis of UHMWPE and a competing catalyst based on Ti(III) was developed by DSM.[135,136] More recently, catalysts based on Cr(III)[137] and specially designed, highly active Fujita catalyst (by Mitsui Chemicals)[138] have also been employed in the synthesis of UHMWPE (see Figure 2.54).[139]

Due to the attractive material properties that UHMWPE offers, it is being manufactured by multinationals such as DSM, Dow Chemicals, Honeywell, Braskem, Mitsui, and Ticona, to name a few (Table 2.22). UHMWPE serves a completely different market sector than the other (HDPE, LDPE, LLDPE, etc.) polyethylenes. UHMWPE fibers are used in manufacturing body armors, in particular, personal armor and on occasion as vehicle armor, cut-resistant gloves, bow strings, climbing

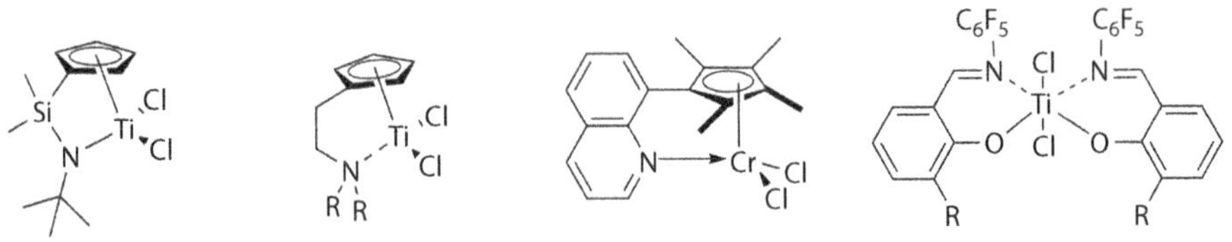

Ti(IV) Dow   Ti(III) Lovacat DSM   Cr(III) Enders   FI Catalyst

FIGURE 2.54   Representative catalytic systems for the production of UHMWPE.

TABLE 2.22
Manufacturers of Commercial UHMWPE with Respective Trade Names

Entry	Manufacturer	Trade Name	Process
1	DSM	Dyneema	Gel spinning
2	Honeywell	Spectra	Gel spinning
3	Celanese	GUR® UHMWPE	Pressure less sintering etc.
4	Braskem	UTEC	Ram extrusion etc.
5	Mitsui	MIPELON	Not available
6	Ticona	GUR	Not available
7	LyondellBasell	Lupolen UHM 5000	Ram extrusion etc.
8	Quadrant	TIVAR	Ram extrusion etc.

equipment, fishing line, spear lines for spear guns, high-performance sails, suspension lines on sport parachutes and paragliders, rigging in yachting, kites, and kites lines for kites sports. Spun UHMWPE fibers excel as fishing line, as they have less stretch, are more abrasion resistant and are thinner than traditional monofilament line. It is used in skis and snowboards, often in combination with carbon fiber, reinforcing the fiberglass composite material, adding stiffness, and improving its flex characteristics. The UHMWPE is often used as the base layer, which contacts the snow, and includes abrasives to absorb and retain wax. High-performance lines (such as backstays) for sailing and parasailing are made of UHMWPE, due to their low stretch, high strength, and light weight. UHMWPE has over 40 years of clinical history as a successful biomaterial for use in hip, knee, and (since the 1980s) spine implants.[140] Thus, UHMWPE is an engineering PE for specialty applications, unlike the commodity LDPE, HDPE, and LLDPE encountered in earlier sections. The Global market demand for UHMWPE has already exceeded 180 kilo tons in 2014 and it is expected to reach up to 410 kilo tons by 2022.*

However, current industrial processing of UHMWPE is a tedious task with adverse environmental effects. Due to its high molecular weight and high entanglement density, the polymer cannot be processed *via* conventional melt-processing means and more than 95 wt% of solvent is required for solution-based gel-spinning

---

* For a market survey on UHMWPE, see: https://www.grandviewresearch.com/press-release/global-uhmwpe-market

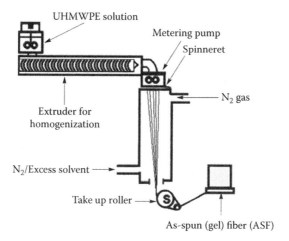

FIGURE 2.55 Pictorial representation of gel-spinning process.

process (Figure 2.55) of the material. The excessive use of solvent makes the process economically and ecologically unattractive. In addition to this, the complete understanding and characterization of UHMWPE, such as quantification of degree of entanglement and so on, are still a sizable challenge.

Recent investigations by Rastogi and coworkers could address these challenges and it appears that UHMWPE can be now melt processed.[141] They were able to tune the polymerization conditions to achieve disentangled UHMWPE using single-site polymerization catalysts.[142] Section 2.9 will present the state of the art developments in disentangled UHMWPE along with commercial success.

2.9 DISENTANGLED ULTRAHIGH MOLECULAR WEIGHT POLYETHYLENE

Ultrahigh molecular weight polyethylene with a very low density of entanglements is referred as *disentangled* ultrahigh molecular weight polyethylene (dUHMWPE/dPE). dUHMWPE is a linear polyethylene with molar mass over 1×10^6 g/mol and it is considered as a high-performance polymer with excellent physical properties such as high toughness, self-lubrication, and abrasion resistance. Because of length and flexibility of such macromolecules, it possesses high melt viscosity. The strong dependence of melt viscosity on molar mass is all because of the presence of intermolecular topological interactions, commonly referred as *entanglements*. These entanglements generally look like cooked spaghetti where long chains are interhooked. A schematic representation of an entanglement is depicted in Figure 2.56.

Such entanglements considerably affect the properties of the resultant polymer in both liquid and solid state. Highly entangled structure determines the course of many processes in which substantial fragments of chains are involved, for example, motion of the (chain mobility) melt becomes extremely slow. This causes difficulty in complete fusion of the polymer particles during conventional processing routes

FIGURE 2.56 Classical model for an entanglement.

and hence melt processing in practice becomes a tedious task. Processing of the UHMWPE with high level of entanglement also leads to nonhomogeneous morphology and causes grain boundaries or fusion defects.[143] The presence of grain boundaries leads to poor mechanical performance of the polymer and affects lifetime, for example, one of the reasons for delamination in the knee prosthesis.[144] Rastogi et al. showed that the grain boundary-free dUHMWPE shows higher toughness and resistance to fatigue in comparison to the sample that is not completely fused.[145] It should also be noted that such entanglements are primarily considered as temporary *physical cross-links* of chains and distinguished from *permanent chemical cross-links* (real networks, e.g., rubbers). Such physical cross linking of the chain can be extricated from entangled chains and the process of releasing the entangled chains is referred as *disentanglement*.

A traditional method to get rid off entanglement is to dissolve less than 5 wt % of the UHMWPE in a high boiling solvent such as decalin. Upon cooling the decalin solution, lamellar UHMWPE crystals could be formed and solvent can be removed, for example, by extraction, and obtained solid UHMWPE should be free of entanglements (Figure 2.57). This disentangled UHMWPE (dUHMWPE) is ductile, easy to draw, and easy to deform into oriented tapes with a high degree of chain alignment.

However, such process demands large volume of solvent, about 95%, which poses additional challenges in terms of process economy and the environment. The above-mentioned procedure to generate disentangled UHMWPE is rather cumbersome but it is being currently practiced for the production of super-strong polyethylene fibers (Table 2.22; Dyneema® by DSM).[146]

In a conceptually different approach, Rastogi and coworkers explored the use of homogeneous polymerization catalysts for the synthesis of dUHMWPE.[147]

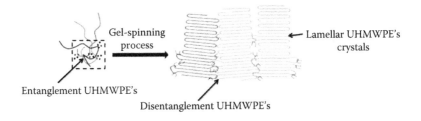

FIGURE 2.57 Solution (gel)-spinning process to obtain disentangled UHMWPE.

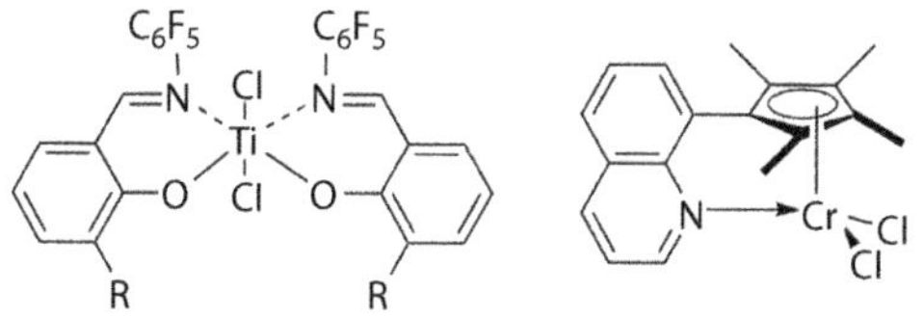

FIGURE 2.58 Successful catalytic systems employed in the synthesis of disentangled UHMWPE.

A methodology to produce dUHMWPE in the form of nascent polymer powder directly in the reactor was demonstrated for the first time. In this method, polymerization conditions were tailored and polymer chains were almost forced to crystallize individually into folded-chain crystals, which form monomolecular crystals, viz one long chain form one crystal. This approach provided an easy pathway to produce highly disentangled linear long-chain polyethylene. The synthesized *disentangled* UHMWPE shows unique features that are of fundamental and technological relevance. The produced polyethylene provided a unique, solvent-free route to obtain high modulus, high-strength tapes where the compressed films can be processed along the uniaxial and biaxial directions in a broad temperature window ranging from 125°C to 145°C. The physical parameters such as high modulus (exceeding 180 GPa) and tensile strength at break (exceeding 4.0 GPa), in the tapes processed via this route, match the high-end solution-spun fibers that are commercially available.

Although dUMHWPE is more of a processing achievement, once again, the catalyst plays the most important role. There have been only two catalytic systems that have been academically explored for the production of dUMHWPE, although there are many more potential candidates. The two catalyst families consist of the FI system and the chromium-based mono-Cp system depicted in Figure 2.58. These catalysts after activation with MAO produce dUMHWPE in high yields and with minimum amount of entanglements. Typical features of these catalytic systems are as follows: (1) The metal center should be highly electrophilic (acidic in nature), (2) there should be almost no or very little β-hydride elimination, (3) the chain-transfer reactions should be reduced to a minimum, (4) the catalyst should retain the activities even at lower temperature, and (5) balanced electronic and steric congestion around the metal center to obtain high molecular weights. These features are not in the exclusive list of parameters but these can simply serve as guiding principles for rational ligand design, although other parameters can also significantly affect the catalyst properties.

Majority of the activities in the area of dUHMWPE are still in academic domain. However, industries have started showing some interest in the recent past. Recently, Reliance Industries Limited filed a series of patents on catalyst and process development to produce disentangled UHMWPE.[148] The only commercial producer of dUHMWPE as of now is Teijin. Teijin has commercialized the solvent-free process for dUHMWPE and has named the product as Endumax® (trade name).[149] Teijin claims that on a weight-to-weight basis Edumax is 11 times stronger than steel and is the strongest material in the world. It is remarkably a light-weight material with very high-strength and very high-modulus, which can be melt processed into various forms such as tapes, ropes, fibers, and so on. Based on these properties it offers, Edumax finds applications in bullet-proof vests, helmets, ballistic protection equipments, sails, cargo-containers, and robotics, to name a few.

2.10 FUNCTIONAL OLEFIN (CO)POLYMERIZATION: STATE-OF-THE-ART

Polyethylene is inherently a long-chain of hydrophobic methylene repeat units without any functional group on the backbone. This partly limits the potential application of PE in adhesives, binders, paints, printing ink, dying, and so on. Incorporation of even small amount of functional groups in PE can significantly enhance these material properties and can further broaden the PE application window. However, due to the high oxophilicity of early transition metal-based ZN type catalysts, the functional group on the polar vinyl monomer coordinates to the metal (occupying the vacant site) that generally leads to catalyst poisoning. Therefore, it has been a long cherished dream of organometallic chemists to synthesize functionalized PE in a single-step via insertion (co)polymerization of ethylene with industrially relevant polar vinyl monomers. Despite the significant progress in olefin polymerization, insertion (co)polymerization of functional olefins remained in accessible until recently. The three fundamental challenges in polar vinyl monomer copolymerization have been depicted in Figure 2.59 and are listed as follows: (a) Ethylene being electron-rich olefin as compared to the functional olefins, most of the times only ethylene coordinates to the metal and polyethylene is obtained without any functionality on the backbone; (b) back coordination of the monomer functional group (denoted by X-Figure 2.59) to the metal to form a σ-complex blocks the vacant site on the metal and polymerization is halted; and (c) the third bottleneck has been chelate formation after the insertion of functional olefin. The chelate is usually quite stable as compared to just olefin coordination and once again, it poisons the catalyst. These three parameters, along with other factors, suppress the copolymerization of functional olefins. A paradigm shift in functional olefin polymerization could be achieved if these three could be suitably addressed.

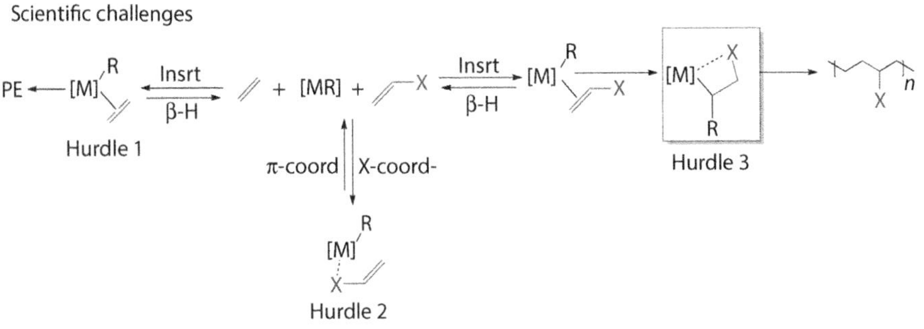

FIGURE 2.59 Three major scientific challenges in polar vinyl monomer copolymerization (X represents functional group).

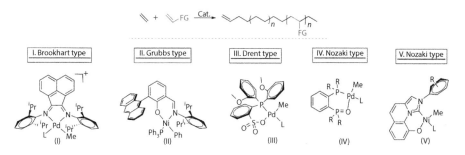

FIGURE 2.60 Overview of recent advances in ancillary ligand and catalysts design for functional olefin copolymerization.

The three fundamental obstacles could be at least partly addressed by (1) choosing a functional group tolerant metal, (2) by designing a ligand that will stabilize the metal center with slight electrophilic character, (3) by playing around with the ethylene pressure or by changing the concentration of the polar vinyl monomer in the polymerization reactor, and (4) by engaging the functional group in other interactions (ionic solvents that will interact with the functional group) and keeping them away from the active metal center. The insertion copolymerization of ethylene with polar vinyl monomers has re-energized the field of olefin polymerization and a large number of publications appeared in the last decade.

Apart from few random attempts, the insertion (co)polymerization of functional olefins mainly explored five ligand classes (Figure 2.60): (a) Brookhart α-diimine ligands,[97] (b) Grubbs-type catalysts,[113] (c) Drent's phosphine-sulfonate ligand system,[150] (d) Nozaki's bisphosphine monoxide,[151] and (e) Nozaki's carbine-based catalytic system.[152] The seminal work of Brookhart led the foundation of polar monomer copolymerization. Palladium and nickel complexes equipped with a bidentate imine ligand were found to catalyze the copolymerization of polar monomers such as methylacrylate with ethylene (Figure 2.60, I).[153] The Brookhart system mainly copolymerizes acrylates with high branching density and the acrylate units are preferentially located at branch ends due to extensive chain walking. Recent ligand modifications suppress the chain-walking process and allow access to linear polyolefins and copolymers with polar functional groups on the backbone.[154] The reduced chain walking is ascribed to the bulky substituents on the two nitrogen atoms.

Although the Grubbs-type phenoxy-imine complex was initially found to catalyze copolymerization of functional olefin with ethylene,[113,155] later applications were mainly in the area of ethylene-α-olefin copolymerization. However, the bimetallic analogs of this catalyst were found to be active in polar monomer copolymerization. Recent reports by Agapie and coworkers revealed that about 1%–3% incorporation of polar monomers could be achieved using the bimetallic nickel catalysts.[156] It is argued that the intramolecular cooperativity between the two metals enables incorporation of polar monomers. It is proposed that the functional group forms a σ-complex with one metal and the olefinic double bond from the same polar monomer coordinates to the next metal in the same complex, which is suitably placed. Thus, the incoming functional group on the polar monomer is kept away from the polymerization metal center and the polymer chain grows to produce polyethylenes with functional groups. To demonstrate the same concept of bimetallic cooperativity, Takeuchi and Osakada reported double-decker type bimetallic phenoxy-imine nickel complexes on a Xanthene backbone.[157] The two nickel centers were close enough (4.7 Å) to cooperate and perform insertion copolymerization of polar monomers with ethylene. Although the incorporation of polar monomers is limited to only 1.5%, addressing the catalyst deactivation through bimetallic cooperativity by suppressing the σ-coordination represents a conceptual advancement.

The above-mentioned two systems could partly address the three challenges and enabled at least some polar monomer incorporation. The most successful catalytic system to date is Drent's phosphine-sulfonate ligated palladium complexes of type III (Figure 2.60). These complexes display broad functional group tolerance, and various polar vinyl monomers such as acrylates, acrylonitrile, vinyl acetate, vinyl ethers, acrylic acid, and vinyl chloride could be incorporated, and the percentage incorporation varies between 1%–52% polar monomer incorporation. In the due course, phosphine-sulfonate ligand family has been tailored to achieve better incorporations and higher molecular weights. The copolymerization experiments are typically carried out in toluene or in neat polar monomer under suitable ethylene pressure and the most significant results from the literature are summarized in Table 2.23. In the first report, Drent et al. disclosed that palladium phosphine-sulfonate complexes catalyze insertion copolymerization of acrylates with ethylene and reported 17% incorporation of methyl acrylate in a linear polyethylene chain.[150] A similar complex with additional methyl substituent on the benzene backbone was investigated by Jordan and coworker in the insertion copolymerization of vinyl ethers. It was found that the catalyst tolerates vinyl ether and revealed 6.9% incorporation of vinyl-butyl ether.[158] Almost parallel to this development, Nozaki and coworkers reported insertion copolymerization of ethylene with acrylonitrile, a monomer that has direct industrial relevance (Table 2.23, run 3).[159] Under optimized conditions, 2%–9% acrylonitrile incorporation was observed. Functional group tolerance toward vinyl ketones was investigated by Sen and coworkers. It was observed that the catalyst tolerates vinyl ketones with 7.7% incorporation of methyl vinyl ketone.[160] Claverie and coworkers investigated the copolymerization of N-Vinyl-2-pyrrolidnone (NVP) and N-Isopropylacrylamide (NIPAM) using the phosphine-sulfonate palladium complex. Upto 9.7% incorporation of NVP and 1.6% incorporation of NIPAM were observed.[161] Similarly, up to 1.9% vinyl acetate incorporation was observed by Nozaki and coworkers.[162] Among the family of phosphine-sulfonate-palladium complexes, the highest polymerization activities were reported when dimethyl sulfoxide (DMSO) was used as a weakly coordinating donor (L = DMSO). The DMSO-ligated complex displayed very high methylacrylate incorporation of 52%, although with a number average molecular weight of 1800 g/mol.[163] Not only the esters, but even acidic groups could be tolerated. Direct insertion copolymerization of acrylic acid with ethylene led to 9.6% acrylic acid incorporation along with reasonable molecular weight.[164] Such acid functionalized polyethylene self-assembles into micelle nanoparticles that were demonstrated to be thermoresponsive.[165] The amphiphilic behavior opens up new application window for functional PE. Apart from these, other accessible polar monomer includes vinyl sulfones,[166] vinyl halides,[167]

TABLE 2.23

Insertion Copolymerization of Ethylene with Polar Monomers to Produce Functional Polyethylene

Entry	FG/Monomer	R	Incorporation (%)	Mol. Wt. (g/mol)
1	COOMe	OMe	17	6400
2	OBu	OMe	6.9	3100
3	CN	OMe	2–9	12300–2900
4	COOMe	OMe	52	1800
5	COMe	OMe	7.7	9600
6	Pyrrolidinone	OMe	9.7	5000
7	NIPAM	OMe	1.6	1230
8	OAc	OMe	1.9	5800
9	COOH	OMe	9.6	6100
10	SO$_2$Me	OMe	2.1	3000
11	F	OMe	3.6	1600
12	Cl	OMe	0.4	3900
13	DMAA	OMe	3.5	1500
14	MA	OMe	9	2800
15	CF$_3$	OMe	8.9	5100
16	CH$_2$OAc	OMe	7.9	4400
17	OCH$_2$OCHCH$_2$	OMe	12.5	1200
18	CH$_2$OAc	Men	0.6	177000
19	ECA	OMe	6.5	5800

DMAA: Dimethyl acrylamide; MA: Maleic anhydride; Men: Menthyl; and ECA: Ethylcyanoacrylate.

functionalized norbornene,[168] acrylamides,[169] anhydrides,[170] trifluoropropene,[171] acrylic anhydride,[172] formal groups,[173] and allylic monomers.[174] Thus the phosphine-sulfonate tolerates a large number of functional groups, but the usual limitation is lower molecular weight. To address this bottleneck and to investigate the catalyst structure–polymer properties correlation, Nozaki and coworkers studied the effect of *ortho*-phenyl substituent in the phosphine-sulfonate ligand system on the polymer molecular weight.[175] It was found that increasing steric bulk on the *ortho*-substituent (R) increases the polymer molecular weight. The most effective substituent found was menthyl group, that led to the copolymerization allyl acetate with ethylene and the molecular weight could be raised to 177000 g/mol along with 0.6% incorporation (Table 2.23, entry 18).

Thus, the aim of the polar monomer copolymerization with ethylene is to increase the polarity of the resultant polyethylene and the investigations so far considered only mono-functional polar monomers. More recently, Chikkali and coworkers reported the insertion copolymerization of difunctional polar vinyl monomers with ethylene using a slightly modified catalyst (L = acetonitrile).[176] It was demonstrated that the phosphine-sulfonate-palladium complex is capable of incorporating highly polar 1,1-disubstituted polar monomers such as ethyl cyanoacrylate (ECA) or super glue. Upto 6.5% ECA incorporation was observed under optimized polymerization conditions. An added advantage of copolymerizing difunctional olefin is that, even if the percent incorporation remains same as a mono-functional monomer, the functional group density will be increased by two folds, making the resultant copolymer highly polar in nature.

Excellent performance by phosphine-sulfonate ligand family in functional olefin copolymerization motivated researchers to investigate secrete behind its success.[177] In their attempts to understand the characteristic features of this ligand, Nozaki and coworkers performed DFT calculations,[178] along with others.[179] Thus, the unique features of phosphine-sulfonate system can be compared on the basis of two parameters: charge effects and orbital interactions. (a) As compared to a cationic complex (such as Brookhart type I), the neutral phosphine-sulfonate palladium complexes are less electrophilic and therefore more functional group tolerant. (b) In the insertion copolymerization of polar monomers, ethylene insertion occurs from a *trans* π-complex (Figure 2.61), which is in equilibrium with a nonproductive *cis* σ-complex that cannot produce any polymer. Therefore, ligands that can destabilize *cis* σ-complex and stabilize *trans* π-complex will produce copolymers. In this situation, the intrinsic design of *ortho*-phosphinebenzenesulfonate stabilizes *trans* π-complex through two types of orbital interactions: (1) It has been understood that the sulfonate (which is *trans* to the olefin) oxygen lone pair repulses the filled d-orbitals on palladium, which in turn are back donated to the unfilled π*-orbitals of the olefin. Such dπ–pπ back bonding stabilizes the *trans* π-complex (Figure 2.62), paving the way for insertion and subsequent polymerization. (2) Second, formation of alkyl-palladium bond is favored when the sulfonate is *trans* to the olefin, due to weaker *trans*-influence of sulfonate compared to phosphine. Thus, due to these two effects the concentration of *trans* π-complex is increased in the reaction, which is a prerequisite for the next insertion and polymerization. The unique features associated with *ortho*-phosphinebenzenesulfonate make this ligand stand out and it is the best ligand to date for polar olefin copolymerization.

cis σ-complex   trans π-complex

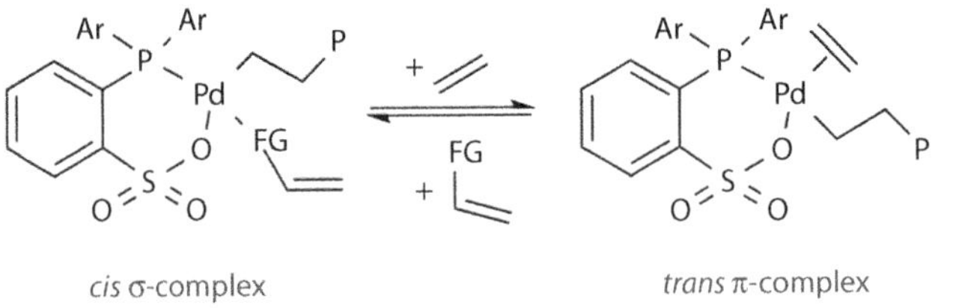

FIGURE 2.61 The decisive equilibrium between a nonproductive *cis* σ-complex and the productive *trans* π-complex.

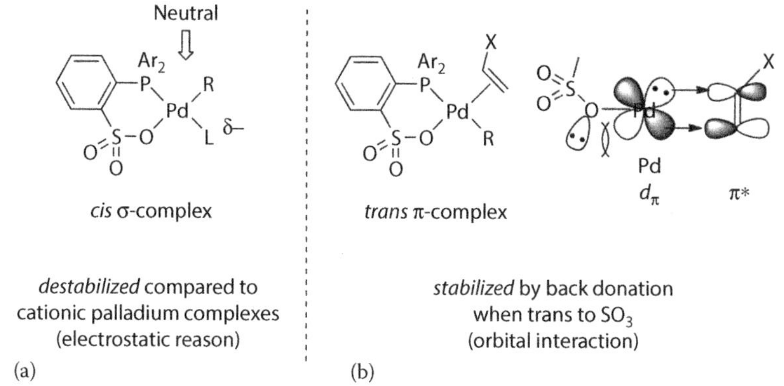

cis σ-complex

trans π-complex

Pd
d_π π*

destabilized compared to
cationic palladium complexes
(electrostatic reason)

(a)

stabilized by back donation
when trans to SO₃
(orbital interaction)

(b)

FIGURE 2.62 Unique features of ortho-phosphinebenzenesulfonate palladium complex, net neutral charge (a) and stabilization of *trans* π-complex through dπ–pπ interactions (b).

Inspired by the excellent performance of *ortho*-phosphinebenzenesulfonate the next generation of ligand design focused on replacing either the phosphine or the sulfonate arm. In one such attempt, Nozaki and coworkers reported the synthesis of bisphosphine monoxide (BPMO) (Figure 2.60, type IV) ligand that is capable of incorporating a wide range of polar monomers in the ethylene backbone.[180] The phosphine arm was untouched, whereas the sulfonate arm was replaced by phosphineoxide in BPMO ligand system. The cationic palladium complexes derived from P-O chelating BPMO ligand were found to catalyze insertion copolymerization of industrially relevant polar monomers such as vinyl acetate, acetonitrile, and vinyl ethers. In their attempts to mimic the *ortho*-phosphinebenzenesulfonate system, Jordan and coworkers reported *N*-heterocyclic carbene-based ligand systems. Thus, the phosphine arm in *ortho*-phosphinebenzenesulfonate was now replaced by an isoelectronic carbene donor and the anionic sulfonate donor was not touched.[181] However, this system was not active in insertion polymerization and was found to rapidly decompose under polymerization conditions. Undeterred by this failure, Nozaki and coworkers designed a catalytic system based on *N*-heterocyclic carbene and phenolate ligand system (Figure 2.60, Nozaki type-V). The designing principles of imidazo[1,5-a]quinolin-9-olate-1-ylidene (IzQO) ligands include the following features (Figure 2.63): (a) A strong σ-donor *N*-heterocyclic carbene (NHC) motif, which is isoelectronic to the phosphine donor in Drent-type complex. (b) A monoanionic oxygen will serve as a weakly coordinating heteroatom donor, which was proved to be essential in both phosphine-sulfonate system and Grubbs system. (c) The two donors are connected via a rigid aromatic backbone, which is quintessential to maintain the coplanarity of the two donors in the metal complex. (d) The coplanarity will avoid the possible overlap of empty p-orbitals of carbene and metal-hydrides in *cis*-configuration and will prevent the catalysts decomposition via reductive elimination. (e) The substituent (Figure 2.63, 5: substituent-R) on NHC provides an additional handle to regulate the steric bulk around the metal center, which can potentially direct the insertion of functional olefin. Equipped with these special features the resultant palladium complexes derived from IzQO were found to be capable

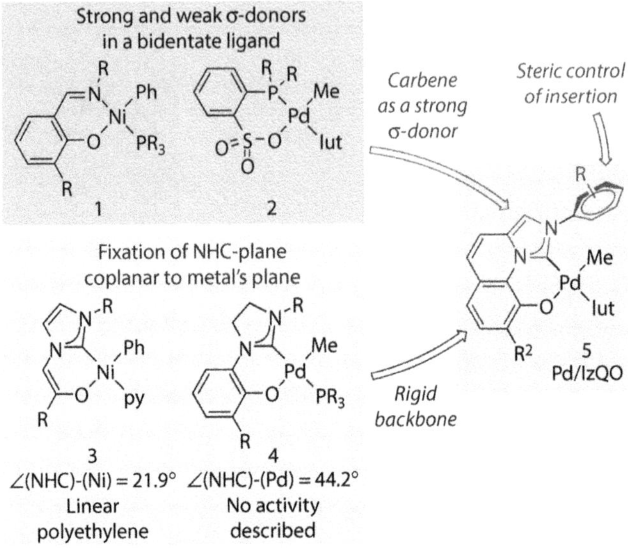

FIGURE 2.63   Rational designing of IzQO ligand system based on catalysts 1–4.

of initiating the insertion copolymerization of propylene with functional polar mono-mers for the first time.[152] Note that the Pd/IzQO catalysts with the smallest dihedral angles between the NHC plane and the palladium square plane performed better (in terms of activity) than those with larger bite angle. The corresponding nickel com-plexes were found to be active in polar monomer copolymerization, although they suffer from poor polar monomer incorporation.

The functional polyolefins discussed in this section are still in the academic domain. Increasing the polymer molecular weight and percentage incorporation of the functional olefin are the two major challenges that the academic world has to address before these materials can see the light of the day. However, there exist other methods to manufacture functional polyolefins via radical polymerization (high pressure, high temperature) or postfunctionalization or reactive functionalization.[182]

2.11   POLYOLEFINS AND THEIR APPLICATION SEGMENTS

Applications of various polyolefins have been summarized under each section. The purpose of the section is to give a broader overview of the polyolefin industry and differentiate various classes of polyolefins based on the market and size of the market (Figure 2.64). Polyolefin-including PE, PP, poly(1-butene), poly(4-methyl-1-pentene), ethylene–propylene elastomer (EPR), and EPDM are the most widely used com-mercial polymers, with over 180 million metric tons global annual consumption, or close to 60% of the total polymer produced in year 2014.* The major types of

---

* Based on the market data base the current production of polyolefins is quoted. The following link may confirm this number, but might not be always available in due course: https://www.statista.com/statistics/282732/global-production-of-plastics-since-1950/.

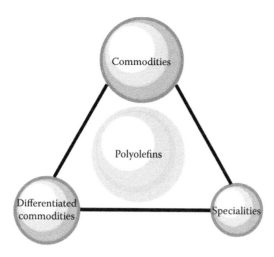

FIGURE 2.64 Classification of polyolefins based on application volume and the market they serve.

polyolefins include PP, HDPE, LLDPE, LDPE, metallocene polyethylene (mPE), and various copolymers and elastomers. The polyolefin family of products serves a wide variety of end-use markets in the major sectors of packaging, automotive, construction, medical, wire and cable, and others. By nature, an industry of this size is considered a commodity producer. However, almost all of the major polyolefin producers consider polyolefins as a specialty product with plenty of opportunities for value addition and creation through technological innovations. Unlike the incremental technological developments more common in other polymers, polyolefin technology developments are significant and routinely leapfrog the existing ones. These tremendous developments in technology impact the whole industry as a unit as well as the high profit sectors, often causing confusion regarding expected impact on profitability and classification of specialty versus commodity. As a case in point, metallocenes introduced in the early 1990s were initially positioned in the market as specialties with high expectations for profitability, but are settling down as *differentiated commodities*. The family of polyolefins can be classified into the following three classes based on the market size they serve: (a) commodities, (b) differentiated commodities, and (c) specialties based on the *specialty index* which is a combination of various parameters such as profitability, demand, number of players, price, and technical barriers as the criteria.

2.11.1 COMMODITY POLYOLEFINS

Commodities are the very large volume products that are sold by a large number of suppliers to large number of customers, with less product differentiation and price as the major criteria. Thus, Ziegler Natta catalysts ($MgCl_2$ supported $TiCl_4$) have been used for the commercial manufacturing of various polymeric materials such as plastics, elastomers, and rubbers since 1956. These collectively fall under commodities segment.

Examples are ZN-derived polyolefins such as HDPE, LDPE, LLDPE, PP, and so on.

2.11.2 Differentiated Commodity Polyolefins

Differentiated commodities are large volume products that are sold by selected suppliers to a large number of customers, differentiating product based on performance and price.

Metallocenes revolutionized the last decade by developing products that have improved characteristics compared to traditional ZN catalyst-based products. Metallocene and single-site catalyst-based products provide

1. Narrower molecular weight distribution
2. Better comonomer incorporation
3. Lower densities compared to conventional ZN-based products

These characteristics provided several advantages at the end-user level including (1) impact strength, (2) clarity, (3) organoleptic properties, (4) heat-seal characteristics, and (5) most importantly, an opportunity to down gage. Metallocenes have been looked at from two points of view: (1) a natural evolution of polyolefin technology—LDPE to LLDPE to metallocenes and (2) as new polymers that provided new opportunities, especially at lower densities and applications requiring down gaging. Considering the technological barriers involved and a few number of companies involved in the manufacturing of metallocene-grade polyolefins, they can be classified as differentiated commodities. If we look at metallocenes as yet another step in polyolefins evolution, we have to accept that in a few more years, they will reach or exceed LLDPE/LDPE to become the next commodities. The only factor at present that prevents metallocenes from becoming a commodity is the sunk research and development costs. Excluding the R&D costs, metallocenes are poised to become commodities because they, for the most part, provide the same functions as LLDPEs—new and improved.

Incorporation of low molecular weight alpha olefin into the polyethylene backbone provides impact and clarity properties to the product independent of the process. The major limitation is that octene is harder to incorporate in nongas-phase processes because of its molecular weight limitations. Octene copolymers have definite performance advantages in processing but are limited by technologies and suppliers, making them unique and can be considered as differentiated commodities. Examples are metallocene-derived polyolefins such as metallocene-grade LLDPE (mLLDPE), mPP, and so on.

2.11.3 Specialties Polyolefins

Large-volume specialties are large to medium volume products that are sold based on technology, performance, and meeting end-use requirements as the major criteria. Specialties are small-volume products supplied by limited suppliers with high profit margins and high barriers to entry based on technology. Examples of speciality polyolefins include PE/PP foam, syndiotactic PP (sPP), polybutene-1, polyolefin ionomers, and so on.

Plastomers/elastomers are essentially extensions of traditional polyolefins into elastomeric applications via metallocene's capability to produce lower density materials. They are at present considered specialties poised to become differentiated commodities in the future.

Vinyl acetate up to 4% to 6% is included in most LDPEs without any perceptible impact on the performance. In most cases, their presence is not identified. The definition of high ethylene-vinyl acetate (EVA) copolymers is reserved for VA content of 11% and more. High EVA copolymers of 11% up to 16% provide improved strength and barrier performance properties to the LDPE film applications and are positioned in the industry as high value-added LDPEs. High EVA copolymers of 18% to 22% affect the crystallinity of the polymer, making them amorphous and more suitable for adhesives and sealants applications. Beyond 22%, VA content materials are used as specialty sealants. The value addition of VA is a reactor process, and thus all the value added is kept at the manufacturer level. The value added to high EVA at the end-user level has to pay for the cost of VA comonomer as well as the reduction of reactor capacity utilization costs, in addition to the expected additional profit. High EVAs represent an excellent example of specialty products with process and comonomer as the differentiation points.

REFERENCES

1. Kamnisky, W. 2012. Discovery of Methylaluminoxane as Cocatalyst for Olefin Polymerization. *Macromolecules* 45: 3289–3295.
2. Ziegler, K., Gellert, H. G. 1950. Addition von lithium alkylen an aethylen. *Justus Liebigs Ann. Chem.* 567: 195–203.
3. Fischer, K., Jonas, K., Misbach, P., Stabba, R., Wilke, G. 1973. The "nickel effect". *Angew. Chem. Int. Ed.* 12: 943–953.
4. Ziegler, K., Holzkamp, E., Breil, H., Martin, H. 1955. Das Muelheimer normaldruck-polyaethylen-verfahren. *Angew. Chem.* 67: 541–547.
5. Boor, J. Jr. *Ziegler-Natta Catalysts and Polymerizations.* Academic Press, New York. 1979.
6. Ziegler, K., Breil, H., Holzkamp, E., Martin, H. 1953. Verfahren zur Herstellung von hochmolekularen Polyaethylenen. German patent number DE 973626.
7. Natta, G., Pino, P., Corradini, P., Danusso, F., Mantica, E., Mazzanti, G., Moraglio, G. 1955. Crystalline high polymers of α-olefins. *J. Am. Chem. Soc.* 77: 1708–1710.
8. Natta, G. 1963. Macromolecular chemistry: From the stereospecific polymerization to the asymmetric autocatalytic synthesis of macromolecules. *Science* 147: 261–272.
9. Beredjick, N., Schuerch, C. 1958. Stereospecific vinyl polymerization by asymmetric induction. *J. Am. Chem. Soc.* 80: 1933–1938.
10. Coates, G. W. 2000. Precise control of polyolefin stereochemistry using single site metal catalysts. *Chem. Rev.* 100: 1223–1252.
11. Natta, G., Corradini, P., Sianesi, D., Morero, D. 1962. Isomorphism phenomena in macromolecules. *J. Polym. Sci.* 51: 527–539.
12. Galli, P., Vecellio, G. 2001. Technology: Driving force behind innovation and growth of polyolefins. *Prog. Polym. Sci.* 26: 1287–1336.
13. Natta, G. 1955. Polymeres isotactiques. *Makromol. Chem.* 16: 213–237.
14. Natta, G., Mazzanti, G. 1960. Organometallic complexes as catalysts in ionic polymerization. *Tetrahedron* 8: 86–100.
15. Cossee, P. 1964. Ziegler-Natta catalysis I. Mechanism of polymerization of α-olefins with Ziegler-Natta catalyst. *J. Catal.* 3: 80–88.
16. Arlman, E. J. 1964. Ziegler-Natta catalysis II. Surface structure of layer-lattice transition metal chlorides. *J. Catal.* 3: 89–98.

17. Arlman, E. J., Cossee, P. 1964. Ziegler-Natta catalysis III. Stereospecific polymerization of propene with the catalyst system TiCl₃-AlEt₃. *J. Catal.* 3: 99–104.

18. Sacchi, M. C., Forlini, F., Tritto, I., Locatelli, P. 1995. A study for distinguishing mono- and bimetallic mechanisms in heterogeneous Ziegler-Natta catalysis. *Macromol. Chem. Phys.* 196: 2881–2890.

19. Patat, F., Sinn, H. 1958. Zum Ablauf der Niederdruckpolymerisation der α-olefine. Komplexpolymerisation I. *Angew. Chem.* 70: 496–500.

20. Arlman, E. J. 1962. Chlorine vacancies in transition metal chlorides and stereospecific polymerization of α-olefins. *J. Polym. Sci.* 62: S30–S33.

21. Wilkinson, G., Pauson, P. L., Birmingham, J. M., Cotton, F. A. 1953. Bis-cyclopentadienly derivatives of some transition elements. *J. Am. Chem. Soc.* 75: 1011–1012.

22. Fischer, E. O. 1955. Metallverbindungen des cyclopentadiens und indens. *Angew. Chem.* 67: 475–482.

23. Natta, G., Pino, P., Mazzanti, G., Giannini, U. 1957. A crystallizable organometallic complex containing titanium and aluminum. *J. Am. Chem. Soc.* 79: 2975–2976.

24. Kamnisky, W., Sinn, H. 1975. Mehrfach druch metalle substituierte aethane. *Liebigs Ann. Chem.* 424–437.

25. Anderson, A., Cordes, H. G., Herwig, J., Kamnisky, W., Merck, A., Mottweiler, R., Pein, J., Sinn, H., Vollmer, H. J. 1976. Halogen free soluble Ziegler catalysts for the polymerization of ethylene. Control of molecular weight by choice of temperature. *Angew. Chem. Int. Ed. Engl.* 15: 630–632.

26. Reichert, K. H., Meyer, K. R. 1973. Zur kinetik der niederdruckpolymerisvon aethylen mit loeslichen Ziegler-katalysatorn. *Macromol. Chem.* 169: 163–176.

27. Long, W. P., Breslow, D. S. 1975. Der einfluss von wasser auf die katalytische aktivitaet von bis(π-cyclopentadienyl)titandichlorid-dimethylaluminiumchlorid zur polymerisation von aethylen. *Justus Liebigs Ann. Chem.* 463–469.

28. Herwig, J., Kamnisky, W. 1983. Halogen-free soluble Ziegler catalysts with methylalumoxan as catalyst. *Polym. Bull.* 9: 464–469.

29. Sinn, H. 1995. Proposals for structure and effect of methylalumoxane based on mass balances and phase separation experiments. *Macromol. Symp.* 97: 27–52.

30. Koide, Y., Bott, S. G., Barron, A. R. 1996. Alumoxanes as cocatalysts in the palladium catalyzed copolymerization of carbon monoxide and ethylene: Genesis of structure-activity relationship. *Organometallics* 15: 2213–2226.

31. Kamnisky, W. 1996. New polymers by metallocene catalysis. *Macromol. Chem. Phys.* 197: 3907–3945.

32. Wild, F. R. W. P., Zsolnai, L., Huttner, G., Brintzinger, H. H. 1982. *Ansa*-Metallocene derivatives IV. Synthesis and molecular structures of chiral *ansa*-titanocene derivatives with bridged tetrahydroindenyl ligands. *J. Orgmet. Chem.* 232: 233–247.

33. Kamnisky, W., Kuelper, K., Brintzinger, H. H., Wild, F. R. W. P. 1985. Polymerization of propene and butene with a chiral zirconocene and methylalumoxane as cocatalyst. *Angew. Chem. Int. Ed. Engl.* 24: 507–508.

34. Ewen, J. A. 1984. Mechanism of stereochemical control in propylene polymerization with soluble group 4B metallocene/methylalumoxane catalysts. *J. Am. Chem. Soc.* 106: 6355–6364.

35. Yang, X., Stern, C. L., Marks, T. J. 1991. Models for organometallic molecule-support complexes. Very large counterion modulation of cationic actinide alkyl reactivity. *Organometallics*, 10: 840–842.

36. Sishta, C., Hathorn, R. M., Marks, T. J. 1992. Group IV metallocene-alumoxane olefin polymerization catalysts. CPMAS-NMR spectroscopic observation of "cation like" zirconocene alkyls. *J. Am. Chem. Soc.* 114: 1112–1114.

37. Hlatky, G. G., Turner, H. W., Eckman, R. R. 1989. Ionic, base free zirconocene catalysts for ethylene polymerization. *J. Am. Chem. Soc.* 111: 2728–2729.

38. Bochmann, M., Lancaster, S. J. 1993. Base-free cationic zirconiumbenzyl complexes as highly active polymerization catalysts. *Organometallics* 12: 633–640.
39. Eisch, J. J., Pombrik, S. I., Zheng, G.-X. 1993. Active sites for ethylene polymerization with titanium (IV) catalysts in homogeneous media: Multinuclear NMR study of ion-pair equilibria and their relation to catalyst activity. *Organometallics* 12: 3856–3863.
40. Tait, P. 1988. Ethylene polymerization process with a highly active Ziegler-Natta catalyst-Kinetic studies. In *Transition Metals and Organometallics as Catalysts for Olefin Polymerization*. Eds. Kaminsky, W. and Sinn, H. Springer Press, Berlin, 1988.
41. Jordan, R. F., Dasher, W. E., Echols, S. F. 1986. Reactive cationic dicyclopentadienylzirconium(IV) complexes. *J. Am. Chem. Soc.* 108: 1718–1719.
42. Kaminsky, W., Kuiper, K., Niedoba, S. 1986. Olefin polymerization with highly active soluble zirconium compounds using aluminoxane as co-catalyst. *Makromol. Chem. Macromol. Symp.* 3: 377–387.
43. Giannetti, E., Nicoletti, G., Mazzochi, R. 1985. Homogeneous Ziegler-Natta catalysis. II. Ethylene polymerization by group IVB transition metal complexes/methyl aluminoxane catalyst systems. *J. Polym. Sci., Polym. Chem. Ed.* 23: 2117–2133.
44. Mohring, P. C., Coville, N. 1994. Homogeneous group 4 metallocene Ziegler-Natta catalysts: The influence of cyclodpentadienyl-ring substituents. *J. Organomet. Chem.* 479: 1–29.
45. Tait, P. J. T., Booth, B. L., Jejelowo, M. O. 1992. *Rate of ethylene polymerization with catalyst system* $(\eta^5\text{-}RC_5H_4)_2ZrCl_2$—*Methylaluminoxane*. Catalysis in polymer science, Chapter 6: 78–87.
46. Brintzinger, H. H., Fischer, D., Mülhaupt, R., Rieger, B., Waymouth, R. M. 1995. Stereospecific olefin polymerization with chiral metallocene catalysts. *Angew. Chem. Int. Ed. Engl.* 34: 1143–1170.
47. Kaminsky, W., Arndt, M. 1997. Metallocenes for polymer catalysis. *Advances in polymer science,* Springer 127: 143–187.
48. Suhm, J., Heinemann, J., Wörner, C., Müller, P., Stricker, F., Kressler, J., Okuda, J., Mülhaupt, R. 1998. Novel polyolefin materials via catalysis and reactive processing. *Macromol. Symp.* 129: 1–28.
49. Alt, H. G., Jung, M., Milius, W. 1998. Verbrueckte indenyliden cyclopentadienyliden komplexe des typs $(C_9H_9CH_2Ph\text{-}X\text{-}C_5H_4)MCl_2$ (X = CMe$_2$, SiMe$_2$; M = Zr, Hf) at metallokatalysatoren fuer die ethylenpolymerisation. Die molekuelstrukturen von $(C_9H_9CH_2Ph\text{-}X\text{-}C_5H_4)MCl_2$ (X = CMe$_2$, SiMe$_2$; M = Zr, Hf). *J. Organomet. Chem.* 558: 111–121.
50. Kaminsky, W. 1998. Highly active metallocene catalysts for olefin polymerization. *J. Chem. Soc. Dalton Trans.* 1413–1418.
51. Peifer, B., Welch, M. B., Alt, H. G. 1997. Synthese und charakterisierung C1- und C2-verbrueckten bis(fluorenyl)komplexes des zirkoniums und hafniums und deren unwendung bei der katalytischen olefinpolymerisation. *J. Organomet. Chem.* 544: 115–119.
52. Brookhart, M., Green, M. L. H. 1983. Carbon-hydrogen-transition metal bonds. *J. Organomet. Chem.* 250: 395–408.
53. Krauledat, H., Brintzinger, H. H. 1990. Isotope effects associated with α-olefin insertion in zirconocene based polymerisation catalysts: Evidences for an α-agostic transition state. *Angew. Chem. Int. Ed. Engl.* 29: 1412–1413.
54. Piers, W. E., Bercaw, J. E. 1990. α "Agostic" assistance in Ziegler–Natta polymerization of olefins. Deuterium isotopic perturbation of stereochemistry indicating coordination of an α C–H bond in chain propagation. *J. Am. Chem. Soc.* 112: 9406–9407.
55. Xu, Z., Vanka, K., Ziegler, T. 2004. Influence of the counter MeB(C$_6$F$_5$)$^{3-}$ and solvent effects on ethylene polymerization catalyzed by [(CpSiMe$_2$NR)TiMe]$^+$: A combined density functional theory and molecular mechanism study. *Organometallics* 23: 104–116.

56. Margl, P., Deng, L., Ziegler, T. 1999. A unified view of ethylene polymerization by d^0 and d^0f^n transition metals. 3. Termination of the growing polymer chain. *J. Am. Chem. Soc.* 121: 154–162.

57. Naga, N., Mizunuma, K. 1998. Chain transfer reaction by trialkylaluminum (AlR_3) in the stereospecific polymerization of propylene with metallocene $AlR_3/Ph_3CB(C_6F_5)_4$. *Polymer* 39: 5059–5067.

58. Reddy, S. S., Sivaram. S. 1995. Homogeneous metallocene-methylaluminoxane catalyst systems for ethylene polymerization. *Prog. Polym. Sci.* 20: 309–367.

59. Quijada, R., Rojas, R., Alzamora, L., Retuert, J., Rabagliati, F. M. 1997. Study of metallocene supported on porous and non-porous silica for the polymerization of ethylene. *Catal. Lett.* 46: 107–112.

60. Alonso, C., Antinolo, A., Carrillo-Hermosilla, F., Carrion, P., Otero, A., Sancho, J., Villasenor, E. 2004. Modified silicas as supports for single site zirconocene catalysts. *J. Mol. Catal. A: Chem.* 220: 285–295.

61. Ketloy, C., Jongsomjit, B., Praserthdam, P. 2007. Characterization and catalytic properties of [t-BuNSiMe$_2$Flu]TiMe$_2$/dMMAO catalyst dispersed on various supports towards ethylene/1-octene copolymerization. *Appl. Catal. A: Gen.* 327: 270–277.

62. Severn, J. R., Chadwick, J. C., Duchateau, R., Friederichs, N. 2005. "Bound but not gagged" -immobilizing single-site α-olefin polymerization catalysts. *Chem. Rev.* 105: 4073–4147.

63. Brambilla, R., Radtke, C., Stedile, F. C., Santos, J. H. Z. D., Miranda, M. S. L. 2010. Metallocene catalyst supported on silica-magnesia xerogels for ethylene polymerization. *Appl. Catal. A: Gen.* 382: 106–114.

64. Silveira, F., Loureiro, S. R., Galland, G. B., Stedile, F. C., Santos, J. H. Z., Teranishi, T. 2003. Hybrid zirconocene supported catalysts. *J. Mol. Catal. A: Chem.* 206: 389–398.

65. Silveira, F., Petry, C. F., Pozebon, D., Pergher, S. B., Detoni, C., Stedile, F. C., Santos, J. H. Z. 2007. Supported metallocene on mesoporous materials. *Appl. Catal. A: Gen.* 333: 96–106.

66. Resconi, L., Cavallo, L., Fait, A., Piemontesi, F. 2000. Selectivity in propene polymerization with metallocene catalysts. *Chem. Rev.* 100: 1253–1346.

67. Ewen, J. A., Elder, M. J., Jones, R. L., Curtis, S., Cheng, H. N. Syndiospecific Propylene Polymerizations with iPr [CpFlu] ZrCl$_2$. In *Catalytic Olefin Polymerization, Studies in Surface Science and Catalysis.* Eds. Keii, T., Soga, K. Elsevier, New York, 1990, p. 439.

68. Ewen, J. A., Haspeslagh, L., Elder, M. J., Atwood, J. L., Zhang, H., Cheng, H. N. In *Transition Metals and Organometallics as Catalysts for Olefin Polymerization.* Eds. Kaminsky, W., Sinn, H. Springer-Verlag, Berlin, 1988, p. 281.

69. Yano, A., Yamada, S., and Akimoto, A. 1999. Propylene polymerization with dimethylsilylbis(3-methylcyclopentadienyl)MCl$_2$ [M = Ti, Zr, Hf] in combination with methylaluminoxane. *Macromol. Chem. Phy.* 200: 1356–1362.

70. Collins, S., Gauthier, W. J., Holden, D. A., Kuntz, B. A., Taylor, N. J., Ward, D. G. 1991. Variation of poly(propylene) microtacticity by catalyst selection. *Organometallics* 10: 2061–2068.

71. Resconi, L., Piemontesi, F., Camurati, I., Sudmeijer, O., Nifant'ev, I. E., Ivchenko, P. V., Kuz'mina, L. G. 1998. Highly regiospecific zirconocene catalysts for the isospecific polymerization of propene. *J. Am. Chem. Soc.* 120: 2308–2321.

72. Resconi, L., Balboni, D., Baruzzi, G., Fiori, C., Guidotti, S. 2000. *rac*-[Methylene (3-tert-butyl-1-indenyl)$_2$ZrCl$_2$]: A simple high-performance zirconocenes catalyst for isotactic polypropylene. *Organometallics* 19: 420–429.

73. Miyake, S., Okumura, Y., Inazawa, S. 1995. Highly isospecific polymerization of propylene with unsymmetrical metallocene catalysts. *Macromolecules* 28: 3074–3079.

74. Spaleck, W., Antberg, M., Aulbach, M., Bachmann, B., Dolle, V., Haftka, S., Kueber, F., Rohrmann, J., Inter, A. New Isotactic Polypropylenes via Metallocene Catalysts. In *Ziegler Catalysts*. Eds. Fink, G., Muelhaupt, R., Brintzinger, H.-H. Springer-Verlag, Berlin, 1995, p. 83.

75. Ewen, J. A., Haspeslagh, L., Elder, M. J., Atwood, J. L., Zhang, H., Cheng, H. N. Propylene Polymerizations with Group 4 Metallocene/Alumoxane Systems. In *Transition Metals and Organometallics as Catalysts for Olefin Polymerization*. Eds. Kaminsky, W., Sinn, H. Springer-Verlag, Berlin, 1988, p. 281.

76. Hart, J. A., Rappé, A. K. 1993. Predicted structure selectivity trends: Propylene polymerization with substituted [*rac*(1,2-Ethylenbis(η^5-indenyl))zirconium(IV)] catalysts. *J. Am. Chem. Soc.* 115: 6159–6164.

77. Yoshida, T., Koga, N., Morokuma, K. 1996. A combined *ab* Initio MO–MM study on isotacticity control in propylene polymerization with silylene-bridged group 4 metallocenes. C_2 symmetrical and asymmetrical catalysts. *Organometallics* 15: 766–777.

78. Ewen, J. A., Jones, R. L., Razavi, A., Ferrara, J. D. 1988. Syndiospecific propylene polymerization with group 4 metallocenes. *J. Am. Chem. Soc.* 110: 6255–6256.

79. Grisi, F., Longo, P., Zambelli, A., Ewen, J. A. 1999. Group 4 C_s symmetric catalysts and 1-olefin polymerization. *J. Mol. Catal. A: Chem.* 140: 225–233.

80. Winter, A., Antberg, M., Bachmann, B., Dolle, V., Kueber, F., Rohrmann, J., Spaleck, W. 1994. Verfahren zur Herstellung von polyolefinen. Assigned to Hoechst AG, EP0584609A2.

81. Resconi, L., Jones, R. L., Rheingold, A. L., Yap, G. P. A. 1996. High molecular weight atactic polypropylene from metallocene catalysts. 1. $Me_2Si(9\text{-}Flu)_2\text{-}ZrX_2$ (X = Cl, Me). *Organometallics* 15: 998–1005.

82. Giardello, M. A., Eisen, M. S., Stern, C. L., Marks, T. J. 1993. Chiral, non-C_2 symmetric zirconocene complexes as catalysts for stereoregular α-olefin polymerization. *J. Am. Chem. Soc.* 115: 3326–3327.

83. Giardello, M. A., Eisen, M. S., Stern, C. L., Marks, T. J. 1995. Chiral C_1-symmetric group 4 metallocenes as catalysts for stereoregular α-olefin polymerization. Metal, ancillary ligand and counteranion effect *J. Am. Chem. Soc.* 117: 12114–12129.

84. Coates, G. W., Waymouth, R. M. 1995. Oscillating stereocontrol: A strategy for the synthesis of thermoplastic elastomeric polypropylene. *Science* 267: 217–219.

85. Hauptman, E., Waymouth, R. M., Ziller, J. W. 1995. Stereoblock polypropylene: Ligand effects on the stereospecificity of 2-arylindene zirconocene catalyst. *J. Am. Chem. Soc.* 117: 11586–11587.

86. Bruce, M. D., Coates, G. W., Hauptman, E., Waymouth, R. M., Ziller, J. W. 1997. Effect of metal on the stereospecificity of 2-arylindene catalyst for elastomeric polypropylene. *J. Am. Chem. Soc.* 119: 11173–11182.

87. Kravchenko, R. L., Masood, A., Waymouth, R. M., Myers, C. L. 1998. Strategies for the synthesis of elastomeric polypropylene: Fluxional metallocenes with C_1 symmetry. *J. Am. Chem. Soc.* 120: 2039–2046.

88. Hoffmann, R. 1982. Building bridges between inorganic and organic chemistry. *Angew. Chem. Int. Ed.* 21: 711–724.

89. Braunschweig, H., Breitling, F. M. 2006. Constrained geometry complexes—synthesis and applications. *Coord. Chem. Rev.* 250: 2691–2720.

90. Johnson, L. K., Killian, C. M., Brookhart, M. S. 1995. New Pd(II)- and Ni(II)-based catalysts for polymerization of ethylene and α-olefins. *J. Am. Chem. Soc.* 117: 6414–6415.

91. Johnson, L. K., Mecking, S., Brookhart, M. S. 1996. Copolymerization of ethylene and propylene with functionalized vinyl monomers by Pd(II) catalysts. *J. Am. Chem. Soc.* 118: 267–268.

92. Killian, C. M., Temple, D. J., Johnson, L. K., Brookhart, M. S. 1996. Living polymerization of α-olefins using Ni(II)-α-diimine catalysts. Synthesis of new block polymers based on α-olefins. *J. Am. Chem. Soc.* 118: 11664–11665.
93. Small, B. L., Brookhart, M., Bennett, A. M. 1998. Highly active iron and cobalt catalysts for the polymerization of ethylene. *J. Am. Chem. Soc.* 120: 4049–4050.
94. Britovsek, G. J. P., Gibson, V. C., Kimberley, B. S., Maddox, J., McTavish, S. J., Solan, G. A., White, A. P., Williams, D. 1998. Novel olefin polymerization catalysts based on iron and cobalt. *Chem. Commun.* 849–850.
95. Wang, C., Friedrich, S. K., Younkin, T. R., Li, R. T., Grubbs, R. H., Bansleben, D. A., Day, M. W. 1998. Neutral Ni(II)-based catalysts for ethylene polymerization. *Organometallics* 17: 3149–3151.
96. Matsui, S., Tohi, Y., Mitani, M., Saito, J., Makio, H., Tanaka, H., Nitabaru, M., Nakano, T., Fujita, T. 1999. New bis(salicylaldiminato)titanium complexes for ethylene polymerization. *Chem. Lett.* 28: 1065–1066.
97. Ittel, S. D., Johnson, L. K., Brookhart, M. 2000. Late metal catalysts for ethylene homo- and co-polymerization. *Chem. Rev.* 100: 1169–1203.
98. Stevens, P. N., Timmers, F. J., Wilson, D. R., Schmidt, G. F., Nickias, P. N., Rosen, R. K., Knight, G. W., Lai, S. 1991. Constrained geometry addition polymerization catalysts, processes for their preparation, precursors therefor, methods of use, and novel polymers formed therewith. *Eur. Pat. Appl.* EP-416815-A2.
99. McKnight, A. L., Waymouth, R. 1998. Group 4 ansa-cyclopentadienyl-amido catalysts for olefin polymerization. *Chem. Rev.* 98: 2587–2598.
100. Shapiro, P. J., Bunel, E., Schaefer, W. P., Bercaw, J. E. 1990. [{(η^5-C$_5$Me$_4$)Me$_2$Si(η^1-NCMe$_3$)} (PMe$_3$)ScH]$_2$: A unique example of a single component α-olefin polymerization catalyst. *Organometallics* 9: 867–869.
101. Okuda, J. 1990. Functionalized cyclopentadienyl ligands, IV. Synthesis and complexation of linked cyclopentadienyl-amido ligands. *Chem. Ber.* 123: 1649–1651.
102. Baier, M. C., Zuideveld, M. A., Mecking, S. 2014. Post-metallocenes in the industrial production of polyolefins. *Angew. Chem. Int. Ed.* 53: 9722–9744.
103. Klosin, J., Fontaine, P. P., Figuerao, R. 2015. Development of group IV molecular catalysts for high temperature ethylene-α-olefin copolymerization reactions. *Acc. Chem. Res.* 48: 2004–2016.
104. Pellecchia, C., Zambelli, A. 1996. Syndiotactic-specific polymerization of propene with a Ni-based catalyst. *Macromol. Rapid Commun.* 17: 333–338.
105. McLain, S. J., Feldman, J., McCord, E. F., Gardner, K. H., Teasley, M. F., Coughlin, E. B., Sweetmann, K. J., Johnson, L. K., Brookhart, M. 1998. Addition polymerization of cyclopentene with nickel and palladium catalysts. *Macromolecules* 31: 6705–6707.
106. Small, B. L., Brookhart, M. 1999. Polymerization of propylene by a new generation of iron catalysts: Mechanism of chain initiation, propagation and termination. *Macromolecules* 32: 2120–2130.
107. Gibson, V. C., Spitzmesser, S. K. 2003. Advances in non-metallocene olefin polymerization catalysis. *Chem. Rev.* 103: 283–316.
108. Pincer ligands have been recently reviewed, see: Zell, T., Milstein, D. 2015. Hydrogenation and dehydrogenation iron pincer catalysts capable of metal ligand cooperation by aromatization and de-aromatization. *Acc. Chem. Res.* 48: 1979–1994.
109. Gibson, V. C., Redshaw, C., Solan, G. A. 2007. Bis(imino)pyridines: Surprisingly reactive ligands and a gateway to new families of catalysts. *Chem. Rev.* 107: 1745–1776.
110. Ma, J., Feng, C., Wang, S., Zhao, K.-Q., Sun, W.-H., Redshaw, C., Solan, G. A. 2014. Bi- and tri-dentate imino-based iron and cobalt pre-catalysts for ethylene oilgo-/polymerization. *Inorg. Chem. Front.* 1: 14–34.

111. Saito, J., Mitani, M., Matsui, S., Sugi, M., Tohi, Y., Tsutsui, T., Fujita, T., Nitabaru, M., Makio, H. 1997. Olefin polymerization catalysts, transition metal compounds, processes for olefin polymerization, and alpha-olefin/conjugated diene copolymers (Assigned to Mitsui Chemicals Inc.) EP-0874005.

112. Keim, W., Kowaldt, F. H., Goddard, R., Krueger, C. 1978. Novel coordination of (Benzoylmethylene)triphenylphosphorane in a nickel oligomerization catalyst. *Angew. Chem. Int. Eg. Engl.* 17: 466–467.

113. Younkin, T. R., Conner, E. F., Henderson, J. I., Friedrich, S. K., Grubbs, R. H., Bansleben, D. A. 2000. Neitral, single-component Ni(II) polyolefin catalysts that tolerate heteroatoms. *Science* 287: 460–462.

114. Osichow, A., Rabe, C., Vogtt, K., Narayanan, T., Harnau, L., Drechsler, M., Ballauff, M., Mecking, S. 2013. Ideal polyethylene nanocrystals. *J. Am. Chem. Soc.* 135: 11645–11650.

115. Mu, H., Pan, L., Song, D., Li, Y. 2015. Neutral nickel catalysts for olefin homo- and copolymerization: Relationship between catalyst structures and catalytic properties. *Chem. Rev.* 115: 12091–12137.

116. Saito, J., Mitani, M., Mohri, J., Yoshida, Y., Matsui, S., Ishii, S., Kojoh, S., Kashiwa, N., Fujita, T. 2001. Living polymerization of ethylene with a titanium complex containing two phenoxy-imine chelate ligands. *Angew. Chem. Int. Ed.* 40: 2918–2920.

117. Makio, H., Fujita, T. 2009. Development and application of FI catalysts for olefin polymerization: Unique catalysis and distinctive polymer formation. *Acc. Chem. Res.* 42: 1532–1544.

118. Mitani, M., Saito, J., Ishii, S., Nakayama, Y., Makio, H., Matsukawa, N., Matsui, S., Mohri, J., Furuyama, R., Terao, H., Bando, H., Tanaka, H., Fujita, T. 2004. FI catalysts: New olefin polymerization catalysts for the creation of value-added polymers. *Chem. Rec.* 4: 137–158.

119. Furuyama, R., Saito, J., Ishii, S., Mitani, M., Matsui, S., Tohi, Y., Makio, H., Matsukawa, N., Tanaka, H., Fujita, T. 2003. Ethylene and propylene polymerization behaviour of a series of bis(phenoxy-imine)titanium complexes. *J. Mol. Catal. A: Chem.* 200: 31–42.

120. Mitani, M., Mohri, J., Yoshida, Y., Saito, J., Ishii, S., Tsuru, K., Matsui, S., Furuyama, R., Nakano, T., Tanaka, H., Kojoh, S., Matsugi, T., Kashiwa, N., Fujita, T. 2002. Living polymerization of ethylene catalyzed by titanium complexes having fluorine-containing phenoxy-imine chelate ligands. *J. Am. Chem. Soc.* 124: 3327–3336.

121. Mitani, M., Furuyama, R., Mohri, J., Saito, J., Ishii, S., Terao, H., Kashiwa, N., Fujita, T. 2002. Fluorine-trimethylsilyl-containing phenoxy-imine Ti complex for highly syndiotactic living propylenes with extremely high melting temperatures. *J. Am. Chem. Soc.* 124: 7888–7889.

122. Khaubunsongserm, S., Jongsomjit, B., Praserthdam, P. 2013. Bis[N-(3-tertbutylsalicylidene)cyclooctylamine]titanium dichloride activated with MAO for ethylene polymerization. *Eur. Polym. J.* 49: 1753–1759.

123. Khaubunsongserm, S., Hormnirun, P., Nanok, T., Jongsomjit, B., Praserthdam., P. 2013. Fluorinated bis-(phenoxy-imine) titanium complexes with methylaluminoxane for the synthesis of ultrahigh molecular weight polyethylene. *Polymer* 54: 3217–3222.

124. Li, T., Kong, F. W., Liu, R., Li, Z. Y., Zhu, F. M. 2011. The effect of cocatalysts on ethylene polymerization with fluorinated bis-phenoxyimine titanium as a catalyst. *J. Appl. Polym. Sci.* 119: 572–576.

125. Matsugi, T., Fujita, T. 2008. High-performance olefin polymerization catalysts discovered on the basis of a new catalyst design concept. *Chem. Soc. Rev.* 37: 1264–1277.

126. Kurtz, S. M. in *The UHMWPE Handbook*, Elsevier Academic Press, 2004.

127. Jones, R. L., Armoush, 2009. Catalysts for UHMWPE. *Macromol. Symp.* 283–284: 88–95.

128. Kaminsky, W. 2016. Production of polyolefins by metallocene catalysts and their recycling by pyrolysis. *Macromol. Symp.* 360: 10–22.

129. Makio, H., Kashiwa, N., Fujita, T. 2002. FI catalysts: A new family of high performance catalysts for olefin polymerization. *Adv. Synth. Catal.* 344: 477–493.

130. Mark, S., Kurek, A., Muelhaupt, R., Xu, R., Klatt, G., Koeppel, H., Enders, M. 2010. Hydridoboranes as modifiers for single-site organochromium catalysts: From low- to ultrahigh-molecular weight polyethylene. *Angew. Chem. Int. Ed.* 49: 8751–8754.

131. Zhang, L., Gao, W., Tao, X., Wu, Q., Mu, Y., Ye, L. 2011. New chromium (III) complexes with imine-cyclopentadienyl ligands: Synthesis, characterization and catalytic properties for ethylene polymerization. *Organometallics* 30: 433–440.

132. Kurek, A., Mark, S., Enders, M., Stuerzel, M., Muelhaupt, R. 2014. *J. Mol. Catal. A: Chem.* 383–384: 53–57.

133. Wang, Y., Fan, H., Jie, S., Li, B. G. 2014. Synthesis and characterization of titanium (IV) complexes bearing end functionalized biphenyl: Efficient catalysts for synthesizing high molecular weight polyethylene. *Inorg. Chem. Commun.* 41: 68–71.

134. Diamond, G. M., Leclerc, M. K., Zhu, G. 2010. Methods for producing very high or ultra-high molecular weight polyethylene. Assigned to Symyx solutions Inc. USA, WO 2010/078164A1.

135. Stevens, J. C., Timmers, F. J., Rosen, G. W., Knight, G. W., Lai, S. Y. 1991. Constrained geometry addition polymerization catalysts, processes for their preparation, precursor therefor, methods of use, and novel polymers formed therewith. European patent assigned to *Dow*, EP416815.

136. van Beek, J. A. M., van Doremaele, G. H. J., Gruter, G. J. M., Arts, H. J., Eggels, G. H. M. R. 1996. Catalyst composition and process for the polymerization of an olefin. Assigned to *DSM*. WO9613529.

137. Mihan, S., Lilge, D., De Lange, P., Schweier, G., Schneider, M., Rief, U., Hack, J., Enders, M., Ludwig, G., Rudolph, R. 2001. Monocyclopentadienyl complexes of chromium, molybdenum or tungsten with a donor bridge. Assigned to BASF, Ludwigshafen, Germany. WO 01/12641A1.

138. Mitani, M., Matsukawa, N., Saito, J., Nitabara, M., Tsuru, K., Matsui, S., Fujita, T. 2000. Olefin polymerization catalysts and olefin polymerization methods using same. Assigned to *Mitsui Chemicals, Japan*. JP 2000230010A.

139. Makio, H., Terao, H., Iwashita, A., Fujita, T. 2011. FI catalysts for olefin polymerization-A comprehensive treatment. *Chem. Rev.* 111: 2363–2449.

140. Lewis, G. 2001. Properties of crosslinked ultra-high-molecular-weight polyethylene. *Biomaterials* 22: 371–401.

141. Rastogi, S., Yao, Y., Ronca, S., Bos, J., van der Eem, J. 2011. Unprecedented high-modulus high-strength tapes and films of ultrahigh molecular weight polyethylene via solvent free route. *Macromolecules*, 44: 5558–5568 and the references therein.

142. a) Yao, Y. F., Graf, R., Spiess, H. W., Rastogi, S. 2008. Restricted segmental mobility can facilitate medium-range chain diffusion: A NMR study of morphological influence on chain dynamics of polyethylene. *Macromolecules*, 41: 2514–2519; b) Yao, Y., Jiang, S., Rastogi, S. 2014. ^{13}C solid state NMR characterization of structure and orientation development in the narrow and broad molecular mass disentangled UHMWPE. *Macromolecules*, 47: 1371–1382.

143. Jenkins, H., Keller, A. 1975. Radiation induced changes in physical properties of bulk polyethylene. I. Effect of crystallization conditions. *J. Macromol. Sci. Part: B, Phys.* 11: 301–323.

144. Bartczak, Z. 2010. Effect of chain entanglement on plastic deformation behaviour of ultra-high molecular weight polyethylene. *J. Polym. Sci. Part B: Polym. Phys.* 48: 276–285.

145. Rastogi, S., Kurelec, L., Lippits, D., Cuijpers, J., Wimmer, M. Lemstra, P. J. 2005. Novel route to fatigue-resistant fully sintered ultrahigh molecular weight polyethylene for knee prosthesis. *Biomacromolecules* 6: 942–947.

146. Smith, P., Lemstra, P. J. 1980. Ultra-high-strength polyethylene filaments by solution spinning/drawing. *J. Mater. Sci.* 15: 505–514.

147. a) Rastogi, S., Lippits, D., Peter, G. W. M., Graf, R., Yao, Y.-F., Spies, H.W. 2005. Heterogeneity in polymer melts from melting of polymer crystals. *Nat. Mater.* 4: 635–641; b) Rastogi, S., Lippits, D., Talebi, S., Bailley, C. 2006. Formation of entanglements in initially disentangled polymer metls. *Macromolecules* 39: 8882–8885; c) Talebi, S., Duchateau, R., Rastogi, S., Kaschta, J., Peters, G. W. M., Lemstra, P. J. 2010. Molar mass and molecular weight distribution determination of UHMWPE synthesized using a living homogeneous catalyst. *Macromolecules* 43: 2780–2788; d) Ronca, S., Romano. D., Forte, G., Andablo-reyes, E., Rastogi, S. 2012. Improving the performance of a catalytic system for the synthesis of ultrahigh molecular weight polyethylene with a reduced number of entanglements. *Adv. Polym. Technol.* 31: 193–204; e) Ronca, S., Forte, G., Tjaden, H., Yao, Y., Rastogi, S. 2013. Tailoring molecular structure via nanoparticles for solvent free processing of ultra-high molecular weight polyethylene composites. *Polymer* 53: 2897–2907; f) Ronca, S., Forte, G., Ailianou, A., Kornfield, J. A., Rastogi, S. 2012. Direct route to colloidal UHMWPE by including LLDPE in solution during homogeneous polymerization of ethylene. *ACS Macro Lett.* 1: 1116–1120; g) Romano, D., Ronca, S., Rastogi, S. 2015. A hemi-metallocene chromium catalyst with trimethylaluminium-free methylaluminoxane for the synthesis of disentangled ultra-high molecular weight polyethylene. *Macromol. Rapid Commun.* 36: 327–331.

148. a) Mathur, A. B., Gandham, S. S. R., Satpathy, U. S., Sarma, K. R., Patil, Y. P., Patel, N. F., Mehta, G. M., Jasra, R. V. 2014. High strength polyethylene products and a process for preparation thereof. Assigned to Reliance Industries Limited, WO 2014192025 A3; b) Gandham, S. S. R., Mathur, A. B., Satpathy, U. S., Mehta, G. M., Patel, N. F., Sarma, K. R., Jasra, R. V. 2015. A process for preparing a disentangled uhmwpe product. Assigned to Reliance Industries Limited, WO 2015140681 A1; c) Mathur, A. B., Gandham, S. S. R., Satpathy, U. S., Sarma, K. R., Jasra, R. V. 2015. Disentangled ultrahigh molecular weight polyethylene graft co-polymers and a process for preparation thereof. Assigned to Reliance Industries Limited, US 20150240017 A1; d) Satpathy, U. S., Gandham, S. S. R., Mathur, A. B., Jasra, R. V., Sarma, K. R., Shah, A. K. P., Amin, Y. M., Mehta, G. M., Patel, N. F., Patel, V. K. 2015. Polyethylene fibers, ultrahigh molecular weight high strength, high modulus. Assigned to Reliance Industries Limited, WO 2015125064 A1.

149. a) De Weijer, A. P., Van de Hee, H., Peters, G. W. M., Rastogi, S., Wang, B. 2009. Polyethylene film with high tensile strength and high tensile energy to break. Assigned to Teijin Aramid B.V., WO 2009007045 A1; b) Eem van der, J., De Weijer, A. P., Rastogi., S. 2010. Polyethylene film with high tensile strength and high tensile energy to break. Assigned to Teijin Aramid B.V., WO 2010079172 A1; c) Romano, D., Ronca, S., Bos, J., Rastogi, S. 2014. Process for the synthesis of dis-entangled UHMW-PE using a Cr or Ti catalyst. Jointly assigned to Teijin Aramid B.V. and Loughborough University, UK. WO 2014060252 A1; d) Teijin can be accessed at: http://www.teijinendumax.com/products/products/.

150. Drent, E., van Dijk, R., van Ginkel, R., van Ort, B., Pugh, R. I. 2002. Palladium catalyzed copolymerization of ethene with alkylacrylates: Polar comonomer built into the linear polymer chain. *Chem. Commun.* 744–745.

151. Carrow, B. P., Nozaki, K. 2012. Synthesis of functional polyolefins using cationic bisphosphine monoxide-palladium complexes. *J. Am. Chem. Soc.* 134: 8802–8805.

152. Nakano, R., Nozaki, K. 2015. Copolymerization of propylene and polar monomers using Pd/IzQO catalysts. *J. Am. Chem. Soc.* 137: 10934–10937.

153. Mu, H., Pan, L., Song, D., Li, Y. 2015. Neutral nickel catalysts for homo- and copolymerization: Relationship between catalyst structure and catalytic properties. *Chem. Rev.* 115: 12091–12137.

154. a) Long, B. K., Eagan, J. M., Mulzer, M., Coates, G. W. 2016. Semi-crystalline polar polyethylene: Ester functionalized linear polyolefins enabled by a functional-group tolerant, cationic nickel catalyst. *Angew. Chem. Int. Ed.* 55: 7106–7110; b) Dai, S., Chen, C. 2016. Direct synthesis of functionalized high molecular weight polyethylene by copolymerization of ethylene with polar monomers. *Angew. Chem. Int. Ed.* 55: 13281–13285.

155. Connor, E. F., Younkin, T. R., Henderson, J. I., Hwang, S., Grubbs, R. H., Roberts, W. P., Litzau, J. J. 2002. Linear functionalized polyethylene prepared with highly active neutral Ni(II) complexes. *J. Polym. Sci. Part A: Polym. Chem.* 40: 2842–2854.

156. a) Radlauer, M. R., Day, M. W., Agapie, T. 2012. Dinickel bisphenoxyiminato complexes for the polymerization of ethylene and α-olefins. *Organometallics* 31: 2231–2243; b) Radlauer, M. R., Day, M. W., Agapie, T. 2012. Bimetallic effects on ethylene polymerization in the presence of amines: Inhibition of the deactivation by Lewis bases. *J. Am. Chem. Soc.* 134: 1478–1481; c) Radlauer, M. R., Buckley, A. K., Henling, L. M., Agapie, T. 2013. Bimetallic coordination insertion polymerization of unprotected polar monomers: Copolymerization of amino olefins and ethylene by dinickel bisphenoxyiminato catalysts. *J. Am. Chem. Soc.* 135: 3784–3787.

157. Takeuchi, D., Chiba, Y., Takano, S., Osakada, K. 2013. Double-decker-type dinuclear nickel catalyst for olefin polymerization: Efficient incorporation of functional co-monomers. *Angew. Chem. Int. Ed.* 52: 12536–12540.

158. Luo, S., Vela, J., Lief, G. R., Jordan, R. F. 2007. Copolymerization of ethylene and alkyl vinyl ethers by a (Phosphine-sulfonate)PdMe catalyst. *J. Am. Chem. Soc.* 129: 8946–8947.

159. Kochi, T., Noda, S., Yoshimura, K., Nozaki, K. 2007. Formation of linear copolymers of ethylene and acrylonitrile catalyzed by phosphine sulfonate palladium complexes. *J. Am. Chem. Soc.* 129: 8948–8949.

160. Borkar, S., Newsham, D. K., Sen, A. 2008. Copolymerization of ethene with styrene derivatives, vinyl ketone and vinylcyclohexane using a (phosphine-sulfonate) palladium(II) system: Unusual functionality and solvent tolerance. *Organometallics* 27: 3331–3334.

161. Skupov, K. M., Piche, L., Claverie, J. P. 2008. Linear polyethylene with tunable surface properties by catalytic copolymerization of ethylene with *N*-Vinyl-2-pyrrolidone (NVP) and *N*-Isopropylacrylamide. *Macromolecules* 41: 2309–2310.

162. Ito, S., Munakata, K., Nakamura, A., Nozaki, K. 2009. Copolymerization of vinyl acetate with ethylene by palladium/alkylphosphine-sulfonate catalysts. *J. Am. Chem. Soc.* 131: 14606–14607.

163. Guironnet, D., Roesle, P., Ruenzi, T., Goettker-Schnetmann, I., Mecking, S. 2009. Insertion polymerization of acrylate. *J. Am. Chem. Soc.* 131: 422–423.

164. Ruenzi, T., Froelich, D., Mecking, S. 2010. Direct synthesis of ethylene-acrylic acid copolymers by insertion polymerization. *J. Am. Chem. Soc.* 132: 17690–17691.

165. Kryuchkov, V. A., Daigle, J.-C., Skupov, K. M., Claverie, J. P., Winnik, F. M. 2010. Amphiphilic polyethylenes leading to surfactant free thermo-responsive nanoparticles. *J. Am. Chem. Soc.* 132: 15573–15579.

166. Bouilhac, C., Ruenzi, T., Mecking, S. 2010. Catalytic copolymerization of ethylene with vinyl sulfones. *Macromolecules* 43: 3589–3590.

167. a) Shen, Z., Jordan, R. F. 2010. Copolymerization of ethylene and vinyl fluoride by (Phosphine-bis(arenesulfonate))PdMe(pyridine) catalysts: Insights into inhibition mechanism. *Macromolecules* 43: 8706–8708; b) Leicht, H., Goettker-Schnetmann, I., Mecking, S. 2013. Incorporation of vinyl chloride in insertion polymerization. *Angew. Chem. Int. Ed.* 52: 3963–3966.

168. Liu, S., Borkar, S., Newsham, D., Yennawar, H., Sen, A. 2007. Synthesis of palladium complexes with an anionic P~O chelate and their use in copolymerization of ethene with functionalized norbornene derivatives: Unusual functionality tolerance. *Organometallics* 26: 210–216.

169. Friedberger, T., Wucher, P., Mecking, S. 2012. Mechanistic insights into polar monomer insertion polymerization from acrylamides. *J. Am. Chem. Soc.* 134: 1010–1018.

170. Daigle, J.-C., Piche, L., Claverie, J. P. 2011. Preparation of functional polyethylenes by catalytic copolymerization. *Macromolecules* 44: 1760–1762.

171. Lanzinger, D., Giuman, M. M., Anselment, T. M. J., Rieger, B. 2014. Copolymerization of ethylene and 3, 3, 3-trifluoropropene using (phosphinesulfonate) Pd(Me)(DMSO) as catalyst. *ACS Macro Lett.* 3: 931–934.

172. Jian, Z., Baier, M. C., Mecking, S. 2015. Suppression of chain transfer in catalytic acrylate polymerization via rapid and selective secondary interactions. *J. Am. Chem. Soc.* 137: 2836–2839.

173. Jian, Z., Mecking, S. 2016. Insertion polymerization of divinyl formal. *Macromolecules* 49: 4395–4403.

174. Ito, S., Kanazawa, M., Munakata, K., Kuroda, J., Okumura, Y., Nozaki, K. 2011. Coordination-insertion copolymerization of allyl monomers with ethylene. *J. Am. Chem. Soc.* 133: 1232–1235.

175. Ota, Y., Ito, S., Kuroda, J., Okumura, Y., Nozaki, K. 2014. Quantification of the steric influence of alkylphosphine-sulfonate ligands on polymerization, leading to high-molecular-weight copolymers of ethylene and polar monomers. *J. Am. Chem. Soc.* 136: 11898–11901.

176. Gaikwad, S. R., Deshmukh, S. S., Gonnade, R. G., Rajamohanan, P. R., Chikkali, S. H. 2015, Insertion copolymerization of difunctional polar vinyl monomers with ethylene. *ACS Macro Lett.* 4: 933–937.

177. For reviews on the success of phosphine-sulfonate ligand systems, see: a) Nakamura, S., Ito, S., Nozaki, K. 2009. Coordination-insertion copolymerization of fundamental polar monomers. *Chem. Rev.* 109: 5215–5244; b) Nakamura, A., Anselment, T. M. J., Claverie, J., Goodall, Jordan, R., Mecking, S., B., Rieger, B., Sen, A., Van Leeuwen, P. W. N. M., Nozaki, K. 2013. ortho-Phosphinebenzenesulfonate: A superb ligand for palladium-catalyzed coordination-insertion copolymerization of polar vinyl monomers. *Acc. Chem. Res.* 46: 1438–1449; c) Gaikwad, S., Deshmukh, S., Chikkali, S. H. 2014. Pd-phosphinesulfonate bravely battles the vinyl halide insertion copolymerization barricade. *J. Polym. Sci. Part A: Polym. Chem.* 52: 1–6.

178. Nozaki, K., Kusumoto, S., Noda, S., Kochi, T., Chung, L. W., Morokuma, K. 2010. Why did incorporation of acrylonitrile to a linear polyethylene become possible? Comparison of phosphine-sulfonate ligand with diphosphine and imine-phenolate ligands in the Pd-catalyzed ethylene/acrylonitrile copolymerization. *J. Am. Chem. Soc.* 132: 16030–16042.

179. Neuwald, B., Caporaso, L., Cavallo, L., Mecking, S. 2013. Concepts for stereoselective acrylate insertion. *J. Am. Chem. Soc.* 135: 1026–1036.

180. a) Carrow, B. P., Nozaki, K. 2012. Synthesis of functional polyolefins using cationic bisphosphine monoxide-palladium complexes. *J. Am. Chem. Soc.* 134: 8802–8805; b) Mitsushige, Y., Carrow, B. P., Ito, S., Nozaki, K. 2016. Ligand controlled insertion regioselectivity accelerates copolymerization of ethylene with methylacrylate by cationic bisphosphine monoxide-palladium catalysts. *Chem. Sci.* 7: 737–744.

181. Zhou, X., Jordan, R. F. 2011. Synthesis, cis-trans isomerization, and reactivity of palladium alkyl complexes that contain a chelating N-heterocylic-carbene sulfonate ligand. *Organometallics* 30: 4632–4642.

182. Mike Chung, T. C. 2013. Functional polyolefins for energy applications. *Macromolecules* 46: 6671–6698.

3 Carbene or C1 Polymerization

Bas de Bruin and Samir H. Chikkali

CONTENTS

3.1 INTRODUCTION TO CARBENE POLYMERIZATION

As noted in Chapter 2, metal-catalyzed polymerization of carbon–carbon double bond (C=C) containing monomers is most widely known and today we produce roughly 180 million tons of polyolefins every year.[1] In this so-called *olefin polymerization* or *vinyl insertion polymerization* method, each monomer delivers two carbon (C2) atoms in each propagation step. Insertion polymerization of monomers delivering only one carbon unit in each chain growth step is named as *C1 or carbene polymerization* (see Figure 3.1).[2] In the recent past, C1 polymerization techniques are being viewed as a valuable alternative to the classical C2 polymerization methods. Not only this but also C1 polymerization offer distinct advantages over C2 polymerization and thus opens up new avenues for the development of new material with polymer properties that are very difficult to achieve using traditional C2 polymerization methods. In this chapter we will discuss the significance of C1 polymerization and the different methods used for C1 polymerization, will summarize the recent developments, and will highlight the elementary steps involved in it.

The current source of C2 monomers is mainly crude oil, which is currently available in ample quantities and hence the monomers are cheaper at this moment.

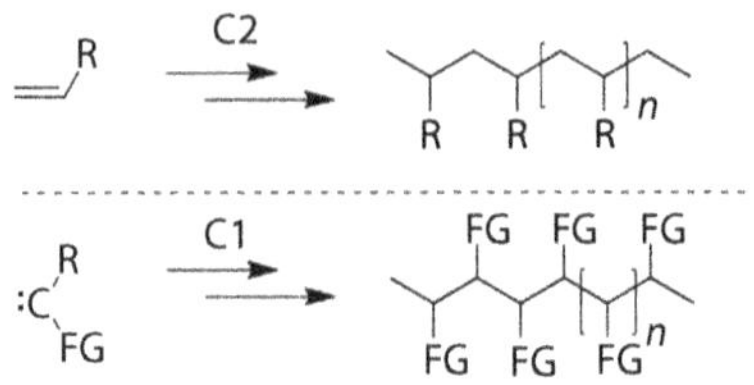

FIGURE 3.1   C2 versus C1 polymerization (R = H, Me, Ph; FG = Functional Group).

However, the supply of crude oil is finite and unsustainable, which calls for long-term sustainable strategies for predictable and uninterrupted supply of monomers. Apart from the monomer supply, another important synthetic limitation of the C2 polymerization methods is their inability to polymerize functional olefins. As presented in Chapter 2, although recent exploration in the field has produced some breakthrough, the functional group incorporation is limited to only few percent at the best with random incorporation. Therefore, producing well-defined highly functionalized and stereoregular (which can be achieved only with metal catalysts) polymers is a significant challenge, even today. The challenge becomes even more pronounced if we attempt metal-catalyzed polymerization of C2 monomers with two or more functional groups. There is only one report on metal-catalyzed copolymerization difunctional olefin with ethylene.[3] If these simple polar functionalities can pose such a huge problem, even larger synthetic challenges are anticipated if more complex functionalities such as energy- or electron-transporting functionalities have to be incorporated. Thus, a polymerization method that can tolerate a wide range of functional groups and incorporate them in a controllable amount and at designated positions will be highly sought. C1 polymerization can be one such approach, which is reported to tolerate functional groups and insert them in a stereoregular fashion.

The C1 polymerization techniques use carbenoid monomers with one or two functional groups per monomer. Such carbenoid monomers can be accessed from diazo-compounds or ylides of sulfoxides, and ylides of phosphonium. Figure 3.2 depicts

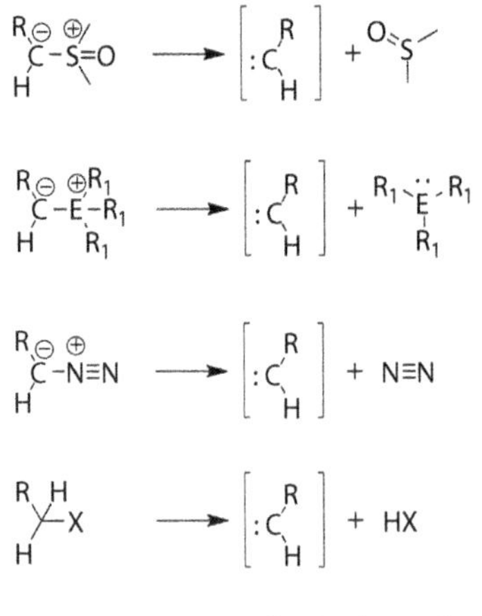

FIGURE 3.2   Accessible monomer precursors for *in situ* carbene (C1) polymerization.

the accessible range of C1 monomer precursor and the monomers in square bracket. There are various methods to prepare these monomers for *in situ* polymerization. The most common transformations used for the *in situ* generation of C1 monomers are deprotonation, elimination, and thermal decomposition. The most reported C1 polymerization methods make use of Lewis acid (LA) or transition metal (TM) catalysts, baring random reports on anionic polymerization[4] or Mg-mediated polymerization[5] of C1 monomers.

3.2 LEWIS ACID-MEDIATED POLYMERIZATION

As depicted in Figure 3.2, the *in situ*-prepared monomers carry a lone pair on the carbon atom and therefore the respective carbon is nucleophilic in character. Due to the nucleophilicity of these C1 monomers, LA compounds can easily react with the lone pair of these carbene precursors. Use of these types of diazo-compounds or ylides as a monomer precursor involves insertion of their carbene fragment into the growing polymer chain. Aluminum and in particular boron are well-known mediators for C1 polymerization of diazoalkanes and ylides. Although boron is considered to be a metalloid (metal-like, not a metal), it does play an important role in C1 polymerization and hence is included here for completeness. A summary of the boron and aluminum-mediated C1 polymerization is presented in this section. A detailed overview of LA-mediated C1 polymerization has been presented.[6]

3.2.1 Boron-Mediated C1 Polymerization

Boron-mediated C1 polymerization represents one of the oldest C1 polymerization methods. A variety of boron compounds such as organoboranes, haloboranes, and borohydrides are known to polymerize C1 monomers. Diazomethane has been intensively investigated as monomer in the boron-mediated/catalyzed C1 polymerization to produce highly crystalline and highly linear polymethylene. Table 3.1 summarizes boron-mediated polymerization of diazomethanes. C1 polymerization of diazomethane by BF_3 represents one of the most successful boron-mediated polymerization, and molecular weight up to 3 million Da (Dalton or gm/mol) could be obtained.[7] However, the higher analogs of diazomethane seem to polymerize to

TABLE 3.1
Boron-Mediated C1 Polymerization of Diazo-Precursors

Run	Boron Compound	Monomer Precursor	Molecular Weight (Da)
1	BF_3	Diazomethane	3×10^6
2	BF_3	Diazoethane	5000
3	BF_3	(2-Diazo ethyl)benzene	3000
4	$B(\alpha\text{-naphtyl})_3$	Diazomethane + diazoethane	70000
5	$B(\alpha\text{-naphtyl})_3$	Diazomethane + diazo-*iso*-butane	1700
6	$BF_3.Et_2O$	4-Diazomethyl[2.2]paracyclophanes	1200
7	BH_3	Dethyl diazoacetate	500

much lower molecular weight polymers/oligomers. BF_3-catalyzed C1 polymerization of (2-diazo ethyl)benzene produced a polymer with only 3000 Da molecular weight. A similar trend was observed in organoborane-catalyzed polymerization of diazomethanes (Table 3.1 run 4 and 5). Boron hydride was found to be a very poor catalyst under the given conditions and produced only oligomers (molecular weight of 500 Da). The lower molecular weight of the polymers can be ascribed to hydride or alkyl group transfer to the electrophilic carbon bound to boron. This proposition is further supported by the unsaturated chain ends in the polymer, pointing to a hydride/alkyl transfer mechanism for chain transfer. Apart from diazoalkane, other diazo-compounds such as diazo esters or diazoketones were evaluated in boron-mediated C1 polymerization. However, with these polar-substituted monomers, only one monomer was found to insert in the B-C bonds of the trialkyl boron reagent, and the subsequent insertions are largely blocked.[8,9] It is most likely that the diazo ester monomers chelate through the oxygen lone pair of boron and form stable O-boron enolate intermediates, which hamper subsequent monomer insertions.[10]

Two types of mechanisms as depicted in Figure 3.3a and b have been proposed to be operative, which might heavily depend on the boron reagent. Davies et al. proposed an ionic mechanism that involves external nucleophilic attack of the diazo-compound on the polarized carbon atom of the growing chain (Figure 3.3b),[11] while repetitive intramolecular nucleophilic attack of the growing chain on the diazo-methane adduct via a migratory insertion mechanism is favored by Matthies and coworkers (Figure 3.3a).[12]

The next class of monomer precursors used in C1 polymerization concerns sulfur ylides. The reactivity of sulfur ylide (i.e., sulfoxide ylides) was investigated as early as 1966 by Tufariello and Lee.[13] It was reported that sulfoxide ylide reacts with an organoborane compound and the carbon chains on organoborane are elongated by one methylene (CH_2) unit. This process of elongation of the three carbon-chains on boranes by one carbon each is termed as *homologation*. Thus, all the three chains (substituents) on boron grow in length to produce long-chain trialkyl boranes. After the chain growth, nucleophilic reagents (such as –OH) are added to displace the alkyl chains from boron. The mechanism of *homologation* was found to be slightly different than those proposed in Figure 3.3a and b. Nucleophilic attack of the carbene-carbon (in the sulfonxide ylide) on to the electrophilic boron

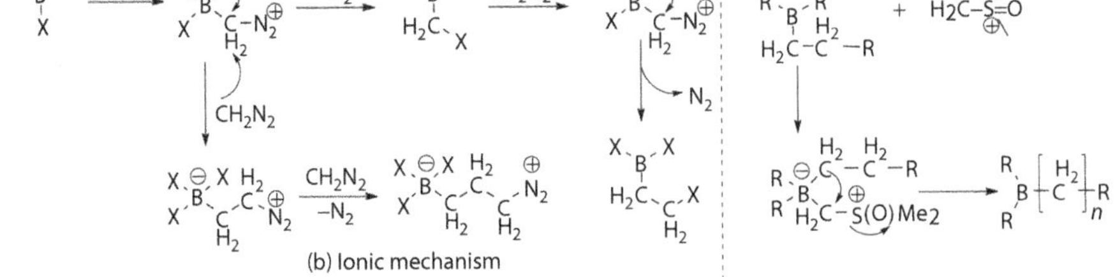

FIGURE 3.3 Proposed mechanisms for boron-mediated polymerization of C1 monomers: (a) Migration, (b) ionic, and (c) polyhomologation.

center followed by 1,2-migration leads to insertion of one carbon atom in the boron-alkyl bond (Figure 3.3c). After a gap of three decades, Shea and coworkers succeeded in converting the *homologation* reaction into *polyhomologation* process and reported the synthesis of linear polymethylene in 1997.[14] The polyhomologation involves repetitive homologation of ylide in the presence of excess ylide to long-chain alkylboranes, which after basic workup yield hydroxyl-terminated polymethylene. The reaction was found to be living in nature and the polymerization displayed characteristic features of a *living polymerization* method. Thus, the ylide/borane ratio determines the chain length of the resultant polymer, and polydispersitites as low as 1.01 could be achieved.[15] Further, polyhomologation enables access to very high molecular weight polymers of 500,000 Da. The reaction kinetics indicated that the polyhomologation is of zero order in ylide and first order in borane. Mechanistic investigations suggested a similar mechanism as proposed by Tufariello (Figure 3.3c).

Subsequent to these initial leads, a very general and robust methodology has been developed and polymethylene with an array of functional groups has been prepared.[16] A variety of terminal functionalized polymethylenes, with different architectures,[17,18] telechelic polymers,[19] with different functional chain ends,[20] and gradient methylidene-ethylidene copolymers[21] have been prepared, which would not have been possible using conventional vinyl insertion polymerization methods.[22,23] However, with the exception of one example,[24] polyhomologation of functionalized (containing hetero-atoms) sulfur ylides was largely unsuccessful, and functional polymethylene could not be produced using C1 polymerization of sulfur ylides.[25–27]

3.2.2 ALUMINUM-MEDIATED C1 POLYMERIZATION

Alkyl aluminum compounds have been scarcely investigated in C1 polymerization, but a series of organoaluminum compounds such as trimethyl aluminum, triethyl aluminum, di-isobutyl aluminum hydride, and so on were tested in the C1 polymerization of diazoketones by Ihara and coworkers.[28] Rather uncontrolled polymerization was observed producing polymers/oligomers with polydispersity index (PDI) values of 1.5 and low molecular weights. Quite remarkably, the resultant polymers revealed the presence of azo functionalities, along with –CHMe units. The incorporation of azo-units in the polymer was proposed to involve attack of alkyl-aluminum on the nitrogen fragment of the diazo-compounds. The formation of ethylidene units was momentarily explained by reductive cleavage of the acryl group in the polymerization reaction. As for boron, the aluminum-mediated C1 polymerization fails to incorporate functional groups.

3.3   TRANSITION METAL-CATALYZED C1 POLYMERIZATION

Although high molecular weight (co)polymers could be prepared using boron-mediated C1 polymerization, the LA-mediated C1 polymerization suffers from following limitations: (1) Polymerization of functional C1 monomers could not be realized; (2) stereoselective synthesis of a polymer is highly unlikely and remains a synthetic challenge; (3) only certain boron reagents could perform well in

C1 polymerization, whereas other boron reagents or other LAs produced only low molecular weight oligomers/polymers. Therefore, the scope of the LA-mediated C1 polymerization is limited.

TM-catalyzed polymerization of C1 monomers can potentially address the above-mentioned limitations and might offer additional advantages over the LA-mediated C1 polymerization methods. Indeed, a significant amount of literature deals with TM-catalyzed carbene polymerization, which has been reviewed by Imoto and Nakaya.[29] An added advantage of TM catalysis is the modularity of the associated ligand, which can sometimes completely shut down certain reaction to enforce a different course of reactivity. We will begin this section with early strides in this area before we focus on the state of the art developments in metal-catalyzed C1 polymerization.

3.3.1 COPPER-, GOLD-, NICKEL-, AND PALLADIUM-CATALYZED C1 POLYMERIZATION

During the early days, reactivity of ligand-free copper salts with diazoalkanes was investigated and it was noted that copper salts polymerize diazoalkanes to polymethylenes. Although Cu(II) salts were employed as catalysts, Cu(I) salts were proposed to be the active species. One of the Cu(II) salts, that is, copper stearate was found to catalyze the carbene polymerization of diazoethane and diazobutane to produce polymers in high (92%) yield.[30] Polymers with reasonable molecular weight (20000 Da) could be obtained, although the polymerization mechanism could not be fully established. Several mechanistic pathways have been proposed: (1) Cu(I)-initiated radical polymerization, (2) repetitive migratory insertion of *in situ*-generated carbene into Cu(I)-alkyl chain, and (3) cationic polymerization.[31]

Although copper salts do catalyze the polymerization of diazoalkanes, they failed to address the limitations of borane-mediated C1 polymerization. Hence, in attempts to find better catalysts, Nasini et al. investigated carbene polymerization of diazoalkanes using another coinage metal: Gold (AuCl$_3$).[32] AuCl$_3$ was found to catalyze the polymerization of diazoalkane in quantitative yields and polymer molecular weights as high as 50,000 Da could be achieved. Detailed investigation indicated that AuCl$_3$ reduces to metallic gold upon addition of diazoalkane and the metallic gold is the actual active catalyst. Higher diazoalkanes were found to be less reactive and afforded the polymers in about 10% yield. Thus, the simple metal salts of copper and gold, without any ancillary ligand, were found to be active in the polymerization of diazomethane, with their own limitations and advantages.

The crucial observation that gold nanoparticles can catalyze the reaction inspired Nasini et al. to expose the surfaces of various metals to diazomethane, which extend the scope of this reaction to heterogeneous catalysis. A broad range of metals such as Cu, Ti, Fe, Mg, W, Ni, V, Mn, Ta, Pt, Co, Zn, Cd, Au, Cr, Al, and Mo were evaluated and their activity was found to decrease from Cu to Mo in the above-mentioned order. Even though the performance of Au(III) was much better than Cu(I/II) in the homogeneous systems, metallic surfaces of copper were found to be the most active heterogeneous systems. On the other hand, although less active than copper, gold surfaces produced stereoregular polyethylidene from diazo-*ethane*. This is the first

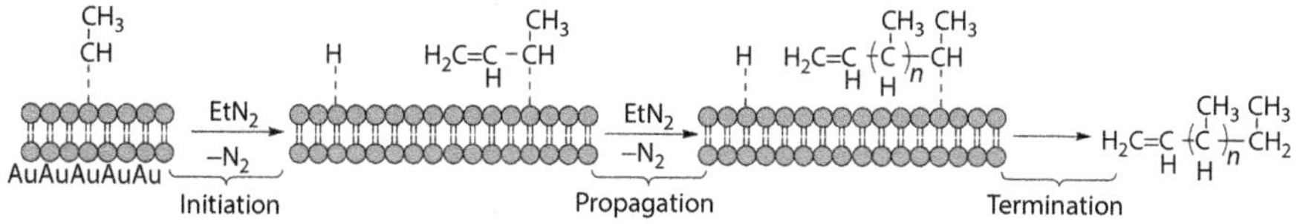

FIGURE 3.4 Plausible mechanisms for the C1 polymerization of diazoethane on gold (Au) surface.

example of stereoregular polyalkylidene, though the actual tacticity of the polymer could not be fully established. The authors proposed multiple initiating pathways of which the most likely mechanism is depicted in Figure 3.4. Decomposition of diazoethane to carbene which is then held on the metal surface was considered to be the first step. Subsequent addition of diazoethane results into hydrogen transfer to the metal surface with liberation of nitrogen gas. Repetitive migratory insertions of the metal-bound diazoethane into the metal surface-bound alkyl unit led to propagation and formation of polyethylidene. Termination by reductive elimination of the metal surface-bound alkyl unit with the hydride was proposed to be the last event.

The above-mentioned findings were very cleverly utilized by Tao and Allara to prepare polymethylene-coated gold surfaces.[33] Exposing the gold surface to a solution of diazomethane produced ultrathin nanometer scale coating on the gold surfaces. In the early stage, the polymerization is initiated at the isolated defect sites of the gold surface and the polymer grows to reach a thickness of about 20 nm. At this stage, the polymerization spreads over the gold surface and eventually covers the entire gold surface with the polymethylene film. The thus-obtained polymethylene is packed in a crystalline manner over the heterogeneous film surface and displays characteristic features of bulk polyethylene, including the melting temperature (T_m). A free radical mechanism was evoked to explain the observation of a –CH_3 group in the polymer and radical disproportionation was proposed to be the preferred mode of chain termination.

To overcome the problems associated with heterogeneous film surface coverage in the above-mentioned gold-catalyzed C1 polymerization reaction, Guo and Jennings came up with a modified gold catalyst.[34] Modification of the gold surface with copper or silver tailors the properties of the gold surface and altered the kinetics and final properties of the polymethylene produced. The authors could successfully demonstrate that on a copper-modified gold surface polymethylene grows almost linearly with time, unlike the rapid termination observed in unmodified gold surface reported by Tao and Allara. It is worth noting that the activity of copper-modified gold surfaces was much higher than individual copper or gold surfaces. In a sharp contrast to copper, increasing the coverage of silver on the gold surface dramatically inhibits the polymethylene film growth. The reduced polymethylene formation was explained by assuming that increasing the silver coverage can potentially block the gold active sites. The same group expanded the scope of this reaction to functional carbene precursors such as ethyl diazoacetate (EDA) in the hope of obtaining functionalized polymethylenes. Homopolymerization of EDA largely failed, but copolymerization of diazomethane and EDA could be achieved. Thus, exposing the gold surface to diazomethane and EDA produced a copolymer with 0%–5% of EDA incorporation.[35] The rate of copolymer film growth was found to be linear with time over a period of 24 hours. Detailed experiments revealed that the polymer growth does not initiate at the metal-surface-solvent interface, but rather propagation occurs at the film-metal junction. This observation prompted the authors to propose an insertion mechanism wherein a carbene is inserted between the gold metal and the growing chain, which pushes the outer termini of the polymer further away from the gold surface. Addition of radical inhibiters does not halt the polymerization reaction and externally added alkenes were not built in the polymer. Both these experiments further favored the insertion mechanism and rule out the possibility of radical or cationic polymerization proposed by Tao and Allara. A hypothetical mechanism with adsorbed gold-carbene (Au = $CHCO_2Et$; derived from EDA) species was proposed to be the first step. Due to the electron-withdrawing nature of the ester group, the gold center becomes electron deficient and that makes the gold center a more active catalyst. Therefore, it was suggested that EDA acts as a cocatalyst and enhances the rate of polymerization, although with low incorporation. The activity of the catalyst and copolymer properties could be manipulated by varying the concentrations of EDA and diazomethane (DM), allowing formation of a copolymer film with several hundred nanometers.

In their endeavor to tune the reactivity of the metal center, Werner and Richards reported the first organometallic complex, nickelocene as an active catalyst for the polymerization of diazomethane.[36] Exposing the solution of diazomethane to nickelocene at −78°C instantaneously released nitrogen gas and afforded a flaky white precipitate of polymethylene. Careful polymer analysis ruled out the presence of any nitrogen or any norcarane fragments in the polymethylene. Even addition of other alkenes does not produce any side product and the nickelocene could be recovered. Moreover, the rate of nickelocene-catalyzed polymerization of diazomethane was found to be much higher than the rate of polymerization by other metal (copper or gold) catalysts. The existence of radical intermediates was ruled out, as the addition of radical scavengers did not influence the performance of nickelocene. Addition of analogous metal complexes such as ferrocene or chromocene does not affect the nickelocene-catalyzed polymerization of diazomethane.

Higher diazo-compounds reduced the activity of catalyst. Changing substituents on the diazo-carbon allowed the authors to conclude that as the steric bulk and degree of substitution at the diazo-carbon increase, the reactivity of nickelocene decreases. An attempt to understand the underlying mechanism was made and based on their observations the authors proposed a mechanism as depicted in Figure 3.5. Formation of nickelocene–methylene complex was proposed to be the first step, followed by proton transfer by the next diazomethane molecule to generate a rather uncommon cationic Ni(IV) species. The formation of highly acidic Ni(IV) is considered to be the most important step in the polymerization mechanism. Failing to generate this species would completely block the polymerization. The author's claim is based on their observation that diazo-compounds (such as diphenyldiazomethane, without any proton on the carbeniod carbon) that cannot generate $-CHRN_2$ will not be able to produce the cationic intermediate and thus fail to polymerize. Reaction of diazomethane with this highly acidic intermediate generates nickel-alkyl-carbene species. In the subsequent step, the alkyl group migrates to the carbene and a cationic nickel-alkyl

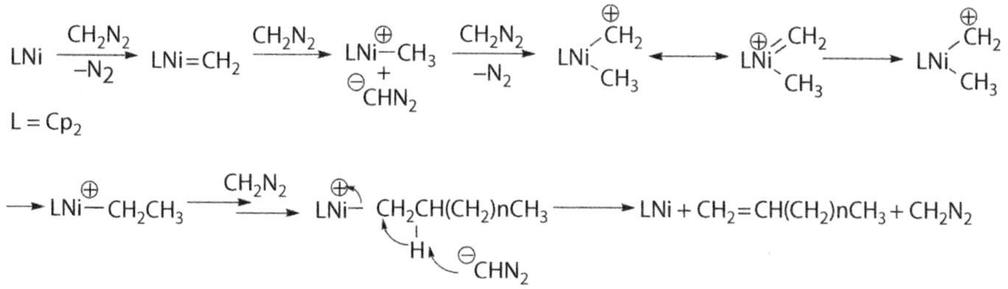

FIGURE 3.5 Proposed mechanisms for nickelocene (LNi)-catalyzed polymerization of diazomethane.

species is generated. Repetition of these sequences leads to chain growth and production of polymethylene.

Thus, nickelocene catalyzes the polymerization of diazomethane to polymethylene with quantitative conversion. However, various disubstituted diazo-compounds or functionalized diazo-compounds such as ethyl diazoacetate could not be polymerized using nickelocene.

Ihara et al. tested a palladium precursor [$Pd_2(dba)_3(CHCl_3)$/pyridine] in the C1 polymerization of diazacarbonyl compounds, only to find low molecular weight polymers.[37] Although a substantial amount (25%) incorporation of azo-groups was observed, the yields were very poor (only about 20%). Analysis of these polymers indicated the presence of azo-groups or acyl-groups, which indicates the existence of multiple insertion modes and the lack of a controlled reaction.

3.3.2 Transition Metal-Catalyzed Oligomerization/Polymerization of Functionalized Diazo-Compounds

Functional group tolerance has been the limiting element for many metal-catalyzed transformations and C1 polymerization is not an exception to this limitation. Early efforts in C1 polymerization of functionalized diazo-compounds could not yield desired polymer products but mainly produced dimers, trimers, and higher oligomers.[38] However, almost after a century, Liu and coworkers landed into an unexpected product while they were investigating the copper-mediated cyclopropanation polymerization of allyl diazoacetate.[39] Detailed structural characterization of the resultant product revealed the formation of poly(carballyloxycarbene), instead of anticipated cyclopropanation polymerization. The authors claimed a carbene polymerization of allyl diazoacetate with a molecular weight of 3000 Da and PDI of 1.2.

Along the same line, Ihara and coworkers reported C1 polymerization of a variety of diazo-carbonyl compounds. A series of novel oligomers/polymers containing various functional groups were prepared by the so-called *poly(substituted methylene) synthesis* procedure. Mainly palladium(II)-catalyzed homo- or copolymerization of diazo-compounds such as diazoacetates,[40] aliphatic/aromatic diazoketones,[41] cyclic diazoketones,[42] and diazoacetamides[43] was investigated. Simple palladium salts such as $PdCl_2$ or solvated $PdCl_2$ compounds were employed as catalysts. However, the

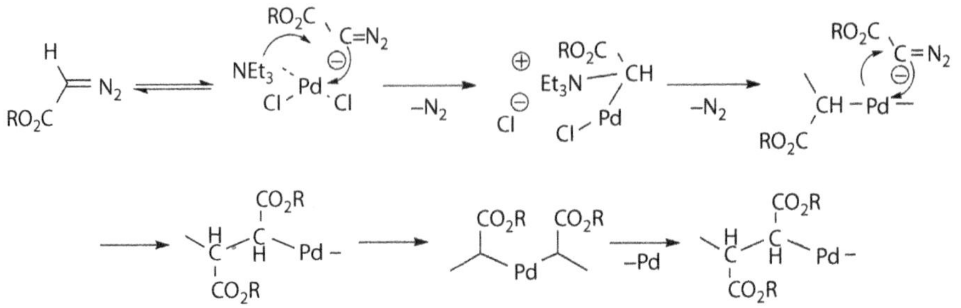

FIGURE 3.6 Hypothetical mechanism for palladium-catalyzed polymerization of functional carbenes.

majority of the monomers could not be polymerized to high molecular weight polymers, but rather low molecular weight oligomers were obtained and polydispersities in the range of 1.1–1.9 were obtained. The authors proposed a triethyl amine-mediated mechanism. The reaction is proposed to be initiated by nucleophilic attack of triethyl amine on the α-carbon atom of the diazoacetate, followed by palladium-carbon bond formation (see Figure 3.6). Direct migration of the growing chain to the carbon atom of the diazoacetate was considered to be the propagation step. The termination was considered to be a reductive elimination step as each palladium center will have two growing chains. However, the above-mentioned mechanism contrasts with the carbene insertion mechanism, which has been well investigated and many mechanistic aspects have been reported. In the later studies, polymerization took place even without the addition of amine. This indicates that the reaction is not initiated by amine as proposed in Figure 3.6 but requires the presence of weak nucleophiles such as acetonitrile, water, or the monomer itself. The reaction was found to be initiated by even Pd(0) precursors, but the authors proposed that the Pd(0) precursors were oxidized to Pd(II) *in situ*, which then activates the polymerization. However, this proposal seems to contrast the recent revelation by Ortuno and coworkers.[44] The authors demonstrated that *in situ*-formed Pd(0) nanoparticles are capable of catalyzing the polymerization of diazomethane.

After the diazoalkanes and diazoacetates, the authors evaluated various palladium complexes in the polymerization of different diazoketones and diazo-carbonyl compounds. Both diazoketones and diazo-carbonyl compounds posed considerable challenge and only low molecular weight oligomers with very low yields could be obtained. Polymerization of these monomers frequently led to incorporation of diazo-group in the polymer backbone with varying degrees of incorporation. Thus, the polymerization of these functionalized monomers proved to be more difficult and produced rather ill-defined, random (co)polymers. Similar behavior was observed in the case of cyclic diazoketones and diazoacetamides. In fact, some of the cyclic diazoketones could not be homopolymerized, whereas others produced oligomers. Although there could be various reasons for this behavior, the two primary reasons seem to be functional group intolerance by the catalyst (catalyst poisoning by the functional group on the monomer precursor) and the steric crowding around the carbene center. Thus, a very general palladium-mediated polymerization of functionalized diazo-carbonyl compounds was introduced by Ihara and coworkers for a broad range of monomers. However, polymerization of majority of the monomers led to only oligomers with ill-defined compositions.

3.3.3 METAL-MEDIATED POLYMERIZATION OF FUNCTIONALIZED CARBENE PRECURSORS TO HIGH MOLECULAR WEIGHT AND STEREOREGULAR (CO)POLYMERS

As discussed in Section 3.3.2, metal-catalyzed polymerization of functionalized carbene precursor gave, at the best atactic, higher molecular weight oligomers, which are of limited practical significance. More recently, de Bruin and coworkers reported rhodium-mediated C1 polymerization of diazo-esters to very high molecular weight polymers, interestingly with high stereoregularity.[45] Performance of rhodium complexes of chelating bidentate ligands, such as pyridine-carboxylic acid, proline, or derivatives of these, was evaluated in the C1 polymerization of functionalized diazo-compounds. Treating the rhodium complexes with EDA produced polymers with a molecular weight of 120 000 Da to 540 000 Da. Quite remarkably, the resultant polymers revealed sharp nuclear magnetic resonance (NMR) spectroscopy resonances both in solution as well as in solid state, which indicates formation of a defined, regular polymeric structure. The authors considered this polymer to be syndiotactic in nature, though the exact structure could not be fully established during these early investigations. Subsequent studies comparing the ^{13}C NMR resonances of the polymer with those of known reference materials confirmed the syndiotactic assignment, which for C1-polymers means that in a regular zig-zag arrangement of the polymer backbone, all substituents point to the *same* side (in contrast to e.g., PP). Unlike the atactic polymers and oligomers obtained via palladium-mediated polymerization of EDA or via radical polymerization of diethyl fumarate, which were isolated as viscous brown oils, the new stereoregular polymers could be isolated as highly crystalline solids. Table 3.2 summarizes some of the most significant results in the rhodium-mediated C1 polymerization of EDA. Although it is a common observation

TABLE 3.2
Rhodium-Mediated Polymerization of EDA

Entry	Catalyst	Yield (%)	M_w (Da)	PDI
1	3.1	11	130 000	3.7
2	3.2	11	140 000	3.7
3	3.3	30	150 000	3.4
4	3.4	15	150 000	3.7
5	3.5	50	150 000	3.6
6	3.6	25	120 000	2.9
7	3.7	5	350 000	15.4
8	3.8	30	540 000	2.0

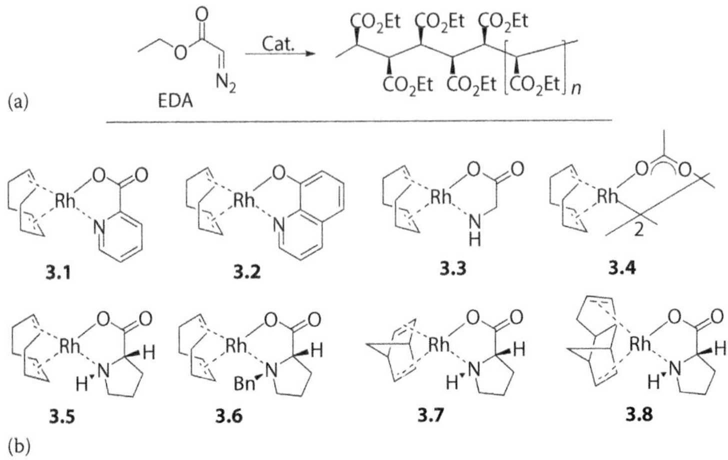

FIGURE 3.7 Rhodium-catalyzed C1 polymerization of EDA (a) and various rhodium complexes used in the EDA polymerization reaction (b).

encountered in stereoregular polymerization, it is important to note that even achiral catalysts such as **3.1**, **3.2**, and **3.3** produced stereoregular polymers (Figure 3.7). This observation suggests a chain-end controlled polymerization mechanism (as discussed in Chapter 1 and Chapter 2), wherein auxiliary ligands hardly modulate the stereoregularity. The same class of catalysts was found to polymerize other diazoesters such as n-butyl diazoacetate, 3-butenyl diazoacetate, and so on.[46] Apart from the isolated polymers, formation of dimers and oligomers was observed.[47] The broad molecular weight distribution coupled with oligomerization suggests the existence of multiple active species in the polymerization. Involvement of three active species was proposed: one producing dimers, another responsible for oligomers, and the third producing desired polymers.[48] Irrespective of the auxiliary ligands, most of the complexes produced high molecular weight polymers with high stereoregularity. This observation indicates that in a catalytic cycle, the N,O-donor ligands are most likely displaced by the monomers. MALDI-ToF-MS analysis of these polymers confirmed that carbene moiety is indeed the repeating unit in these polymers. The scope of this reaction was extended to iridium complexes and the iridium analogs of complex **3.5** and **3.6** were tested in EDA polymerization. Though one of the iridium complexes initiated the polymerization of EDA, only low (12000 Da) molecular weight polymers were obtained. Similar Ir(I) complexes were tested by Buchmeiser et al. in the C1 polymerization of EDA.[49] However, these complexes failed to polymerize EDA, although similar rhodium complexes were found to be active.

Unlike the monoanionic ligand that does not interfere with the polymerization, the diene ligands were found to alter the reactivity of the rhodium center. For example, complexes **3.7** and **3.8** with same monoanionic ligands but different diene ligands produced polymers with significantly higher molecular weight (entry 7–8 versus entry 5–6). Kinetic measurements indicate that, of the 2–3 mol% catalyst used, only about 1%–5% catalyst is involved in the polymerization reaction.

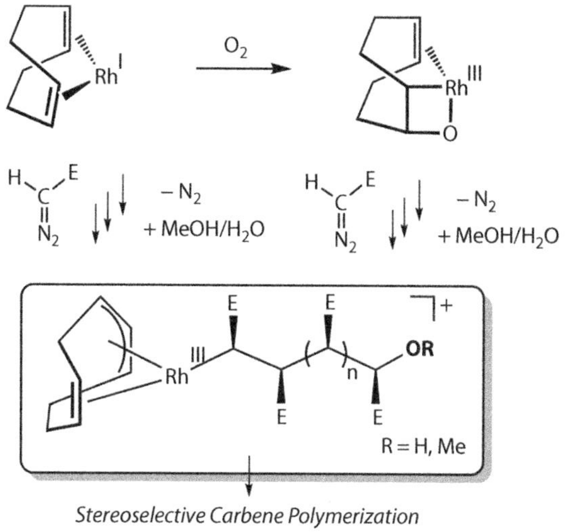

Stereoselective Carbene Polymerization

FIGURE 3.8 Detection of cationic [(allyl-cod)Rh^{III}(polymeryl)]$^+$ species as the active species responsible for polymer chain growth (E = COOEt).

The kinetic investigations also indicated that there is a linear relation between the molecular weight of the polymer and the polymer yield. Shorter polymers were obtained in the presence of alcohols or water in the reaction medium.[50] The chain-transfer process mainly involves nucleophilic protonolysis, with the rate of chain transfer varying with the nucleophilicity of the alcohol (or water). These observations suggest a near-living character with relative chain transfer involving alcohols or water when present in the reaction medium.

The mechanism of these reactions was thus far unclear, but recently the scientists were able to unravel the mechanism in detail through a combination of catalyst screening and high-resolution electron spray ionization mass spectrometry (ESI-MS) measurements. These studies revealed that the Rh^I(diene) species get converted to [(allyl)Rh^{III}(polymeryl-OR)]$^+$ species under the applied reaction conditions, which are the actual polymer-forming species (Figure 3.8). Remarkably, oxygenated [(*diene-O*) Rh^{III}] complexes produce polymers in higher yields (up to 85%) and with higher molecular weights in the C1 polymerization of EDA as compared to their parent nonoxidized Rh^I complex (maximum yield was 50%).[51] The reason for the enhanced polymer yields for the air-oxidized complexes is a better initiation efficiency of the oxygenated rhodium complexes. The same [(allyl)Rh^{III}(polymeryl-OR)]$^+$ species were found to be responsible for chain growth when starting from the oxygenated catalysts (Figure 3.8).

Based on the polymer analysis and density functional theory (DFT) investigations, the authors proposed formation of [Rh-OH] species, which initiates the C1 polymerization of EDA and a carbene inserts between the rhodium and the hydroxy group. In agreement with this proposal, the polymer chain ends were found to be either –OH groups or OMe (derived from solvent or quenching agent) groups. The proposed reaction mechanism is depicted in Figure 3.9.[52–54] The proposed propagation

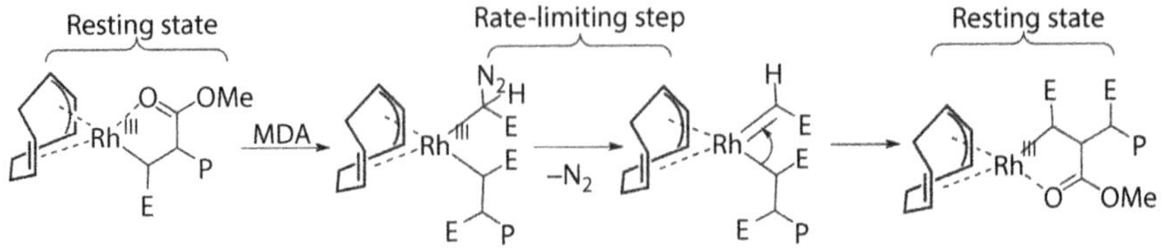

FIGURE 3.9 Proposed mechanism for rhodium-mediated C1 polymerization of MDA (E = COOMe, P = polymer).

mechanism depicted in Figure 3.8 is based on mass spectrometric detection of the growing-chain species,[51] DFT calculations,[55] and some additional experimental findings.[56] It is believed that the resting state is stabilized by the carbonyl oxygen atom of the beta-ester group of the growing polymer chain (Figure 3.9). Substitution of this carbonyl by the diazo-compound produces a five-coordinated rhodium(III) complex. Such dissociation of carbonyl compounds has been commonly observed and has been separately investigated in model compounds.

Elimination of nitrogen and formation of rhodium-carbene complex was found to be the rate-limiting step in this reaction. Repetitive migratory insertions of *cis*-carbene moieties into the Rh-polymeryl bond lead to chain growth, and production of high molecular weight polymer is realized. Alcohol coordination and proton transfer from the coordinated alcohol to the growing polymer chain are proposed as the chain-transfer mechanism.

The scope of the reaction was extended to sulfoxonium ylides and diazo-esters.[57] The rhodium (similar catalysts in Figure 3.8) complexes catalyze the copolymerization of sulfoxonium ylides with diazoesters to produce corresponding copolymers with high molecular weights ranging up to 791000 Da. The authors were successful in demonstrating the synthesis of block copolymers by controlling the addition sequence of the two monomers. The resultant copolymers revealed syndiotactic microstructure and the incorporation of the ester functionality could be tuned to change the ratio of polar and nonpolar fragments in the resultant copolymer. Along the same line, rhodium-mediated C1 polymerization of diazomethane with EDA was reported.[58] The resultant copolymers displayed blocky microstructure with tunable functional group incorporation.

In a new proof of concept investigations, de Bruin and coworkers reported the insertion copolymerization of ethylene with EDA using the same rhodium complexes as in Figure 3.7.[59] This became the first example of a combined C2 and C1 polymerization method and can potentially open up new avenues for the synthesis of functionalized polyethylene. Remarkably high molecular weight (1022000 Da) copolymers of ethylene with EDA were produced with reasonable polydispersities and low ethylene content (up to 11%). Attempts to increase the ethylene content by increasing the ethylene pressure were not very successful but rather revealed low incorporation. This was explained by assuming similar differences in the activities of active rhodium complexes in a polymerization reaction under the given

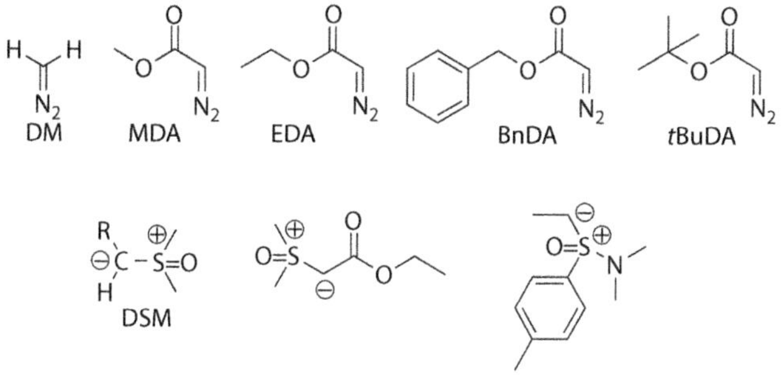

FIGURE 3.10 Various monomers used in the rhodium-mediated C1 polymerization.

polymerization conditions. The resultant block copolymers revealed highly syndiotactic polymer microstructures.

In an attempt to employ cost-effective metals, C1 polymerization using palladium precursor was reported. However, changing the metal center to palladium drastically reduced the activity and only oligomers were obtained.[60] It is likely that the two types of active sites responsible for insertion (olefin) polymerization and carbene polymerization are incompatible and do not yield the desired copolymers.

Thus, the rhodium-mediated polymerization and copolymerization of various functionalized C1 monomers were investigated. Rhodium complexes with diene and monoanionic ligand provided the first breakthrough to produce C1 polymers with considerably higher molecular weight and highly stereoregular polymers were prepared. The oxygenated rhodium complexes were found to be even more active and provided polymers with better overall yields. However, the catalyst efficiencies are still considerably low and only a few percentage of the total catalyst is actually active in the polymerization. So far the monomers/substrates shown in Figure 3.10 have been investigated in rhodium-mediated C1 polymerization, leaving ample room to expand the scope.

3.4 APPLICATIONS

Although nonfunctionalized polyolefins such as polyethylene, polypropylene, and polystyrene dominate the field of polymer science and technology, the demand for functionalized polyolefin was picked up in the past two decades. Introducing small amount of polar functionalities in the backbone of these polymers has beneficial effects on their surface properties. The thus-introduced functional groups enhance the adhesion, stickiness, and binding properties of the resultant material. Potential applications of these functionalized polyolefins include but are not limited to the following set: (1) They can serve as a binder for various binding applications such as bookbinding. (2) If sufficient functional groups are incorporated, they can be used as glue or adhesive. (3) Along the same line, functional polyolefins can be used in

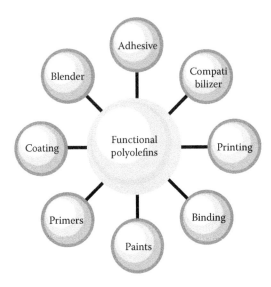

FIGURE 3.11   Applications of functionalized polyolefins.

printing ink. (4) Another innovative application will be to use the functional polyolefins as a compatiblizing agent between a hydrophilic and a hydrophobic polymer.
(5) Further, they can be used in various applications as depicted in Figure 3.11.
At present, the functionalized polyolefins are only prepared for research and development purpose. Industrial production of these materials for daily applications is
yet to be realized.

Another remarkable feature of the stereoregular polymers produced by
Rh-mediated polymerization of EDA leading to possible applications is the fact that
they are liquid crystalline (LC). This is rather unexpected, given the flexible nature
of the flexible sp³-carbon backbone of the polymer chains. Interestingly, the LC
properties of these polymers are in fact a result of their *tertiary structure*, which has
a remarkable and unexpected influence on these materials.[61]

Although the tertiary structure of polypeptides is well known to determine
important functional properties of proteins, a similar influence of supramolecular
aggregation of synthetic polymers on their material properties is much less established. A combination of studies recently revealed that the surprising LC behavior
of these polymers is due to self-aggregation of the stereoregular polymer chains
into triple-helix aggregates, reflecting the importance of the *tertiary structure*
of synthetic polymers on their material properties (Figure 3.12). The observed
liquid crystallinity is a direct consequence of the increased rigidity of the chain
induced by self-assembly of syndiotactic polycarbene polymer chains into van der
Waals-stabilized LC aggregates (rigid-rod behavior).[61] Such unusual behavior is
not observed for more common C2 polymers such as polyethylene (PE), polypropylene (PP), or polystyrene (PS), thus providing ample opportunities for *de novo*
material design.

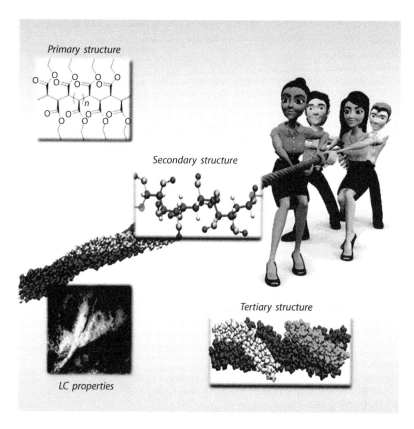

Primary structure

Secondary structure

Tertiary structure

LC properties

FIGURE 3.12 Self-aggregation into a triple-helix tertiary structure leading to LC behavior.

REFERENCES

1. Kamnisky, W. 2012. Discovery of methylaluminoxane as cocatalyst for olefin polymerization. *Macromolecules* 45: 3289–3297.
2. Jellema, E., Jongerius, A. L., Reek, J. N. H., de Bruin, B. 2010. C1 polymerization and related C-C bond forming "Carbene insertion" reactions. *Chem. Soc. Rev.* 39: 1706–1723.
3. Gaikwad, S. R., Deshmukh, S. S., Gonnade, R. G., Rajamohanan, P. R., Chikkali, S. H. 2015. Insertion copolymerization of difunctional polar vinyl monomers with ethylene. *ACS Macro Lett.* 4: 933–937.
4. Maruoka, K., Oishi, M., Yamamoto, H. 1996. Novel anionic oligomerization by a new sequential generation of organolithium compounds. *Macromolecules* 29: 3328–3329.
5. Ihara, E., Kobayashi, K., Wake, T., Mokume, N., Itoh, T., Inoue, K. 2008. Mg-mediated polycondensation of α,α-dibromotoluene with bifunctional electrophiles. *Polym. Bull.* 60: 211–218.
6. Luo, J., Shea, K. J. 2010. Polyhomologation. A living C1 polymerization. *Acc. Chem. Res.* 43: 1420–1433.
7. Kantor, S. W., Osthoff, R. C. 1953. High molecular weight polymethylene. *J. Am. Chem. Soc.* 75: 931–932.

8. Hooz, J., Gunn, D. M. 1969. The reaction of β-vinylic and β-alkyl-9-borabicyclo[3.3.1] nonane derivatives with ethyl diazoacetate and diazoacetone. *Tetrahedron Lett.* 10: 3455–3458.

9. Hooz, J., Linke, S. 1968. Alkylation of diazoacetonitrile and ethyl diazoacetate by organoboranes. Synthesis of nitriles and esters. *J. Am. Chem. Soc.* 90: 6891–6892.

10. Pasto, D. J., Wojtkowski, P. W. 1970. Transfer reactions involving boron. XXI intermediates formed in the alkylation of diazocompounds and dimethylsulfonium phenacylide via organoboranes. *Tetrahedron Lett.* 11: 215–218.

11. Davis, A. G., Hare, D. G., Khan, O. R., Sikora, J. 1963. Monomethylation and polymethylenation by diazomethane in presence of boron compounds. *J. Chem. Soc.* 4461–4471.

12. Bawn, C. E. H., Ledwith, A., Matthies, P. 1959. The mechanism of the polymerization of diazoalkanes catalyzed by boron compounds. *J. Polym. Sci.* 34: 93–108.

13. Tufariello, J. J., Lee, L. T. C. 1966. The reaction of trialkylboranes with dimethyloxosulfonium methylide. *J. Am. Chem. Soc.* 88: 4757–4759.

14. Shea, K. J., Walker, J. M., Zhu, H., Paz, M., Greaves, J. 1997. Polyhomologation. A living polymethylene synthesis. *J. Am. Chem. Soc.* 119: 9049–9050.

15. Busch, B. B., Paz, M. M., Shea, K. J., Staiger, C. L., Stoddard, J. M., Walker, J. R., Zhou, X.-Z., Zhu, H. 2002. The boron catalyzed polymerization of dimethylsulfoxonium methylide. A living polymethylene synthesis. *J. Am. Chem. Soc.* 124: 3636–3646.

16. Shea, K. J. 2000. Polyhomologation: The living polymerization of ylides. *Chem.-Eur. J.* 6: 1113–1119.

17. Shea, K. J., Busch, B. B., Paz, M. M. 1998. Polyhomologation: Synthesis of novel polymethylene architectures by a living polymerization of dimethylsulfoxonium methylide. *Angew. Chem. Int. Ed.* 37: 1391–1393.

18. Shea, K. J., Lee, S. Y., Busch, B. B. 1998. A new strategy for the synthesis of macrocycles. The polyhomologation of boracyclanes. *J. Org. Chem.* 63: 5746–5747.

19. Busch, B. B., Staiger, C. L., Stoddard, J. M., Shea, K. J. 2002. Living polymerization of sulfur ylides. Synthesis of terminally functionalized and telechelic polymethylene. *Macromolecules* 35: 8330–8337.

20. Wagner, C. E., Kim, J.-S., Shea, K. J. 2003. The polyhomologation of 1-boraadamantane: Mapping the migration pathways of a propagating macrotricyclic trialkylborane. *J. Am. Chem. Soc.* 125: 12179–12195.

21. Zhao, R., Shea, K. J. 2015. Gradient methylidene-ethylidene copolymer via C1 polymerization: An ersatz gradient ethylene-propylene copolymer. *ACS Macro Lett.* 4: 584–587.

22. Wagner, C. E., Rodriguez, A. A., Shea, K. J. 2005. Synthesis of linear alpha-olefins via polyhomologation. *Macromolecules* 38: 7286–7291.

23. Luo, J., Lu, F., Shea, K. J. 2012. Hydrocarbon waxes from a salt in water: The C1 polymerization of trimethylsulfoxonium halide. *ACS Macro Lett.* 1: 560–563.

24. Bai, J., Shea, K. J. 2006. Reaction of boranes with TMS diazomethane and dimethylsulfoxonium methylidine. Synthesis of poly(methylidene-co-TMSmethylidene) random copolymers. *Macromol. Rapid Commun.* 27: 1223–1228.

25. Zhou, X.-Z., Shea, K. J. 2000. Ersatz ethylene-propylene copolymers: The synthesis of linear carbon backbone copolymers one carbon atom at a time. *J. Am. Chem. Soc.* 122: 11515–11516.

26. Sulc, R., Zhou, X.-Z., Shea, K. J. 2006. Repetitive sp^3–sp^3 carbon–carbon bond formation copolymerizations of primary and tertiary ylides. Synthesis of substituted carbon backbone polymers: Poly(cyclopropylidine-co-methylidine). *Macromolecules* 39: 4948–4952.

27. Wang, J., Horton, J. H., Liu, G., Lee, S.-Y., Shea, K. J. 2007. Polymethylene-block-poly(dimethyl siloxane)-block-polymethylene nanoaggregates in toluene at room temperature. *Polymer* 48: 4123–4129.

28. Ihara, E., Kida, M., Itoh, T., Inoue, K. 2007. Organoaluminum-mediated polymerization of diazoketones. *J. Polym. Sci. Part A: Polym. Chem.* 45: 5209–5214.
29. Imoto, M., Nakaya, T. 1972. Polymerization by carbenoids, carbenes and nitrenes. *J. Macromol. Sci.* 7: 1–48.
30. Bawn, C. E. H., Rhodes, T. B. 1954. High molecular weight polymethylene. Part 1—The kinetics of the copper salt catalyzed decomposition of diazomethane. *Trans. Faraday Soc.* 50: 934–941.
31. Cowell, G. W., Ledwith, A. 1970. Developments in the chemistry of diazo-alkanes. *Q. Rev. Chem. Soc.* 24: 119–167.
32. Nasini, A. G., Trossarelli, L., Saini, G. 1961. Reactions of diazoalkanes upon metallic surfaces: Polymer formation and a stereoregulating action of gold. *Makromol. Chem.* 44: 550–569.
33. Seshadri, K., Atre, A. V., Tao, Y.-T., Lee, M.-T., Allara, D. L. 1997. Synthesis of crystalline, nanometer-scale, $-(CH_2)_x-$ clusters and films on gold surfaces. *J. Am. Chem. Soc.* 119: 4698–4711.
34. Guo, W., Jennings, G. K. 2002. Use of under potentially deposited metals on gold to affect the surface-catalyzed formation of polymethylene films. *Langmuir* 18: 3123–3126.
35. Bai, D., Gennings, G. K. 2005. Surface-catalyzed growth of polymethylene-rich copolymer films on gold. *J. Am. Chem. Soc.* 127: 3048–3056.
36. Werner, H., Richards, J. H. 1968. Reaction of nickelocene with diazoalkanes. *J. Am. Chem. Soc.* 90: 4976–4982.
37. Ihara, E., Kida, M., Fujioka, M., Haida, N., Itoh, T., Inoue, K. 2007. Palladium mediated copolymerization of diazocarbonyl compounds with phenyldiazomethane. *J. Polym. Sci. Part A: Polym. Chem.* 45: 1536–1545.
38. Lorey, K. 1929. Ueber molekuelverbindungen des diazoessigesters mit anorganischen salzen und oxyden. *J. Prakt. Chem.* 124: 185–190.
39. Liu, L., Song, Y., Li, H. 2002. Carbene polymerization: Characterization of poly(carballyloxycarbene). *Polym. Int.* 51: 1047–1049.
40. Ihara, E., Haida, N., Iio, M., Inoue, K. 2003. Palladium mediated polymerization of alkyl diazoacetates to afford poly(alkoxycarbonylmethylene)s. First synthesis of polymethylenes bearing polar substitutents. *Macromolecules* 36: 36–41.
41. Ihara, E., Fujioka, M., Haida, N., Itoh, T., Inoue, K. 2005. First synthesis of poly(acylmethylene)s via palladium-mediated polymerization of diazoketones. *Macromolecules* 38: 2101–2108.
42. Ihara, E., Hiraren, T., Itoh, T., Inoue, K. 2008. Palladium-mediated polymerization of cyclic diazoketones. *J. Polym. Sci. Part A: Polym.* Chem. 46: 1638–1648.
43. Ihara, E., Hiraren, T., Itoh, T., Inoue, K. 2008. Palladium-mediated polymerization of diazoacetamides. *Polym. J.* 40: 1094–1098.
44. Illa, O., Rodriguez-Garcia, C., Acosta-Silva, C., Favier, I., Picurelli, D., Oliva, A., Gomez, M., Branchadell, V., Ortuno, R. M. 2007. Cyclopropanation of cyclohexanone by diazomethane catalyzed by palladium-diacetate: Evidences for the formation of palladium(0) nanoparticles. *Organometallics* 26: 3306–3314.
45. Hetterscheid, D. G. H., Hendriksen, C., Dzik, W. I., Smits, J. M. M., van Eck, E. R. H., Rowan, A. E., Busico, V. et al. 2006. Rhodium-mediated stereoselective polymerization of "carbenes" *J. Am. Chem. Soc.* 128: 9746–9752.
46. Jellema, E., Jongerius, A. L., van Ekenstein, G. A., Mookhoek, S. D., Dingemans, T. J., Reingruber, E. M., Chojnacka, A. et al. 2010. Rhodium-mediated stereospecific carbene polymerization: From homopolymers to random and block copolymers. *Macromolecules* 43: 8892–8903.
47. Jellema, E., Budzelaar, P. H. M., Reek, J. N. H., de Bruin, B. 2007. Rh-mediated polymerization of carbenes: Mechanism and stereoregulation. *J. Am. Chem. Soc.* 129: 11631–11641.

48. Franssen, N. M. G., Finger, M., Reek, J. N. H., de Bruin, B. 2013. Propagation and termination steps in Rh-mediated carbene polymerization using diazomethane. *Dalton Trans.* 42: 4139–4152.

49. Bantu, B., Wurst, K., Buchmeiser, M. R. 2007. *N*-acetaly-*N,N*-dipyrid-2-yl (cycloocta-diene) rhodium (I) and iridium (I) complexes: Synthesis, X-ray structure, their use in hydroformylation and carbonyl hydrosilylation reaction and in the polymerization of diazocompounds. *J. Organomet. Chem.* 692: 5272–5278.

50. Walters, A. J. C., Jellema, E., Finger, M., Aarnoutse, P., Smits, J. M. M., Reek, J. N. H., de Bruin, B. 2012. Rh-mediated carbene polymerization: From multistep catalyst acti-vation to alcohol mediated chain transfer. *ACS Catal.* 2: 246–260.

51. Walters, A. J. C., Troeppner, O., Ivanovic-Burmazovic, I., Tejel, C., Pilar del Rio, M., Reek, J. N. H., de Bruin, B. 2012. Stereospecific carbene polymerization with oxygen-ated Rh(diene) species. *Angew. Chem. Int. Ed.* 51: 5157–5161.

52. Hoover, J. F., Stryker, J. M. 1990. Coupling of a 2-oxacyclopentylidene and phospho-nium ylide ligand at platinum. Migratory insertion of a Fisher carbene into a metal alkyl-bond. *J. Am. Chem. Soc.* 112: 464–465.

53. Mango, F. D., Dvoretzky, I. 1966. The introduction of methylene into an Iridium com-plex. *J. Am. Chem. Soc.* 88: 1654–1657.

54. Casty, G. L., Stryker, J. M. 1997. Platinum mediated coupling of alkynols and α-diazoesters. Pseudocatalytic synthesis of tetrahydrofuranylidene esters via alkyl migration to a coordinated carbene. *Organometallics* 16: 3083–3085.

55. Walters, A. J. C., Reek, J. N. H., de Bruin, B. 2014. Computed propagation and termination steps in [(cyclooca-2,6-dien-1-yl)Rh^{III}(polymeryl)]$^+$ catalyzed carbene polymerization reactions. *ACS Catal.* 4: 1376–1389.

56. Finger, M., Reek, J. N. H., de Bruin, B. 2011. Role of β-hydride elimination in rhodium-mediated carbene insertion polymerization. *Organometallics* 30: 1094–1101.

57. Olivoz Suarez, A. I., Pilar del Rio, M., Remerie, K., Reek, J. N. H., de Bruin, B. 2012. Rh-mediated C1 polymerization: Copolymers from diazo-esters and sulfoxonium ylides. *ACS Catal.* 2: 2046–2059.

58. Franssen, N. M. G., Remerie, K., Macko, T., Reek, J. N. H, de Bruin, B. 2012. Controlled synthesis of functional copolymers with blocky architectures via carbene polymeriza-ton. *Macromolecules* 45: 3711–3721.

59. Franssen, N. M. G., Reek, J. N. H., de Bruin, B. 2013. A different route to functional polyolefins: Olefin-carbene copolymerization. *Dalton Trans.* 42: 9058–9068.

60. Franssen, N. M. G., Reek, J. N. H., de Bruin, B. 2011. Pd-mediated carbene polymeriza-tion: Activity of palladium (II) versus low valent palladium. *Polym. Chem.* 2: 422–431.

61. Franssen, N. M. G., Ensing, B., Hegde, M., Dingemans, T., Norder, B., Picken, S. J., Alberda van Ekenstein, G. O. R. et al. 2013. On the 'Tertiary Structure' of Poly-Carbenes; Self-assembly of sp^3-Carbon based polymers into liquid crystalline aggre-gates. *Chem. Eur. J.* 19: 11577–11589.

4 Ring-Opening Polymerization and Metathesis Polymerizations

Ashootosh V. Ambade

CONTENTS

4.1 RING-OPENING POLYMERIZATION

Aliphatic polyesters are important industrial polymers that have applications as fibers, coatings, bulk packaging materials, and films. Their biodegradable nature makes them an environmentally friendly alternative to the *nondegradable* plastics.[1] Copolymers of lactide and trimethylene carbonate (TMC) are used as thermoplastic elastomers and biomaterials for applications in tissue engineering, drug delivery, as biodegradable devices for bone fracture repair and sutures. The monomer for poly(lactic acid) (PLA), D,L-lactide is a dimer of lactic acid (LA), which is produced from natural sources such as a starch or sugar *via* bacterial fermentation of D-glucose.[2] Hence, production of poly(lactide) is environmentally friendly. Manufacturing of PLA has become profitable over the years. Natureworks LLC, a joint venture between Cargill and Teijen Limited, set up a 300 million pounds per year PLA production plant and sells PLA under the trade name Ingeo™, which is produced in isotactic form by a carbon neutral process. Its physical properties are similar to polyolefins and polystyrene.[3] The polycondensation route for PLA is undesirable since it is difficult to produce high molecular weight polymer, making ring-opening polymerization (ROP) the method of choice. In this chapter, ROP toward sustainable polymers such as PLA and poly(ε-caprolactone) (PCL) only will be discussed. ROP is a living polymerization, that is, it shows fast initiation and minimal termination and transfer reactions, however it follows step-growth kinetics. Polydispersity is usually low but can be influenced by *trans*-esterification reactions and is of concern while making block copolymers. The thermodynamic driving force for the polymerization is

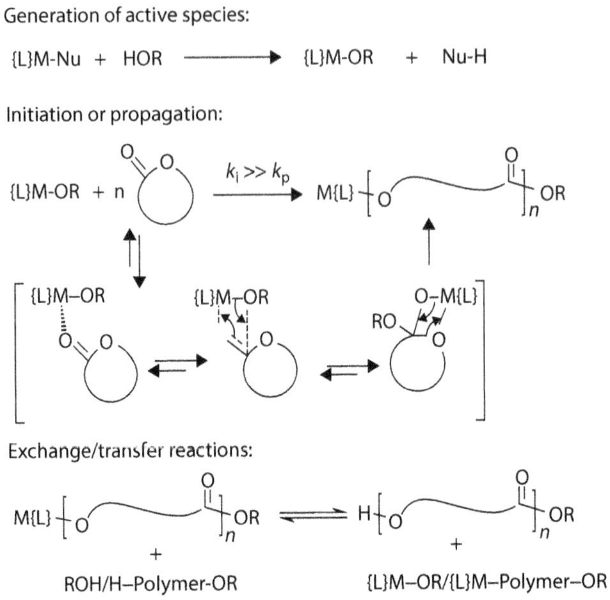

FIGURE 4.1 Coordination–insertion mechanism for ring-opening polymerization. (Adapted from Ajellal, N. et al., *Dalton Trans.*, 39, 8363–8376, 2010.)

the relief of ring strain, which helps to overcome the high entropy values (Lactide: $\Delta S = 25.0$ J mol^{-1} K^{-1}; ε-CL: $\Delta S = 53.9$ J mol^{-1} K^{-1}).[4] ROP performed using metal catalysts that operate through cationic mechanism do not yield high molecular weight polymer desirable for practical applications. Therefore *immortal* ROP that follows *chain-transfer pathway* and involves a catalyst and a nucleophile (either part of the catalyst or externally added) that acts simultaneously as the initiator and chain-transfer agent (CTA), was developed as an efficient alternative to the classical living cationic polymerization.

Two mechanisms are involved in *immortal* ROP. The coordination–insertion mechanism employs metal alkoxides or metal complexes in combination with alcohols. In case of nonalkoxide metal complex (e.g, an alkyl, amido, or borohydrido derivative), it is converted by the alcohol (**ROH**) into the {L}M–OR active species by alcoholysis. The mechanism shown in Figure 4.1 involves the following steps: (1) coordination of the lactide to the Lewis-acid metal center, (2) insertion of the lactide into the metal-alkoxide bond *via* nucleophilic addition, and (3) ring opening of the lactide monomer *via* acyl-oxygen cleavage followed by insertion of next monomer.[5] Polymerization is terminated by hydrolysis of the active chain end and the polymer is isolated.

In the form of active species, {L}M–{O---C(O)}OR, the polymer chain undergoes rapid exchange with other protic species, such as alcohol introduced in excess in the reaction medium, and is converted to α-hydroxy, ω-alkoxycarbonyl polyester chains (H–Pol–OR), which are dormant species. Thus, active and dormant species are converted into each other during the whole polymerization process and can affect propagation and chain transfer, respectively. Control over the polymerization is

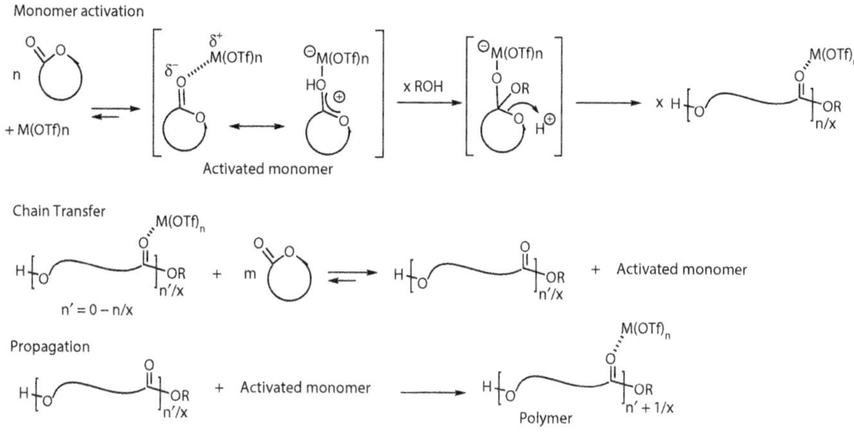

FIGURE 4.2 Activated monomer mechanism for *immortal* ROP. (Adapted from Ajellal, N. et al., *Dalton Trans.*, 39, 8363–8376, 2010.)

governed by the rate of exchange between metal alkoxide species and alcohol (acid–base interaction) as compared to the rate of propagation. Deactivation of the active species into hydroxy-terminated polymer leads to termination of polymerization.[6]

When a simple salt of Lewis acidic metal (e.g., $M(OTf)_n$) is used as catalyst the activated monomer mechanism (AMM) comes into play wherein a protic nucleophilic additive such as an alcohol acts as chain-transfer agent (Figure 4.2). The metal center in the salt coordinates to the carbonyl oxygen in the monomer to form the *activated monomer* species. Polymerization is initiated by ring opening of the monomer *via* oxygen-acyl bond cleavage that occurs due to attack of the protic nucleophile on the carbon atom in carbonyl group. Any source of protons including the alcohol, impurities in the reagents, and terminated polymer chains can act as the nucleophile. Polymer chains formed during the polymerization take part in the chain-transfer equilibria as active as the dormant species as well as dormant species. Similar to the first mechanism, control over polymerization is achieved if the rate of transfer equilibria (k_{tr}) is significantly greater than the rates of initiation and propagation steps.

In *living* polymerizations, each metal center produces only one polymer chain, whereas in *immortal* polymerizations, many polymer chains per catalyst are produced through reversible chain transfer between growing and dormant polymer chain. In *living* ROP the polymer molecular weight (M_n) can be predicted by the monomer/initiator ratio, whereas it is given by the monomer/ROH ratio in *immortal* ROP since one equivalent of alcohol is used to initiate ROP and the remaining excess acts as chain-transfer agent. Concentration of the alcohol remains constant throughout the process and OH-terminated macromonomers (R–polymer–OH) are obtained. A detailed kinetic analysis of this mechanism has been carried out.[7]

A catalyst/initiator for ROP of cyclic esters is typically a single-site catalyst of the general formula of *LMR* or *LMOR*, where M is a central metal atom surrounded by an ancillary ligand L and R/OR is the initiating group.[5] It satisfies the following conditions: (1) The metal should not undergo redox process, (2) it should be tolerant

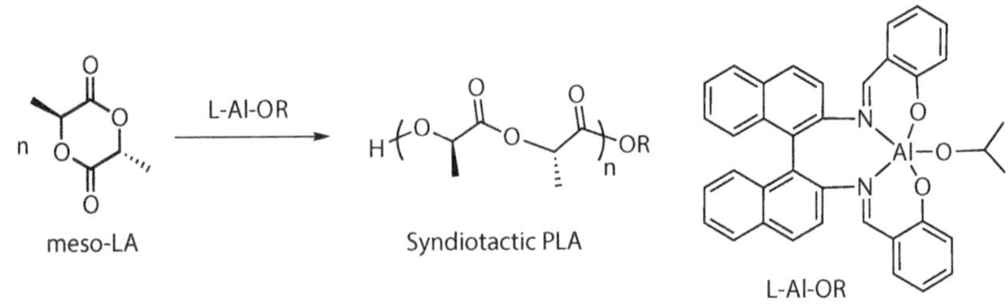

FIGURE 4.3 Synthesis of syndiotactic PLA using aluminum alkoxide complex.

to ligand scrambling, and (3) alkoxide ligand in the *LMOR* complex should be labile to alcohol exchange and insertion reactions with C–X multiple bonds. This allows introduction of functionality into the polymer.[8] An ideal catalyst for ROP should possess characteristics to afford living polymerization and should give stereochemically pure polymer after polymerization of a mixture of monomer stereoisomers. Most of the main group metals offer the advantages of single-site catalysis.

Tetraphenylporphyrin aluminum alkoxides were the earliest discrete metal complex catalysts used for polymerization of D,L-LA. Theses complexes showed control over molecular weight and low polydispersity index (PDI) (<1.25). Later, complexes containing Schiff's base ligands (salen-based framework) were designed that probably operate through equilibrium between dimeric and monomeric species. These catalysts with achiral ligands showed some control over stereochemistry of polymer, however with chiral ligands greater stereoselectivity was observed; for example, syndiotactic polymer with enantiotopic selectivity of 96% was obtained using the chiral catalyst (Figure 4.3).[9] Ligands other than Schiff's bases have also been incorporated into aluminum alkoxide complexes and shown to afford controlled polymerization of ε-caprolactone and δ-valerolactone. To install thioester end groups, aluminum thiolates were synthesized and shown to polymerize ε-caprolactone although low molecular weight polymers were obtained.[10] To improve the slowness of aluminum-based catalysts, complexes of yttrium and lanthanides were designed. To suppress *trans*-esterification reactions by screening polymer chains from active metal center during polymerization, bulky groups like isopropoxy were introduced in rare earth alkoxides.[11] These isopropoxy Ln diethyl acetoacetate (Ln = Y, Nd) catalysts afforded living polymerization of ε-caprolactone so that block copolymers with TMC and D,L-LA could be prepared.

Complexes of metals of low toxicity such as Li, Mg, Ca, Sr, Ba, Zn, and Zr are regarded as suitable catalysts for synthesis of PLA and other biodegradable polyesters for biomedical and food packaging applications. Complexes of Mg(II) and Zn(II) offer advantages such as lack of color, low cost, and low toxicity that are important in biomedical applications. To perform their biological functions these ions have to be kinetically labile, however this is not a desirable attribute for controlled polymerizations. In order to inhibit ligand exchange equilibria and control nuclearity of the complex, sterically hindered multidentate *N*-donor ligands are incorporated.[12] These are efficient catalyst systems for lactide polymerization and polymerization kinetics, and selectivity has been investigated. Polymerizations were first order in lactide and metal complex. Due to higher polarity of the Mg-OR bond and higher electropositivity of metal center, Mg(II) complexes were 100 times faster than Zn(II) complexes, however the latter are more tolerant to air and moisture. Chiral catalysts, with camphor and menthone substituents on ligands, showed preference for polymerization of *meso*-compound from a mixture of *rac*- and *meso*-lactide.[13] Magnesium and zinc complexes afford living polymerizations and low PDIs depending on the structure of the complex.

A highly active class of Zn complexes for the nonstereocontrolled living ROP of rac-LA comprises heteroleptic complexes containing bulky phenolate-based ancillary ligands to stabilize the metal.[14] The Zn(II) alkoxide complex, (LZnOEt)$_2$, where L$_1$ = 2,4-di-*tert*-butyl-6-{[(2'-dimethylaminoethyl)methylamino]methyl}phenolate, catalyzed the polymerization of lactide to afford high molecular weight polymer (M$_n$ = 130,000 g/mol) with low PDI, even at loadings of <0.1%. Detailed kinetic studies were carried out to elucidate empirical second-order rate law $-d[LA]/dt = k_p$ [LZnOEt][LA]. X-ray crystal structure showed that the complex was dimeric in solid state, however NMR and mass spectrometric analyses revealed that it exists predominantly in monomeric form in solution. It was found that a low level (0.07%–0.48% based on [LA]$_0$) of impurity that prevented polymerization below a threshold catalyst concentration (<2.4 mM at 0°C) was present. Based on the analysis of polymer-end groups a coordination–insertion pathway was proposed (Figure 4.4) that also accounted for lower molecular weight than predicted and is consistent with

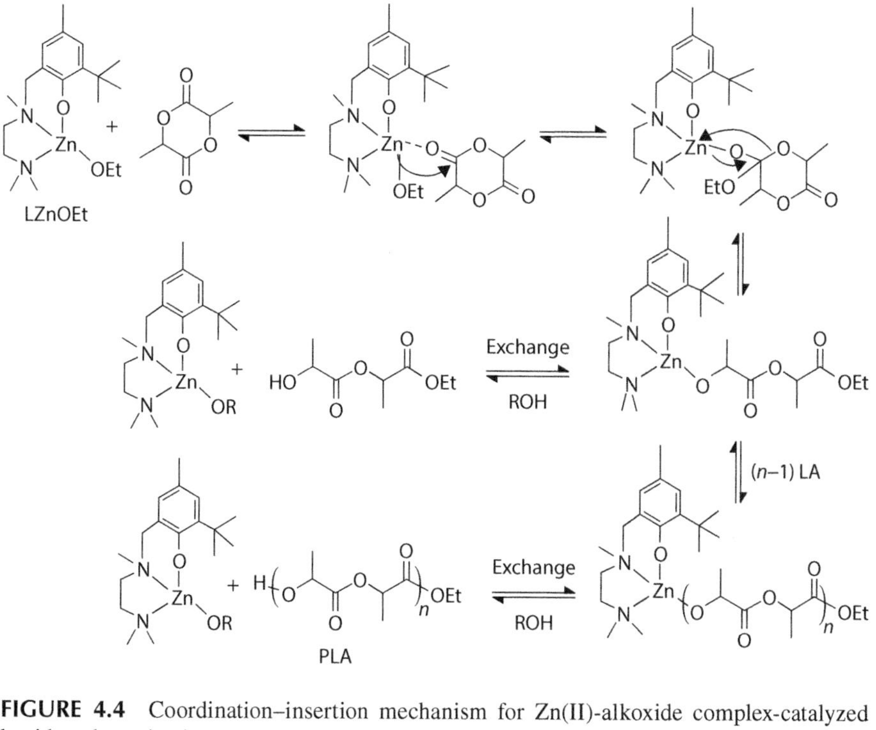

FIGURE 4.4 Coordination–insertion mechanism for Zn(II)-alkoxide complex-catalyzed lactide polymerization.

the second-order rate law. The molecular weight was determined using SEC equipped with light scattering detector.

Another class of highly active Zn complexes that gave stereocontrolled ROP of *rac*-LA comprises bulky β-diketiminate (BDI) bidentate ligands.[15] The complex [(BDI-1)ZnO^{i}Pr]$_2$ [(BDI-1) = 2-((2,6-diisopropylphenyl)amido)-4-((2,6-diisopropylphenyl)imino)-2-pentene] showed high activity and stereoselectivity for the polymerization of *rac*- and *meso*-lactide. The substituents on the β-diiminate ligand influenced both the degree of stereoselectivity and the rate of polymerization; changing the ligand substituents from isopropyl to ethyl groups resulted in a decrease in the heterotacticity ($P_r = 0.79$ from 0.94), whereas with *n*-propyl groups it decreased further ($P_r = 0.76$). (*S,S*)-lactide was polymerized to isotactic PLA without epimerization of the monomer, whereas *rac*-lactide and *meso*-lactide were polymerized to heterotactic PLA ($P_r = 0.94$ at 0°C) and syndiotactic PLA ($P_r = 0.76$ at 0°C), respectively, in a living polymerization as evidenced by the linear nature of M_n versus conversion plots. P_r is the probability of finding racemic diad in the polymer chain. The polymerizations were first order with respect to monomer (*rac*-lactide) and 1.56 order in catalyst as indicated by kinetics experiments. A similar magnesium complex [(BDI-1)MgO^{i}Pr]$_2$ was found to polymerize *rac*-lactide (500 eq.) in 96% yield in less than 5 min at 20°C.

Polymerization of α-methylene-γ-butyrolactone (MBL), a bifunctional cyclic ester with highly reactive exocyclic C=C bond obtained from biomass, follows the vinyl addition pathway (VAP) because the five-membered ring is highly stable under ambient pressure (ΔG_{ROP} is positive), however ring-opening copolymerization with a strained cyclic ester has been successful. Recently, ROP of MBL to exclusively produce the unsaturated polyester homopolymer was reported.[16] A simultaneous propagation by VAP and ROP resulting in crosslinked polymer was also observed. The formation of each type of polymer could be controlled by adjusting the catalyst (La)/initiator (ROH) ratio, which is determined by the chemoselectivity of the La–X (X = OR, NR$_2$, R) group. A homoleptic La complex La[N(SiMe$_3$)$_2$]$_3$ and two heteroleptic Y complexes were employed as catalysts. In the presence of alcohol the lanthanide amide is converted to the alkoxide, which is known to mediate ROP in a controlled manner compared to amides. Thus, ROP of MBL occurred when excess of alcohol was added to the reaction mixture, specifically at the ratio of MBL/La/BnOH = 200:1:3 at −60°C in tetrahydrofuran (THF) polyMBL of $M_n = 10400$ g/mol and $M_w/M_n = 1.28$ was obtained. The polyester was shown to be degradable and functionalizable through the reactive double bond by thiolation and was recycled to the monomer by heating the polymer solution at ≥100°C using the La-silamide catalyst.

Due to its biologically benign nature, calcium also is a metal of interest to prepare biocompatible metal catalysts for synthesis of aliphatic polyesters. Calcium is significantly larger and has different organometallic and coordination chemistry in comparison to Mg and Zn. The earliest example on calcium-based catalyst for ROP contained BDI ligand CH[CMeN(2,6-iPr$_2$C$_6$H$_3$)]$_2$ that had been shown to be an effective spectator ligand in LMX-type complexes (where, X = an amide or alkoxide) of Mg^{2+} and Zn^{2+}.[17] Relative reactivity of Ca, Mg, and Zn complexes of the type [HB(3-^{t}Bupz)$_3$]M was studied and it was found that under identical conditions, less than 90% conversion of LA was observed in 1 min for Ca, 1 h for Mg, and

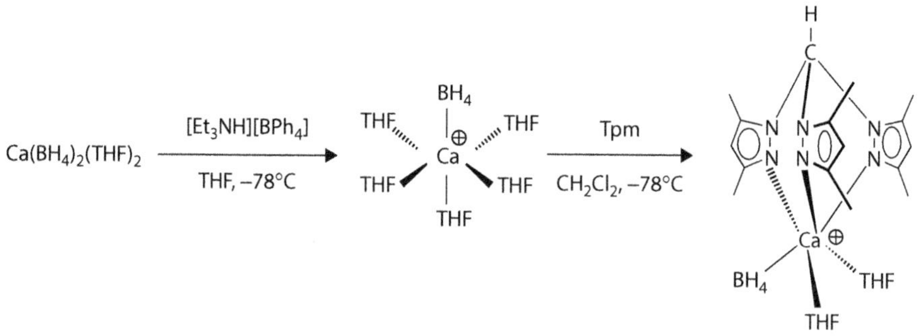

Tpm = Tris(3,5-dimethyl pyrazolyl)methane

FIGURE 4.5 Synthesis of cationic tetrahydroborate complexes of calcium.

six days for Zn. When two of the three metals were present the same order of reactivity, that is, Ca > Mg > Zn was followed with the less reactive metal remaining after the consumption of monomer. Stability of the complexes of these three metals is however generally in the reverse order. In terms of conversion of large quantities of monomers, that is synthesis of high molecular weight polymers, Zn complexes outperform those of Mg and Ca.

Lanthanide tetrahydroborates $[(L)Ln(BH_4)]$ are better controlled than the corresponding amides, and comparable to alkoxides as initiators for ROP.[18,19] Based on this, the first report on main group tetrahydroborate complexes described the cationic $[Ca(BH_4)(THF)_5][BPh_4]$ and the charge neutral $(Tp^{tBu,Me})Ca(BH_4)$ (THF) complexes of calcium (Figure 4.5).[20] These were shown to initiate ROP of *rac*-lactide with the latter complex giving PLA with high heterotactic content due to bulkier ancillary ligands that can induce tacticity through chain-end control. Heterotactically enriched polymer with high P_r values of 0.88–0.90 at −20°C (0.80 at RT) was obtained as determined by tetrad analysis of the CH(Me)O region of the selectively homonuclear decoupled ^1H NMR spectra. The molecular weight determined by GPC was $M_n = 17,600$ g/mol ($M_{n, theo} = 22,820$ g/mol) and polydispersity was low ($M_w/M_n = 1.3$). Calcium complexes containing Schiff base ligands have also been synthesized and found to be tolerant of high monomer/initiator ratios. Homo- and copolymerization of LA and TMC could be carried out using these complexes that gave high molecular weight and low PDI. Bis(trimethylsilyl)amide was used as the initiator.[21]

Limited number of examples exists on well-defined complexes of barium due to its highly ionic nature and large size that leads to reaction with ethereal solvents and makes isolation of complexes difficult. A bulky aminebis(phenolate) ligand was utilized to isolate and characterize a well-defined, trinuclear barium complex. A diamino bis(phenol) ligand was reacted with $Ba\{N(SiMe_3)_2\}_2(THF)_2$ in toluene and allowed to crystallize.[22] The complex obtained was a bis(THF)-solvated trinuclear species in the monoclinic space group C2/c and initiated ROP of *l*-lactide and ε-caprolactone (ε-CL). Lactide was polymerized in bulk at high monomer: initiator ratio of 900:1, however conversions of only 60% were reached due to an increase in the viscosity ($M_n = 25,500$ g/mol; PDI = 1.57). Polymerization of ε-CL in toluene proceeded with first-order kinetics with little or no induction period and initially resulted in bimodal distribution, however prolonged reaction times gave monomodal albeit broad molecular weight distribution. Based on NMR data of the complex in THF it was proposed that the solid structure was probably not intact, however both terminal and bridging aryloxides were present in coordinating solvents. Characterization of the polylactide obtained by using benzyl alcohol (BnOH) as coinitiator by matrix-assisted laser desorption/ionization time-of-flight (MALDI-TOF) mass spectrometry and NMR end-group analysis revealed two kinds of polymer chains: a higher molecular weight initiated by insertion into a Ba-O(L) aryloxide bond and a lower molecular weight initiated via insertion into a Ba-OCH$_2$Ph bond suggesting that active Ba-O(L) sites remain in the presence of a coinitiator and is consistent with the presence of at least two *active site* environments being available within an oligomeric initiating species; this was also supported by NMR spectra of the complex in excess BnOH. Barium complexes with bulky iminopyrrolyl ligand {(THF)$_n$Ba(2-(Ph$_3$CN=CH) C$_4$H$_3$N)$_2$} were synthesized recently, along with those of Ca and Sr, using two methods.[23] The iminopyrrole ligand L-H was directly treated with the metal precursor [Ba{N(SiMe$_3$)$_2$}$_2$(THF)$_n$, n = 3] in 2:1 molar ratio in tetrahydrofuran at ambient temperature. The complex was also prepared using the salt metathesis reaction involving the treatment of the potassium salt [(2-(Ph$_3$CN=CH)C$_4$H$_3$N)K(THF)$_{0.5}$]$_4$ with barium diiodide in 2:1 molar ratio in tetrahydrofuran. The complex was found to possess higher activity and afford narrow polydispersity for ROP of ε-caprolactone compared to that of magnesium complex with iminopyrrolyl ligand. Heteroleptic silylamido complex of barium supported by amino ether phenolate ligand [{LO^3} Ba-N(SiMe$_2$H)$_2$] ({LO^3}$^-$ = 2-[(1,4,7,10-tetraoxa-13-azacyclopentadecan-13-yl) methyl]-4,6-di-tert-butylphenolate) was synthesized and studied for stability and activity toward ROP of L-lactide.[24] Molecular structure of corresponding Ca and Sr complexes determined by X-ray diffraction showed M···H-Si internal β-agostic interactions, which contribute to the stabilization of the complex against ligand redistribution reactions. Immortal ROP of L-LA catalyzed by the Ca, Sr, and Ba complexes in the presence of 10–100 equivalents of propargyl alcohol as a transfer agent was investigated. Propargyl alcohol was used to install alkyne moiety at the chain end of poly(L-lactic acid) (PLLA) to build complex macromolecular architectures using Cu-catalyzed azide-alkyne cycloaddition, a well-known type of click reaction. For the polymerization carried out in toluene at 30°C, the activity decreased in the order Ca < Sr < Ba; the calcium complex was found to be inactive, whereas barium complex at 0.5 mM afforded 49% conversion in 4 min ([L-LA]/[complex]/[propargyl alcohol] = 1000:1:10).

Complexes of iron have been explored due to easy availability of nontoxic metal-containing precursors. For example, bis(imino)pyridine iron bis(alkoxide) complexes were used in the polymerization of (rac)-lactide. A living polymerization that allowed polymer growth up to fifteen sequential additions of lactide monomer was observed. The activities of the catalysts were particularly sensitive to the identity of the initiating alkoxide with more electron-donating alkoxides resulting in faster polymerization rates. The bis(aryloxide) catalysts operated with one alkoxide ligand, whereas the bis(alkylalkoxide) complexes initiated with both alkoxide ligands as observed in the mechanistic studies. Catalysis was shown to be switched on and off upon oxidation and reduction of the iron catalyst, respectively, due to formation of the inactive cationic iron(III) bis-alkoxide complex by oxidation of iron(II) catalyst.[25] Zr-based catalysts were found to give high molecular weight polymers, whereas racemization of polymer was higher for catalysts comprising less electronegative metals.[26]

Tin carboxylates are the most widely used complexes of tin for polymerization of cyclic esters, mainly lactides and it gives atactic polymers. The active species is Sn(II)-alkoxide formed in the presence of protic reagents such as alcohol or LA impurities and traces of water present in monomer or catalyst.[27] Tin(IV) dialkyl dialkoxide complexes and well-defined single-site complexes of tin have also been used. Tin complexes are less active compared to similar zinc complexes due to the lower electrophilicity of the tin center but also produce living polymers with low PDI. Tin (II) complexes {LOx}Sn(X), where {LOx}$^-$ = aminophenolate ancillary, containing amido coligands (X) were studied for copolymerization of lactide with TMC.[28] The catalyst initiated homopolymerization of TMC in the presence of isopropanol, however copolymerization of L-lactide and TMC was not observed. Experimental data and density functional theory (DFT) calculations showed that insertion of L-lactide followed by that of TMC is endothermic by +1.1 kcal/mol compared to insertion of two consecutive lactide units (−10.2 kcal/mol). The drawback with tin complexes is the toxicity of tin, which prevents the use of these polymers in biomedical applications and polymerization using tin catalysts is not considered a sustainable process.

Since the material properties depend strongly on the polymer tacticity, designing single-site catalysts that afford stereoselective polymerization is important. Lactide exists in three isomers—D,L, and meso, however the most commonly used and economically viable monomer is *rac*-lactide. Single-site metal complexes are used to prepare stereoregular polylactide from *rac*- or *meso*-lactide. For example, syndiotactic PLA can only be synthesized from *meso*-lactide using a catalyst that selectively inserts at only one stereocenter of the monomer, whereas heterotactic PLA is obtained from *rac*-lactide by selective insertion of the lactide isomer with opposite configuration to the previously inserted monomer. Preparation of isotactic PLA does not require stereocontrol over polymerization when pure L-lactide is used as the monomer. Two different mechanisms are followed by the catalysts depending on how the configuration of the next inserted monomer in *rac*-lactide polymerization or the cleavage site of the monomer in *meso*-lactide polymerization is determined: (1) a chain-end control mechanism, where it is determined by the stereogenic center in the last repeating unit in a propagating chain and (2) an enantiomorphic site control mechanism, where it is determined by configuration of the surrounding ligand. Thus, only isotactic or syndiotactic PLA can be obtained from *rac*- or *meso*-lactide, respectively, under enantiomorphic control, whereas heterotactic (alternate arrangement of pairs of stereocenters, i.e., -SSRRSSRR-) polymer is obtained in addition to iso- and syndiotactic polymer under chain-end control mechanism.[1] Detailed studies on stereoselectivity of some chiral aluminum complexes have shown that several factors are responsible and it is not straightforward to ascribe the observed stereoselectivity clearly to either of the mechanism.[29] Achiral Schiff base aluminum alkyls have also been shown to afford stereoselective polymerization of *rac*-lactide by introducing substituents (Ph, *t*-Bu) in aromatic rings of Schiff base ligands.[30] Degree of stereoregularity is expressed as coefficients P_m and P_r that give the probability of finding meso or racemic diads, respectively.

Coordination–insertion mechanism discussed earlier gives polymers with alkoxy ester at α-end and hydroxy group at ω-end of the polymer. Introduction of hydroxyl groups at both ends of polyester or polycarbonate opens the way for installing other functional groups that can initiate polymerization of chemically different monomers and preparation of structurally diverse block copolymers (with polypeptides, polymethacrylates) that are important as biodegradable plastics and biomaterials. Synthesis of α,ω-dihydroxytelechelic polyesters has been achieved by the following two strategies: (1) direct synthesis using discrete metal borohydride complexes and (2) *immortal* polymerization using a diol as transfer agent in combination with a catalyst.[31] Polyesters (from lactones) and poly(trimethylene carbonate) have been prepared by using rare earth metal complexes containing borohydride as a ligand. The reactivity of these complexes is due to the hydridic nature of the metal-boron bond, which involves bridging hydrogen atoms.

Variation in macromolecular architecture helps to control polymer properties and makes applications in drug delivery and nanotechnology possible. Star polymer is an architecture that has been explored with aliphatic polyesters and polycarbonates.[32] Star polymers of lactides show lower melting temperatures (T_m), glass transition temperatures (T_g), and crystallization temperatures (T_c) than their linear counterparts. Star polylactides have been synthesized using various polyol cores and stannous octanoate [$Sn(oct)_2$] as the catalyst of choice. The structure of core was shown to have a significant impact on thermal properties of polymers. Cholic acid,[33] β-cyclodextrin,[34] tetra- and hexahydroxy-functionalized perylene,[35] dendrimer,[36] and various polymeric cores have been used. Terminal groups have a significant effect on hydrolytic degradation stability that is important for application in controlled drug release. PLAs with Cl-, -NH_2, and -COOH end groups had higher cold crystallization temperatures than the -OH-terminated PLAs.[37] Polymers with Cl-, -NH_2 groups also had higher thermal stability than -OH-terminated polymers.

4.2   METATHESIS POLYMERIZATIONS

A metathesis reaction is defined as a chemical transformation in which atoms from different functional groups interchange with one another, resulting in the redistribution of functionality yielding similar bonding patterns for both molecules.[38] The term *olefin metathesis* was proposed by Calderon in 1967 and since then it has been applied for the synthesis of complex molecules, particularly cyclic systems to prepare active pharmaceutical ingredients as well as polymers. It is a great advancement in polymer synthesis since functionalized polyolefins are made easily accessible using metathesis polymerization. Two types of metathesis polymerization will be discussed in this chapter—(a) ring-opening metathesis polymerization (ROMP),[39] a chain-growth polymerization and (b) acyclic diene metathesis polymerization (ADMET), a step-growth polymerization.

4.2.1 Ring-Opening Metathesis Polymerization

ROMP, discovered during studies on olefin polymerization, was the first olefin metathesis-based transformation reported. Chauvin proposed the now accepted mechanism (Figure 4.6) in 1971 that is consistent with the experimental evidence gathered later.

In the initiation step, a transition metal alkylidene complex coordinates to a cyclic olefin such as norbornene and undergoes [2+2]-cycloaddition to yield a four-membered metallacyclobutane intermediate, which undergoes a cycloreversion reaction and a new metal alkylidene is formed. The new alkylidene is larger in size due to addition of monomer however its reactivity is similar to the initiator. Hence, more monomer molecules add repeatedly during the propagation stage until all monomer is consumed or the polymerization is terminated. ROMP is considered a living polymerization since the metal alkylidene at the chain end can start the polymerization when fresh batch of monomer is added. ROMP is quenched by deliberate addition of a specialized reagent called chain terminator (CT). The CT reacts with the metal alkylidene to selectively remove and deactivate the transition metal complex from the polymer chain end. Thus, functionalized terminating reagents can be used to install functionality at the chain end of a polymer. Various complex functionalities can be introduced in this manner in ROMP unlike other polymerizations. Typically, aldehydes or ketones are used to terminate ROMP initiated by Ti, Ta, W, and Mo catalysts, whereas vinyl ethers and vinyl lactones[40] are employed to terminate Ru complex-initiated polymerization. A difunctional molecule derived from *cis*-2-butene, called chain-transfer agent (CTA), works through cross-metathesis to install functionalities on both ends of the polymer chain.

The mechanism as shown in Figure 4.3 can also proceed in opposite direction since ROMP reactions are typically reversible. However, they are equilibrium controlled and are driven forward by the release of ring strain in the cyclic olefin accompanied by entropic penalties. Commonly employed ROMP monomers have a ring

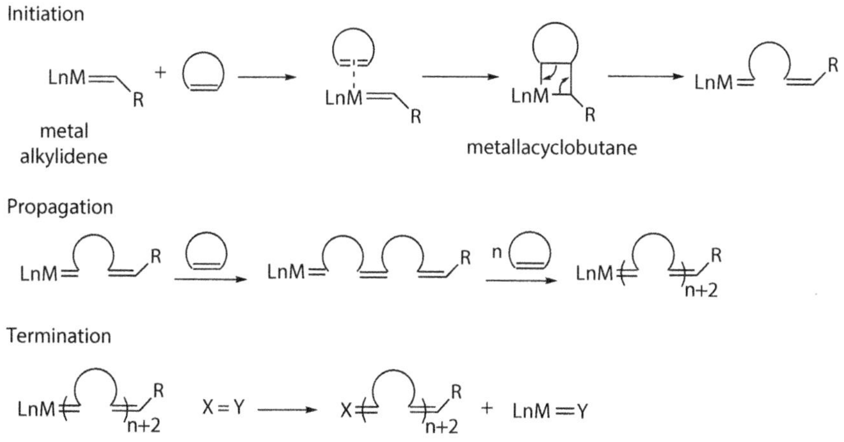

FIGURE 4.6 General mechanism of ROMP.

strain of >5 kcal/mol. Due to these thermodynamic considerations, temperature and concentration strongly influence the polymerization.[40,41] Side reactions (secondary metathesis) such as intermolecular chain-transfer, intramolecular chain-transfer reactions (called *backbiting*), and formation of cyclic oligomers also take place due to the reversible nature of ROMP.

Activity of various metathesis catalysts is given in Table 4.1. Initially, olefin metathesis involved ill-defined catalysts containing metal salts in higher oxidation states and metal oxides as activators. Early on, titanium-based catalysts containing bis(cyclopentadienyl)titanacyclobutane moiety were developed. These were less active but gave polymers with low polydispersity (<1.2) and it was possible to synthesize complex polymer topologies including di- and triblock copolymers.[42] Tantalum-based catalysts designed by Schrock and coworkers were slightly more active than Ti catalysts, however they showed secondary metathesis reactions near completion of polymerization leading to higher polydispersities. This tendency was reduced by the incorporation of bulky, electron-rich diisopropylphenoxide ligands that decreased their electrophilic character. The rates of polymerization with the Ti and Ta catalysts depend on the catalyst and monomer concentration (i.e., second-order). Both of these catalysts are highly Lewis acidic owing to their high oxidation states. In 1986, single-component, Lewis-acid free, W-based catalysts of the type $W(CHR')(N-2,6-C_6H_3-iso-Pr_2)(OR)_2$ were prepared that afforded living ROMP and low polydispersities.[43,44] The catalyst activity could be modulated through the choice of alkoxy ligands.[45] Later, oxo-tungsten-based alkylidene complexes $W(=CH-t-Bu)(O)(PR_3)(OAr)_2$ were designed that showed better functional group tolerance.[46]

For higher functional group tolerance and stability toward air, oxygen, and moisture compared to W-based catalysts, structurally similar molybdenum alkylidenes that tolerated ester, amide, imide, ketal, ether, cyano, trifluoromethyl, and primary halogen-containing functional groups in the monomer were introduced by the Schrock group.[40] Apart from polar norbornene monomers such as endo-5-norbornene-2,3-dicarboximide, other cyclic olefins like cyclopentene could be polymerized.

TABLE 4.1
Activity of Various Metathesis Catalysts

Ti/Ta	W	Mo	Ru	Decreasing Activity
R-COOH	R-COOH	R-COOH	C=C	
R-OH, water	R-OH, water	R-OH, water	R-COOH	
R-CHO	R-CHO	R-CHO	R-OH, water	
$R_2C=O$	$R_2C=O$	C=C	R-CHO	
R-COOR	C=C	$R_2C=O$	$R_2C=O$	
R-CONHR				
C=C	R-COOR	R-COOR	R-COOR	
	R-CONHR	R-CONHR	R-CONHR	

Source: Bielawski, C. W. and Grubbs, R. H., *Prog. Polym. Sci.*, 32, 1–29, 2007.

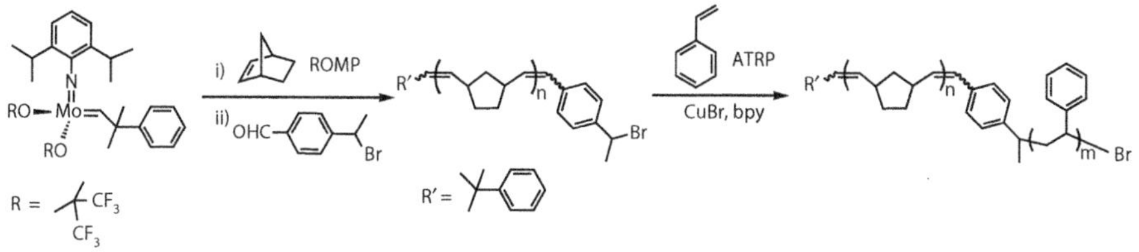

FIGURE 4.7 Synthesis of block copolymer using tandem ROMP and ATRP.

Similar to W-based catalysts a strong donor ligand (PMe$_3$) was required to obtain polymers with controlled molecular weights and low polydispersity.[47] Polymerization was conducted at high monomer concentration (>2.8 M) to obtain high molecular weight polymer with low PDI (1.04). Polymerization of 3,3-disubstituted cyclopropene afforded predominantly *trans* polymer with low polydispersity.[48] Block copolymers from monomers polymerizable by two different techniques were obtained by combining ROMP with atom transfer radical polymerization (ATRP). ROMP of norbornene was terminated with a benzaldehyde that contained initiating group for ATRP and then the polynorbornene was used as macroinitiator for ATRP of styrene or methyl acrylate (Figure 4.7).[49]

Ruthenium-based catalysts designed by Grubbs group are more stable to air and moisture and are tolerant to many functionalities compared to the aforementioned catalysts. The first-generation catalyst, [Ru]-1 is more active than the early Ru-based catalysts due to electron-rich trialkyl phosphine ligands. This can be contrasted with Schrock catalysts wherein activity increased due to electron-withdrawing ligands. Other alkylidenes can be prepared by reacting with a terminal olefin and the trialkyl phosphines (PR$_3$) can be replaced by other trialkyl phosphines (PR'$_3$) to control the catalyst activity. Thus, charged phosphines were incorporated to obtain water-soluble catalysts and ROMP was conducted in aqueous media in the presence of HCl.[50] When one of the PR$_3$ ligands in [Ru]-1 was replaced by *N*-heterocyclic (NHC) carbene the reactivity and functional group tolerance greatly improved, however these complexes catalyze metathesis and olefin isomerization simultaneously (second and third generation [Ru]-2 and [Ru]-3). Structures of the commonly used catalysts are shown in Figure 4.8.

Although ROMP is generally carried out in organic solvents, water has also been used as solvent toward *green chemistry* and for making water-soluble polymers for applications as biomaterials and drug carriers. Initially, miniemulsion

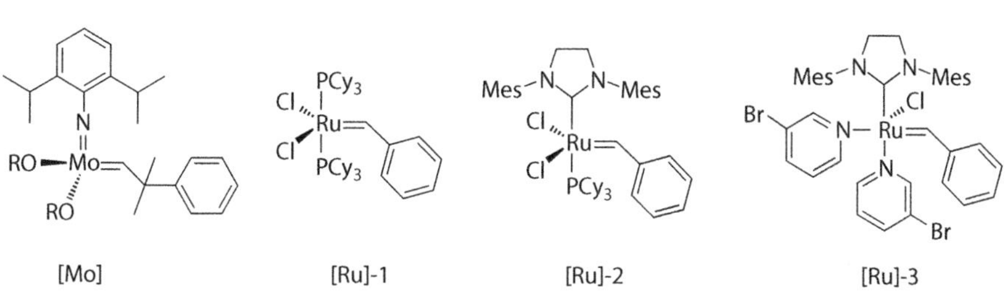

FIGURE 4.8 Commonly used catalysts for metathesis polymerization.

polymerization was carried out with norbornene and cyclooctene, however, higher amounts of organic solvent were required for reasonable yields.[51] Later, water-soluble ruthenium catalysts attached with polyethylene glycol (PEG) chains or quaternary ammonium groups were developed and norbornene monomers containing similar groups were polymerized so that polymers were soluble in water either as single chains or as micelles.[52,53] Norbornene-based polybetaines that contain both cationic and anionic groups and are promising for biomaterials applications have also been made using ROMP.[54] ROMP catalyzed by Ru-complexes has been combined with other controlled radical polymerization (CRP) techniques in order to prepare polymer topologies with various backbone chemistries.[55] ROMP conducted using chain-transfer agent (CTA) containing initiator group for CRP gave ditelechelic polynorbornene, which was then used as macroinitiator for CRP to prepare ABA triblock copolymers.[56,57] Brush copolymers were also obtained by polymerizing a norbornene monomer substituted with ATRP initiator followed by ATRP from these initiating sites along the polymer backbone.[58] A dual ATRP-ROMP initiator was prepared by attaching suitable alkyl halide to the alkylidene carbon in [Ru]-1 and both polymerizations were carried out simultaneously to obtain diblock copolymer in one pot.[59] Functional polymers for various applications have been synthesized through ROMP by direct polymerization of the monomer substituted with oligothiophene derivatives, carbohydrates, metal complexes, and large chromophores containing heteroatoms. Another approach toward functional polymer topologies is the combination of ROMP and efficient coupling reactions called *click chemistry*. Such reactions require azide and alkyne groups that are not compatible with Ru-initiated polymerization and need to be protected. Sulfur-containing functional groups are also not compatible with metal-catalyzed ROMP.

Stereoregular polymers such as polypropylene can crystallize and possess a distinct melting point and hence are desirable for high-end applications. Most of the work on stereoregular polymers utilizes molybdenum-based catalysts. In the ROMP mechanism, exo face of the C=C bond in norbornene adds to the metal-carbene bond to form the cis or trans metallacyclobutane intermediate that leads to cis or trans C=C bonds, respectively, in the final polymer backbone. Isotactic or syndiotactic arrangement is obtained depending on which face of the propagating M=CHR intermediate that the norbornene adds to repeatedly. Initially, *cis, isotactic* polynorbornenes were made using Mo-based catalysts through *enantiomorphic site control*, that is the monomer adds to a given metal center through the same enantiotopic face of the olefin in each step. Later, Mo- and W-based catalysts were designed to yield *cis, syndiotactic* polynorbornenes through a mechanism involving inversion of the stereogenic metal center with each insertion, which is defined by Schrock as *stereogenic metal control*.[60] Polynorbornenes retain their tacticity even after the double bonds in the main chain are hydrogenated. The structures of four possible stereoregular polynorbornenes are shown in Figure 4.9.

Mo-based imido alkylidene catalysts containing chiral biphenolate or binaphtholate ligands (O_2R = biphen, binap) afforded *cis*-polymers with single tacticity[61] but with racemic ligands, polymer with bimodal molecular weight distribution is obtained due to different rates of initiation, propagation, or both from enantiomeric metal centers. *Cis, syndiotactic* polymers were obtained by using monoaryloxide pyrrolide

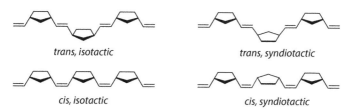

trans, isotactic   trans, syndiotactic

cis, isotactic   cis, syndiotactic

FIGURE 4.9 Structures of possible stereoregular polynorbornenes.

(MAP) imido Mo-alkylidene catalysts containing large aryloxide (OR) ligand such as HIPTO [O-2,6(2,4,6-i-Pr$_3$C$_6$H$_2$)C$_6$H$_3$].[62] In this case, metallacyclobutane intermediate, in which the substituents are pointing away from the bulky OR group is selectively formed and hence the initiator catalyzes Z-selective metathesis reactions. Polymerization of (+)-5,6-dicarbomethoxynorbornene using Mo(NAd)(CHCMe$_2$Ph)-(Pyr)(O-2,6-(2,4,6-i-Pr$_3$C$_6$H$_2$)$_2$C$_6$H$_3$) (Ad = 1-adamantyl; Pyr = pyrrolide) initiator gave a polymer with ~92% of the dyads *trans, isotactic*.[63] Oxo-tungsten-based alkylidene complexes gave >95% *cis, isotactic* polymer when used to initiate ROMP of 2,3-dicarbomethoxynorbornadiene and 2,3-bis(trifluoromethyl)norbornadiene.[46] Stereoregular polymers have also been obtained from other cyclic olefins. 3-Methyl-3-phenylcyclopropene was polymerized using Mo biphenolate and binaphtholate, which gave *cis, isotactic* polymer and using MAP initiators, which gave *cis, syndiotactic* polymer.[48] The two polymers can be distinguished on the basis of ^{13}C NMR spectra using chemical shift of the methyl group. The initiator attacks from the side of the methyl group for steric reasons. This monomer was polymerized with Mo(NAr)(CHCMe$_2$Ph)-(OTPP)(Pyr) (Ar = 2,6-diisopropylphenyl, Pyr = pyrrolide, OTPP = 2,3,4,6-tetraphenylphenoxide) at different temperatures to understand the effect on structural regularity. At −78°C, *cis, syndiotactic* polymer was formed, however as polymerization temperature was increased stepwise to 46°C the regularity was found to steadily decrease. It was proposed that stable *anti* alkylidene insertion product is formed at low temperature giving *cis, syndiotactic* polymer, whereas at higher temperature *anti* alkylidenes are converted to *syn* products and hence the polymer lacks a regular structure. Even when well-defined initiators with known structure are used, unexpected tacticity may result since the lowest energy pathway may depend on the monomer.[60] Ru initiators have also been used to obtain stereoregular polynorbornenes.[64] [RuCl$_2$(p-cymene)]$_2$ catalyst precursor was used in the presence of trimethylsilyldiazomethane (TMSD) as activator to obtain highly *trans, isotactic* polymer from 2,3-dicarboalkoxynorbornadienes.[65]

Polymers for a variety of advanced materials have been made using ROMP. Factors affecting the performance of liquid crystalline polymer-based elastomers and gels, such as microstructure of the polymer backbone, degree of cross-linking, and type and frequency of the mesogen, are difficult to control using conventional polymerization techniques. Side chain liquid crystalline (SCLC) polymers based on ABA-triblock architecture were synthesized using a bimetallic Mo-based initiator.[66] Further, liquid crystalline (LC) polymers containing chiral units such as cholesterol show chiral nematic and smectic phases and hence have applications in optoelectronics such as LCD displays and in biomedicine. Surfaces are functionalized with polymer brushes to impart superhydrophobicity, switchable wettability, and anti-fouling properties. Due to its high functional group tolerance, ROMP has been used to grow functional polymers on surfaces for such applications.[67,68] Self-healing materials have the ability to heal microcracks formed in a polymer composite material and prevent its failure caused by exposure to UV-irradiation, heat, pressure, and so on. ROMP has been used for self-healing materials. A monomer (norbornene) is encapsulated in a urea-formaldehyde microcapsule that ruptures when a crack propagates and the released monomer is polymerized by the initiator present in the matrix thus filling the crack.[69]

A limitation of the ruthenium-catalyzed reactions is the incomplete removal of ruthenium catalyst, which makes the polymers toxic and colored. To overcome this problem, silica particles loaded with functional groups that bind to ruthenium such as amines, thiols, and thiourea have been developed commercially. Silica gel functionalized with isocyanide ligands and several other systems have also been reported toward this end.[70,71]

Polymers made by metathesis polymerizations have found applications in niche areas due to their unique properties. Blends of poly(norbornene) with other elastomers such as natural rubber or styrene-butadiene rubber are being explored to tune the properties of the vulcanizates to fit the desired application. Copolymers of norbornene and other cyclic olefins have found applications in optics, for example, in camera lenses.[72]

4.2.2 ACYCLIC DIENE METATHESIS POLYMERIZATION

ADMET polymerization uses α,ω-dienes as monomers to yield linear polymers with unsaturated backbone that are analogous to polyethylene. The mechanism is shown in Figure 4.10. It is step-growth polymerization in which ethylene is released as the small molecule by-product and same catalysts as in ROMP are used. Polydispersity values are also typical of step-growth polymerization (~2). It is a thermally neutral process and high molecular weight polymer is obtained by removing ethylene under vacuum; up to 80,000 g/mol have been achieved using [Ru] catalysts.[73] A small amount of cyclization common to all polycondensation reactions also occurs in ADMET. Depolymerization can be carried out on ADMET polymers by subjecting the unsaturated bonds in the backbone to ethylene gas.[74]

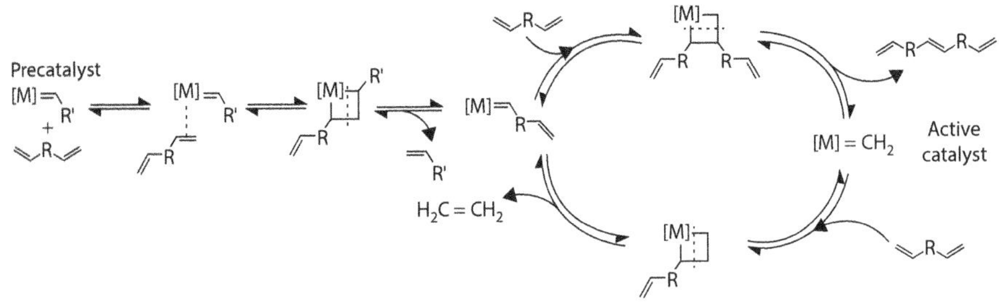

FIGURE 4.10 Catalytic cycle in ADMET polymerization.

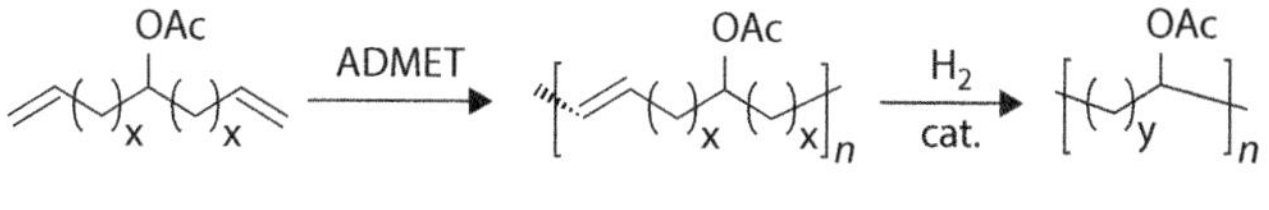

FIGURE 4.11   EVA polymer synthesis using ADMET.

ADMET polymerization followed by hydrogenation affords strictly linear polyethylenes that are suitable for structure–property relationship studies, since entanglement molecular weight for polyethylene is only 1000 g/mol. For certain applications it is necessary to incorporate polar functional groups into polyethylene to enhance adhesive and barrier properties.[75] Linear ethylene-vinyl acetate (EVA) copolymers with wide range of compositions have been obtained using ADMET followed by hydrogenation (Figure 4.11).[76] Polyethylenes with various functionalities such as aromatic groups, acids, silanes, ethers, acetals, ketones, and thioethers have been incorporated into ADMET polymers.[77]

Introduction of branching in a controlled manner into polyethylene backbone is an important tool to tailor its properties. To study the effect of branching on properties of polyethylene, ethylene-propylene copolymers are made using various synthetic methods, however ADMET provides a regularly methyl-branched polymer by polymerization of a single monomer.[78] ADMET also provides model polymers for ethylene-alkene copolymers. A library of polymers with varied branch contents was prepared by varying the number of methylenes between the terminal olefins. While polymerizing the same monomer with [Mo] and [Ru] catalysts, polymers with equal branch content but with M_n values of 72,000 and 17,400 g/mol, respectively, were obtained. Copolymerization of these methyl-branched monomers with different weight percentages of 1,9-decadiene using [Mo] catalyst afforded randomly branched linear polyethylenes.[79] The regularly branched polymer showed a sharp melting transition, whereas the random copolymer showed a broader melting transition owing to additional order in the former as suggested by X-ray scattering data. Effect of spacing between branches on thermal properties was studied by synthesizing a series of polyethylenes with butyl branches. As the spacing between branches increased the melting point also increased.[80] Effect of branch spacing on morphologies was studied by synthesizing 11 different polyethylene copolymer structures with branches precisely located on every 21st carbon.[81] It was found that the smaller branches such as methyl and ethyl were included in the unit cell, whereas the larger branches (propyl, butyl, pentyl, *tert*-butyl, cyclohexyl, adamantyl) were excluded from the crystal. Design of appropriate monomer is the key to obtain polymers with precise branching as well as specific functional groups in ADMET.

Various complex polymer architectures have also been obtained using ADMET. Grubbs and coworkers have synthesized mechanically interlocked polymers such as polyrotaxanes and *daisy-chain* polymers *via* ADMET.[82,83] Hyperbranched polymers are highly branched polymers that possess higher solubility and lower viscosity due to large number of end groups. These are typically synthesized in single-step reaction from AB_n ($n \geq 2$)-type monomers using condensation or CRP chemistry. Selectivity of the cross-coupling of electron-rich and electron-deficient olefins was used to synthesize hyperbranched polymers from an AB_2 monomer containing one terminal olefin and two acrylates using ADMET catalyzed by Grubbs [Ru]-2 catalyst.[84] The acrylate groups at the periphery were reacted with a pyrene derivative using cross-metathesis to modify properties of the polymer. Hyperbranched poly-phosphoester with acrylate end groups was synthesized using a similar approach, and intramolecular crosslinking of endgroups was used to get polymer nanoparticles.[85] Linear-dendritic polyphosphoester copolymer was synthesized using alkene-acrylate reaction in combination with ADMET and thiol-ene reaction.[86]

ADMET polymerization has been used to prepare telechelic oligomers that are used as building blocks in segmented polymers by polymerizing 1,9-decadiene in the presence of 9-decenyl acetate and 1,5-hexadiene in the presence of olefin-containing silane.[87] Reaction of an acrylate with carbon–carbon double bonds in the main chain of an ADMET polymer resulted in depolymerization into oligomers with ester end-groups, that was termed as *insertion metathesis depolymerization*.[88] Selective alkene-acrylate metathesis was also utilized to obtain alternating copolymers by reacting α,ω-diene with α,ω-diacrylate without using high vacuum conditions.[89] Effect of solvent on molecular weight was also studied and dichloromethane was found to afford highest molecular weight polymer ($M_w = 20,000$ g/mol), whereas tetrahydro-furan gave the lowest molecular weight ($M_w = 3300$ g/mol). ADMET polymerization and ATRP have been used in tandem to prepare graft copolymers that contain poly-styrene chains with pendent phosphonic acid groups for proton conduction.[90] The polymers were prepared by two approaches: (1) poly(vinylbenzyl phosphonic acid) (PVBPA) with a diene at one end was synthesized by ATRP and then polymerized by ADMET, and (2) a polymer containing ATRP-initiating site in every repeat unit was synthesized by ADMET followed by functionalized polystyrene synthesis by ATRP.

Characteristic features of conducting organic materials such as flexibility and lightness are being used to develop flexible electronic devices in solar cells, organic light-emitting diodes (OLEDs), supercapacitors, and sensors. To develop a funda-mental understanding of relationship between conjugation length and properties, ADMET polymerization of 2,5-de-heptyloxy-1,4-divinylbenzene with [Mo] catalyst was carried out. It yielded vinyl end-capped poly(p-phenylene vinylene) (PPV) oligo-mers of chain length between 2 and 8. Selective telomerization without reaction with internal double bonds was performed using [Mo] catalyst to obtain cross-metathesis products, which are oligomers of integer multiplicity relative to starting oligomers.[91]

Introduction of amino acids into the main chain of polyolefins renders them chiral. ADMET has been used to synthesize polymers with peptide in the main chain from dienes containing L-valinol and L-leucinol moieties.[92] However, these polymers are hydrolytically less stable than polymers with pendent amino acid groups, which were synthesized from a diene with pendent-protected amino acid using [Ru]-2 catalyst.[93] Silicon-containing polymers are commercially important polymers used in various applications such as biomedical, cosmetics, and electronics due to their unique prop-erties compared to organic polymers. ADMET has been used to incorporate silicon-based groups in hydrocarbon polymers to explore the possibility of making novel materials with enhanced thermal and mechanical properties. As an example, block copolymer of polyoctenamer and poly(dimethylsiloxane) (PDMS) was prepared. 1,9-Decadiene was polymerized in the presence of 4-pentenylchlorodimethylsilane using [Mo] or [Ru]-1 catalyst to obtain telechelic oligomers.[94]

REFERENCES

1. Williams, C. K. Synthesis of functionalized biodegradable polyesters. 2007. *Chem. Soc. Rev.* 36:1573–1580.
2. Vaidya, A. N., Pandey, R. A., Mudliar, S., Kumar, M. S., Chakrabarti, T., Devotta, S. 2005. *Crit. Rev. Environ. Sci. Technol.* 35:429–467.
3. Wheaton, C. A., Hayes, P. G., Ireland, B. J. 2009. Complexes of Mg, Ca and Zn as homogeneous catalysts for lactide polymerization. *Dalton Trans.* 25:4832–4846.
4. Duda, A., Penczek, S. 2000. Thermodynamics, kinetics, and mechanisms of cyclic esters polymerization. In *Polymers from Renewable Resources.* ed. Scholz C., Gross, R. A. ACS Symposium Series 764:160–198, American Chemical Society, Washington DC.
5. Dijkstra, P. J., Du, H., Feijen, J. 2011. Single site catalysts for stereoselective ring-opening polymerization of lactides. *Polym. Chem.* 2:520–527.
6. Ajellal, N., Carpentier, J.-F., Guillaume, C., Guillaume, S. M., Helou, M., Poirier, V., Sarazina Y., Trifono, A. 2010. Metal-catalyzed immortal ring-opening polymerization of lactones, lactides and cyclic carbonates. *Dalton Trans.* 39:8363–76.
7. Wang, L., Poirier, V., Ghiotto, F., Bochmann, M., Cannon, R. D., Carpentier, J.-F., Sarazin, Y. 2014. Kinetic analysis of the immortal ring-opening polymerization of cyclic esters: A case study with tin(II) catalysts. *Macromolecules* 47:2574–84.
8. Wu, J., Yu, T.-L., Chen, C.-T., Lin, C.-C. 2006. Recent developments in main group metal complexes catalyzed/initiated polymerization of lactides and related cyclic esters. *Coord. Chem. Rev.* 250:602–626.
9. Ovitt, T. M., Coates, G. W. 1999. Stereoselective ring-opening polymerization of *meso*-lactide: Synthesis of syndiotactic poly(lactic acid). *J. Am. Chem. Soc.* 121:4072–4073.
10. Huang, C.-H., Wang, F.-C., Ko, B.-T., Yu, T.-L. Lin, C.-C. 2001. Ring-opening polymerization of ε-caprolactone and L-lactide using aluminum thiolates as initiator. *Macromolecules* 34:356–361.
11. Shen, H., Shen, Z., Zhang, Y., Yao, K. 1996. Novel rare earth catalysts for the living polymerization and block copolymerization of ε-caprolactone. *Macromolecules* 29:8289–8295.
12. O'Keefe, B. J., Hillmyer, M. A., Tolman, W. B. 2001. Polymerization of lactide and related cyclic esters by discrete metal complexes. *J. Chem. Soc., Dalton Trans.* 15:2215–2224.
13. Chisholm, M. H., Eilerts, N. W., Huffman, J. C., Iyer, S. S., Pacold, M., Phomphrai, K. 2000. Molecular design of single-site metal alkoxide catalyst precursors for ring-opening polymerization reactions leading to polyoxygenates. 1. Polylactide formation by achiral and chiral magnesium and zinc alkoxides, (η^3-L)MOR, Where L = Trispyrazolyl- and Trisindazolylborate ligands. *J. Am. Chem. Soc.* 122:11845–11854.
14. Williams, C. K., Breyfogle, L. E., Choi, S. K., Nam, W., Young Jr., V. G., Hillmyer, M. A., Tolman, W. B. 2003. A highly active zinc catalyst for the controlled polymerization of lactide. *J. Am. Chem. Soc.* 125:11350–11359.
15. Chamberlain, B. M., Cheng, M., Moore, D. R., Ovitt, T. M., Lobkovsky, E. B., Coates, G. W. 2001. Polymerization of lactide with zinc and magnesium β-diiminate complexes: Stereocontrol and mechanism. *J. Am. Chem. Soc.* 123:3229–3238.
16. Tang, X., Hong, M., Falivene, L., Caporaso, L., Cavallo, L., Chen, E. Y.-X. 2016. Quest for converting biorenewable bifunctional α-methylene-γ-butyrolactone into degradable and recyclable polyester: Controlling vinyl-addition/ring-opening/cross-linking pathways. *J. Am. Chem. Soc.* 138(43):14326–14337. DOI:10.1021/jacs.6b07974.
17. Chisholm, M. H., Gallucci, J., Phomphrai, K. 2003. Lactide polymerization by well-defined calcium coordination complexes: Comparisons with related magnesium and zinc chemistry. *Chem. Commun.* 1:48–49.

18. Dyer, H. D., Huijser, S., Susperregui, N., Bonnet, F., Schwarz, A. D., Duchateau, R., Maron, L., Mountford, P. 2010. Ring-opening polymerization of *rac*-lactide by bis(phenolate)amine-supported samarium borohydride complexes: An experimental and DFT study. *Organometallics* 29:3602–3621.
19. Palard, I., Schappacher, M., Soum, A., Guillaume, S. M. 2006. Ring-opening polymerization of ε-caprolactone initiated by rare earth alkoxides and borohydrides: A comparative study. *Polym. Int.* 55:1132–1137.
20. Cushion, M. G., Mountford, P. 2011. Cationic and charge-neutral calcium tetrahydroborate complexes and their use in the controlled ring-opening polymerisation of *rac*-lactide. *Chem. Commun.* 47:2276–2278.
21. Darensbourg, D. J., Choi, W., Richers, C. P. 2007. Ring-opening polymerization of cyclic monomers by biocompatible metal complexes. Production of poly(lactide), polycarbonates, and their copolymers. *Macromolecules* 40:3521–3523.
22. Davidson, M. G., O'Hara, C. T., Jones, M. D., Keir, C. G., Mahon, M. F., Kociok-Koehn, G. 2007. Synthesis and structure of a molecular barium aminebis(phenolate) and its application as an initiator for ring-opening polymerization of cyclic esters. *Inorg. Chem.* 46:7686–7688.
23. Kottalanka, R. K., Harinath, A., Rej, S., Panda, T. K. 2015. Group 1 and group 2 metal complexes supported by a bidentate bulky iminopyrrolyl ligand: Synthesis, structural diversity, and ε-caprolactone polymerization study. *Dalton Trans.* 44:19865–19879.
24. Liu, B., Roisnel, T., Guegan, J.-P., Carpentier, J.-F. Sarazin, Y. 2012. Heteroleptic silylamido phenolate complexes of calcium and the larger alkaline earth metals: β-Agostic Ae⋯Si-H stabilization and activity in the ring-opening polymerization of L-lactide. *Chem. Eur. J.* 18:6289–6301.
25. Biernesser, A. B., Li, B., Byers, J. A. 2013. Redox-controlled polymerization of lactide catalyzed by bis(imino)pyridine iron bis(alkoxide) complexes. *J. Am. Chem. Soc.* 135:16553–16560.
26. Kundys, A., Plichta, A., Florjanczyk, Z., Frydrych, A., Zurawski, K. 2015. Screening of metal catalysts influence on the synthesis, structure, properties, and biodegradation of PLA-PBA triblock copolymers obtained in melt. *J. Polym. Sci. Part A: Polym. Chem.* 53:1444–1456.
27. Nijenhuis, A. J., Grijpma, D. W., Pennings, A. J. 1992. Lewis acid catalyzed polymerization of L-lactide. Kinetics and mechanism of the bulk polymerization. *Macromolecules* 25:6419–6424.
28. Wang, L., Kefalidis, C. E., Sinbandhit, S., Dorcet, V., Carpentier, J.-F., Maron, L., Sarazin, Y. 2013. Heteroleptic tin(II) initiators for the ring-opening (co)polymerization of lactide and trimethylene carbonate: Mechanistic insights from experiments and computations. *Chem. Eur. J.* 19:13463–13478.
29. Chisholm, M. H., Gallucci, J. C., Quisenberry, K. T., Zhou, Z. P. 2008. Complexities in the ring-opening polymerization of lactide by chiral salen aluminum initiator. *Inorg. Chem.* 47:2613–2624.
30. Nomura, N., Ishii, R., Akakura, M., Aoi, K. 2002. Stereoselective ring-opening polymerization of racemic lactide using aluminum-achiral ligand complexes: Exploration of a chain-end control mechanism *J. Am. Chem. Soc.* 124:5938–5939.
31. Guillaume, S. M. 2013. Recent advances in ring-opening polymerization strategies toward α,ω-hydroxy telechelic polyesters and resulting copolymers. *Eur. Polym. J.* 49:768–779.
32. Cameron, D. J. A., Shaver, M. P. 2011. Aliphatic polyester polymer stars: Synthesis, properties and applications in biomedicine and nanotechnology. *Chem. Soc. Rev.* 40:1761–1776.
33. Fu, H. L., Cheng, S. X., Zhang, X. Z., Zhuo, R. X. 2007. Dendrimer/DNA complexes encapsulated in a water soluble polymer and supported on fast degrading star poly(DL-lactide) for localized gene delivery. *J. Controlled Release* 124:181–188.

34. Adeli, M., Zarnegar, Z., Kabiri, R. 2008. Amphiphilic star copolymers containing cyclodextrin core and their application as nanocarrier. *Eur. Polym. J.* 44:1921–1930.
35. Klok, H. A., Becker, S., Schuch, F., Pakula, T., Mullen, K. 2003. Synthesis and solid state properties of novel fluorescent polyester star polymers. *Macromol. Biosci.*, 3:729–741.
36. Cai, Q., Zhao, Y., Bei, J., Xi, F., Wang, S. 2003. Synthesis and properties of star-shaped polylactide attached to poly(amidoamine) dendrimer. *Biomacromolecules* 4:828–834.
37. Lee, S. H., Kim, S. H., Han, Y. K., Kim, Y. H. 2001. Synthesis and degradation of end-group-functionalized polylactide *J. Polym. Sci. Part A: Polym. Chem.* 39:973–985.
38. Baughman, T. W., Wagener, K. B. 2005. Recent advances in ADMET polymerization. *Adv. Polym. Sci.* 176:1–42.
39. Knall, A.-C., Slugovc, C. 2014. Olefin metathesis polymerization. In *Olefin Metathesis: Theory and Practice.* ed. Grela, K. pp. 269–284. New York: John Wiley & Sons.
40. Bielawski, C. W., Grubbs, R. H. 2007. Living ring-opening metathesis polymerization *Prog. Polym. Sci.* 32:1–29.
41. Hilf, S., Grubbs, R. H., Kilbinger, A. F. M. 2008. End capping ring-opening olefin metathesis polymerization polymers with vinyl lactones. *J. Am. Chem. Soc.* 130:11040–11048.
42. Risse, W., Wheeler, D. R., Cannizzo, L. F., Grubbs R. H. 1989. Di- and tetrafunctional initiators for the living ring-opening olefin metathesis polymerization of strained cyclic olefins. *Macromolecules* 22:3205–3210.
43. Schaverien C. J., Dewan J. C., Schrock R. R. 1986. Multiple metal-carbon bonds. 43. Well-characterized, highly active, Lewis acid free olefin metathesis catalyst. *J. Am. Chem. Soc.* 108:2771–2773.
44. Schrock, R. R., Feldman J., Cannizzo, L. F., Grubbs, R. H. 1987. Ring-opening polymerization of norbornene by a living tungsten alkylidene complex. *Macromolecules* 20:1169–1172.
45. Schrock, R. R., DePue, R. T., Feldman, J., Schaverien, C. J., Dewan, J. C., Liu, A. H. 1988. Preparation and reactivity of several alkylidene complexes of the type W(CHR') (N-2,6-C$_6$H$_3$-iso-Pr$_2$)(OR)$_2$ and related tungstacyclobutane complexes. Controlling metathesis activity through the choice of alkoxide ligands. *J. Am. Chem. Soc.* 110:1423–1435.
46. O'Donoghue, M. B., Schrock, R. R., LaPointe, A. M., Davis, W. M. 1996. Preparation of well-defined, metathetically active oxo alkylidene complexes of tungsten. *Organometallics* 15:1334–1336.
47. Trzaska, S. T., Lee, L.-B. W., Register, R. A. 2000. Synthesis of narrow-distribution "perfect" polyethylene and its block copolymers by polymerization of cyclopentene. *Macromolecules* 33:9215–9221.
48. Singh, R., Czekelius, C., Schrock, R. R. 2006. Living ring-opening metathesis polymerization of cyclopropenes. *Macromolecules* 39:1316–1317.
49. Coca, S., Paik, H., Matyjaszewski, K. 1997. Block copolymers by transformation of living ring-opening metathesis polymerization into controlled/"living" atom transfer Radical polymerization. *Macromolecules* 30:6513–6516.
50. Lynn, D. M., Mohr, B., Grubbs, R. H., Henling, L. M., Day, M. W. 2000. Water-soluble ruthenium alkylidenes: Synthesis, characterization, and application to olefin metathesis in protic solvents. *J. Am. Chem. Soc.* 122:6601–6609.
51. Claverie, J. P., Soula, R. 2003. Catalytic polymerizations in aqueous medium. *Prog. Polym. Sci.* 28:619–662.
52. Leitgeb, A., Wappel, J., Slugovc, C. 2010. The ROMP toolbox upgraded. *Polymer* 51: 2927–2946.
53. Alfred, S. F., Lienkamp, K., Madkour, A. E., Tew, G. N. 2008. *J. Polym. Sci., Part A: Polym. Chem.* 46:6672–666.
54. Colak, S., Tew, G. N. 2008. Synthesis and solution properties of norbornene based polybetaines. *Macromolecules* 41:8436–8434.

55. Nomura, K., Abdellatif, M. M. 2010. Precise synthesis of polymers containing functional end groups by living ring-opening metathesis polymerization (ROMP): Efficient tools for synthesis of block/graft copolymers. *Polymer* 51:1861–1881.

56. Banik, S. M., Monnot, B. L., Weber, R. L., Mahanthappa, M. K. 2011. ROMP-CT/NMP synthesis of multiblock copolymers containing linear poly(ethylene) segments. *Macromolecules* 44:7141–7148.

57. Bielawski, C. W., Morita, T., Grubbs, R. H. 2000. Synthesis of ABA triblock copolymers via a tandem ring-opening metathesis polymerization: Atom transfer radical polymerization approach. *Macromolecules* 33:687–680.

58. Cheng, C., Khoshdel, E., Wooley, K. L. 2006. Facile one-pot synthesis of brush polymers through tandem catalysis using Grubbs' catalyst for both ring-opening metathesis and atom transfer radical polymerizations. *Nano Lett.* 6:1741–1746.

59. Bielawski, C.W., Louie, J., Grubbs, R. H. 2000. Tandem catalysis: Three mechanistically distinct reactions from a single ruthenium complex. *J. Am. Chem. Soc.* 122:12872–12873.

60. Schrock, R. R. 2014. Synthesis of stereoregular polymers through ring-opening metathesis polymerization. *Acc. Chem. Res.* 47:2457–2466.

61. Totland, K. M., Boyd, T. J., Lavoie, G. G., Davis, W. M., Schrock, R. R. 1996. Ring opening metathesis polymerization with binaphtholate or biphenolate complexes of molybdenum. *Macromolecules* 29:6114–6125.

62. Flook, M. M., Ng, V. W. L., Schrock, R. R. 2011. Synthesis of *cis, syndiotactic* ROMP polymers containing alternating enantiomer. *J. Am. Chem. Soc.* 133:1784–1786.

63. Flook, M. M., Börner, J., Kilyanek, S., Gerber, L. C. H., Schrock, R. R. 2012. Five-coordinate rearrangements of metallacyclobutane intermediates during ring-opening metathesis polymerization of 2,3-dicarboalkoxynorbornenes by molybdenum and tungsten monoalkoxide pyrrolide initiators. *Organometallics* 31:6231–6243.

64. Rosebrugh, L. E., Marx, V. M., Keitz, B. K., Grubbs, R. H. 2013. Synthesis of highly cis, syndiotactic polymers via ring-opening metathesis polymerization using ruthenium metathesis catalysts. *J. Am. Chem. Soc.* 135:10032–10035.

65. Delaude, L., Demonceau, A., Noels, A. F. 2003. Probing the stereoselectivity of the ruthenium-catalyzed ring-opening metathesis polymerization of norbornene and norbornadiene diesters. *Macromolecules* 36:1446–1456.

66. Gabert, A. J., Verploegen, E., Hammond, P. T., Schrock R. R. 2006. Synthesis and characterization of ABA triblock copolymers containing smectic C* liquid crystal side chains via ring-opening metathesis polymerization using a bimetallic molybdenum initiator. *Macromolecules* 39:3993–4000.

67. Samanta, S., Locklin, J. 2008. Formation of photochromic spiropyran polymer brushes via surface-initiated, ring-opening metathesis polymerization: Reversible photocontrol of wetting behavior and solvent dependent morphology changes. *Langmuir* 24:9558–9565.

68. Faulkener, C. J., Fischer, R. E., Jennings, G. K. 2010. Surface-initiated polymerization of 5-(perfluoro-n-alkyl)norbornenes from gold substrates. *Macromolecules* 43:1203–1209.

69. White, S. R., Sottos, N. R., Geubelle, P. H., Moore, J. S., Kessler, M. R., Sriram, S. R., Brown, E. N., Viswanathan, S. 2001. Autonomic healing of polymer composites. *Nature* 409:794–797.

70. French, J. M., Caras, C. A., Diver S. T. 2013. Removal of ruthenium using a silica gel supported Reagent. *Org. Lett.* 15:5416–19.

71. Vougioukalakis, G. C. 2012. Removing ruthenium residues from olefin metathesis reaction products. *Chem. Eur. J.* 18:8868–8880.

72. Slugovc, C. 2014. Industrial applications of olefin metathesis polymerization. In *Olefin Metathesis: Theory and Practice.* ed. Grela, K. pp. 329–333. New York: John Wiley & Sons.

73. Hopkins, T. E., Wagener, K. B. 2004. ADMET synthesis of polyolefins targeted for biological applications. *Macromolecules* 37:1180–1189.

74. Watson, M. D., Wagener, K. B. 1999. Acyclic diene metathesis (ADMET) depolymerization: Ethenolysis of 1,4-polybutadiene using a ruthenium complex. *J. Polym. Sci., Part A: Polym. Chem.* 37:1857–1861.

75. Boffa, L. S., Novak, B. M. 2000. Copolymerization of polar monomers with olefins using transition-metal complexes. *Chem. Rev.* 100:1479–1494.

76. Watson M. D., Wagener, K. B. 2000. Ethylene/vinyl acetate copolymers via acyclic diene metathesis polymerization. Examining the effect of "long" precise ethylene run Lengths. *Macromolecules* 33:5411–5417.

77. Schulz, M. D., Wagener, K. B. 2014. Precision polymers through ADMET polymerization. *Macromol. Chem. Phys.* 215:1936–1945.

78. Smith, J. A., Brzezinska, K. R., Valenti, D. J., Wagener, K. B. 2000. Precisely controlled methyl branching in polyethylene via acyclic diene metathesis (ADMET) polymerization. *Macromolecules* 33:3781–3794.

79. Sworen, J. C., Smith, J. A., Wagener, K. B., Baugh, L. S., Rucker, S. P. 2003. Modeling random methyl branching in ethylene/propylene copolymers using metathesis chemistry: Synthesis and thermal behavior. *J. Am. Chem. Soc.* 125:2228–2240.

80. Inci, B., Wagener, K. B. 2011. Decreasing the alkyl branch frequency in precision polyethylene: Pushing the limits toward longer run lengths. *J. Am. Chem. Soc.* 133:11872–11875.

81. Rojas, G., Inci, B., Wei, Y., Wagener, K. B. 2009. Precision polyethylene: Changes in morphology as a function of alkyl branch size. *J. Am. Chem. Soc.* 131:17376–17386.

82. Guidry, E. N., Li, J., Stoddart, J. F., Grubbs, R. H. 2007. Bifunctional [c2]daisy-chains and their incorporation into mechanically interlocked polymers. *J. Am. Chem. Soc.* 129: 8944–8945.

83. Momcilovic, N., Clark, P. G., Boydston, A. J., Grubbs, R. H. 2011. One-pot synthesis of polyrotaxanes via acyclic diene metathesis polymerization of supramolecular monomers. *J. Am. Chem. Soc.* 133:19087–19089.

84. Gorodetskaya, I. A., Choi, T. L., Grubbs, R. H. 2007. Hyperbranched macromolecules via olefin metathesis. *J. Am. Chem. Soc.* 129:12672–12673.

85. Ding, L., Qiu, J., Lu, R., Zheng, X., An, J. 2013. Hyperbranched polyphosphoesters with reactive end groups synthesized via acyclic diene metathesis polymerization and their transformation to crosslinked nanoparticles *J. Polym. Sci. Part A: Polym. Chem.* 51:4331–4340.

86. Ding, L., Qiu, J., Wei, J., Zhu, Z. 2014. Convenient divergent synthesis of linear-dendron block polyphosphoesters via acyclic diene metathesis polymerization. *Polym. Chem.* 5:4285–4292.

87. Marmo, J. C., Wagener, K. B. 1995. ADMET depolymerization. Synthesis of perfectly difunctional (f = 2.0) telechelic polybutadiene oligomers. *Macromolecules* 28:2602–2606.

88. Schulz, M. D., Ford, R. R., Wagener, K. B. 2013. Insertion metathesis depolymerization. *Polym. Chem.* 4:3656–3658.

89. Schulz, M. D., Wagener, K. B. 2012. Solvent effects in alternating ADMET polymerization. *ACS Macro Lett.* 1:449–451.

90. Markova, D., Opper, K. L., Wagner, M., Klapper, M., Wagener, K. B., Müllen, K. 2013. Synthesis of proton conducting phosphonic acid-functionalized polyolefins by the combination of ATRP and ADMET. *Polym. Chem.* 4:1351–1363.

91. Peetz, R., Strachota, A., Thorn-Csanyi, E. 2003. Homologous series of 2,5-diheptyloxy-*p*-phenylene vinylene (DHepO-PV) oligomers with vinyl or 1-butenyl end groups: Synthesis, isolation, and microstructure. *Macromol. Chem. Phys.* 204:1439–1450.

92. Hopkins, T. E., Pawlow, J. H., Koren, D. L., Deters, K. S., Solivan, S. M., Davis, J. A., Gómez, F. J., Wagener, K. B. 2001. Chiral polyolefins bearing amino acids. *Macromolecules* 34:7920–7922.

93. Hopkins, T. E., Wagener, K. B. 2004. ADMET synthesis of polyolefins targeted for biological applications. *Macromolecules* 37:1180–1189.
94. Brzezinska, K. R., Wagener, K. B., Burns, G. T. 1999. Silicon-terminated telechelic oligomers by ADMET chemistry: Synthesis and copolymerization. *J. Polym. Sci., Part A: Polym. Chem.* 37:849–856.

5 Controlled Radical Polymerization

Ashootosh V. Ambade

CONTENTS

5.1 INTRODUCTION

Conventional free radical polymerization (FRP) has been a widely used technique for manufacturing of polymeric materials due to widely available vinyl monomers, tolerance to functional groups, and mild reaction conditions. However, highly reactive radical species cause rapid bimolecular termination reactions that lead to poor control over molecular weight and polydispersity. Since polymer properties and applications depend upon molecular weight distribution, composition, and topology, a polymerization method that affords control over these molecular features was desired. Controlled radical polymerization (CRP) is such a method that is also industrially applicable. IUPAC recommends the term *reversible-deactivation radical polymerization (RDRP)* to describe this technique and hence it will be used in this chapter. Compared to FRP, RDRP minimizes the side reactions and allows synthesis of unprecedented functional polymers with narrow molecular weight distribution while retaining the tolerance of FRP to various reaction conditions.[1] The control over molecular weight is achieved by a dynamic equilibrium between active (growing radical) and dormant (deactivated) species so that active species is present in low concentrations at any moment. The process is characterized by almost quantitative initiation so that rate of propagation is much lower than rate of initiation. Based on the mechanism involved in the deactivation process to maintain the equilibrium, there are three types of RDRP. The first approach involves reversible deactivation

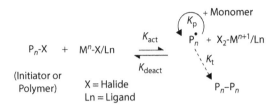

FIGURE 5.1   General mechanism of ATRP.

of propagating radicals to form the dormant species by the coupling between propagating (transient) radical and a *stable* or *persistent* free radical. Due to persistent radical effect[2] the coupling of propagating and stable radical is favored over the self-coupling of propagating radicals, thereby giving controlled polymerization. Examples of this approach include nitroxide-mediated polymerization (NMP)[3] and organometallic-mediated radical polymerization (OMRP)[4] that involves reversible homolytic cleavage of the weak bond between an alkyl group and a transition metal catalyst. In the second type, degenerate transfer (thermodynamically neutral bimolecular exchange) takes place between propagating radicals and dormant species. The degenerative radical transfer polymerization does not involve a stable radical and the kinetics of polymerization is similar to FRP. Examples include reversible addition-fragmentation chain-transfer (RAFT) polymerization,[5] macromolecular architecture design by interchange of xanthates (MADIX),[6] and iodine transfer polymerization.[7] In the third type of CRP, called as atom transfer radical polymerization (ATRP),[8,9] the reversible deactivation is accomplished by atom transfer catalyzed by a transition metal catalyst. This chapter will focus on ATRP since it is a metal-catalyzed CRP. General mechanism of ATRP is presented in Figure 5.1.

5.2   MECHANISM OF ATOM TRANSFER RADICAL POLYMERIZATION

In ATRP, a transition metal complex or halide (M^n-Y/Ligand, where Y may be a ligand or a counter-ion) is used as a catalyst and a halogen compound (e.g., alkyl or acyl halide) is employed as an initiator. The radicals (active species) are generated through a reversible redox process involving abstraction of a halogen atom from a dormant species. The dormant species is the initiator in the initiation step, whereas it is the propagating chain in the propagation step. The generation of radicals is catalyzed by the metal complex, which concomitantly undergoes a one-electron oxidation. Propagation takes place in a manner similar to FRP by addition of monomers to growing radicals. Growth of polymer chains stops after all the monomer is consumed and the halide at the chain end can be reactivated to prepare block copolymers. Termination reactions and other side reactions also occur in ATRP, however, if the polymerization is well controlled, percentage of the terminated polymer chains is very low (<10%). At the most 5% of the total growing polymer chains terminate during the initial, short, nonstationary stage of the polymerization, which generates persistent radicals (X-M^{n+1}) that reduce the stationary concentration of growing polymer radicals and thereby minimize the contribution of termination.

Thus, ATRP shows the characteristics of a RDRP, which are fast initiation and reversible deactivation (slow propagation) with minimal contribution of transfer or termination reactions resulting in almost uniform growth of all the chains to afford low polydispersity.

ATRP can be catalyzed by many redox-active transition metal complexes, however Cu has been the most often used metal and hence discussion in this chapter will focus on polymerizations using Cu catalyst. The rate of polymerization depends on the rate constant of propagation and on the concentrations of monomer and growing radicals, which in turn depends on the equilibrium constant and the concentration of dormant species, activators, and deactivators, as shown in Equation 5.1[8]:

$$R_p = k_p[M][P^*_n] = k_p K_{ATRP}\left(\frac{[P_nX][Cu^I/Ln][M]}{[X - Cu^{II}/Ln]}\right) \qquad (5.1)$$

Lifetime of propagating chains in FRP is short (~1 s) and at any time their concentration is at ppm level, and the concentration of dead chains is very high. However in RDRP, most of the chains are in dormant state that are reactivated for ~1 ms. The current thinking is focused on extending the life of propagating chains to more than 1 day, by inserting dormant periods of ~1 minute between the periods of activity.[8] As mentioned earlier termination in RDRP is very low, however it is important to know how many dead chains exist. For this purpose, dead chain fraction (DCF) is defined as ratio of concentration of terminated chains to initial concentration of initiator. DCF depends on targeted degree of polymerization (DP = $[M]_0/[I]_0$), monomer conversion, propagation and termination rate constants, and reaction time. DCF values are low at slower rates of polymerization, lower monomer conversion, lower targeted DP, and higher initial monomer concentration, and for highly reactive monomers.[8] Formation of radicals from dormant species in a Cu-catalyzed process, considering the mechanism, can take place by two pathways—the outer-sphere electron transfer (OSET) or the inner-sphere electron transfer (ISET). OSET can involve formation of radical anion intermediates in a stepwise or concerted fashion, whereas in ISET transfer of halogen atom from the halide to Cu(I) takes place via a Cu–X–C transition state, which is formally a single electron transfer process. The energy barrier for OSET as obtained by Marcus analysis is ~15 kcal/mol higher than the experimentally determined value, that is, OSET is ~10^{10} times slower than ISET.[10] Therefore, it was concluded that Cu-catalyzed ATRP occurs via concerted homolytic dissociation of alkyl halide via ISET—an atom transfer process.

5.3 INITIATING SYSTEMS AND CATALYSTS

ATRP equilibrium constants and hence control over the polymerization depend on structure of the catalysts, initiator, and the reaction medium. Activation rate constants are influenced by the structure of catalyst and initiator more than deactivation rate constants, which are usually high ($k_{deact} > 10^7$ M^{-1} s^{-1}). Reactivity of the halide (A–X, where X is a halogen) depends on the structure of the A group and the transferable (halogen) group. Obviously, chlorine (Cl), bromine (Br), and iodine(I) are active as the halogen, whereas fluorine (F) is inactive for metal-catalyzed

RDRP. The component A should contain some conjugated or radical-stabilizing groups to facilitate radical generation, such as ester [–C(=O)OR], ketone [–C(=O)R], amide [–C(=O)NR2], cyano (–C≡N), phenyl (-Ar), and so on, similar to *conjugated* monomers for radical polymerization. ATRP is more sensitive than other RDRPs to polar effects rather than to steric effects.[11] Reactivity of alkyl halides follows the order of tertiary > secondary > primary, in agreement with the bond dissociation energy needed for homolytic bond cleavage and also the order I > Br > Cl. The least dissociable chlorine is favored for methyl methacrylate (MMA), which gives a highly stable and sterically conjugated tertiary carbon radical. In contrast, acrylates and styrenes favor Br- or I-initiators because of the less active secondary structure of their radicals with less activity.[1] This knowledge is important for synthesis of block copolymers by ATRP where the order is acrylonitrile > methacrylates > acrylates ≈ styrene > acrylamides. However, this order can be changed by halogen exchange process wherein a methacrylate block can be grown from a Br-terminated acrylate chain by using CuCl catalyst complex.[12] Density functional theory (DFT) calculations were used to derive the bond dissociation energies and free energies for several R–X type initiators (X = Cl, Br, I, N_3, and S_2CNMe_2) that are used for ATRP. The computed energies were comparable with the experimentally determined bond dissociation energy. The ATRP equilibrium constants were calculated using the free energies and were found to agree with those determined from polymerization rates. Thus, DFT computations can predict polymerization rates for new monomers. For example, it was predicted that under same conditions of ATRP, it would take 1 h for methyl acrylate, 1 s for acrylonitrile, 10 h for styrene, and 15 years for vinyl acetate polymerization to reach 90% conversion.[13] The most active initiator is ethyl α-bromophenylacetate, which is more active than 1-phenylethyl bromide and the commonly used initiator, 2-bromopropionate. Commercially available α-halocarboxylic acid chloride/bromide can be reacted with various molecules containing —OH or NH_2 groups to generate a large variety of initiators, which can be used to generate different kinds of polymeric architectures and for conjugation with different scaffolds such as biomolecules. Arylsulfonyl halides have also been used as initiators. They show a fast and quantitative addition of sulfonyl radical to the monomer due to their unstable nature and give high initiation efficiency and controlled molecular weights. All Cl-, Br-, and I-based initiators have been found to be effective.[14,15] Structures of commonly used initiators are shown in Figure 5.2.

Although halogen compounds are generally used as initiators, Matyjaszewski and coworkers have developed halogen-free initiating systems such as alkyl diethyl dithiocarbamates (R-S(C=S)NEt₂) for ATRP.[16] Here, the proposed mechanism involves the exchange of diethyldithiocarbamate group between dormant and active

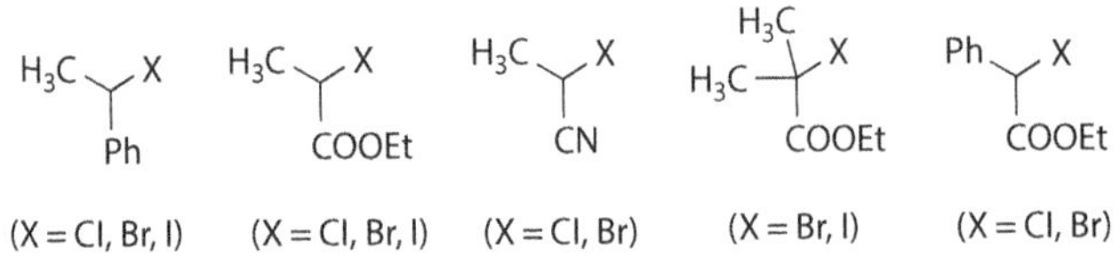

FIGURE 5.2 Chemical structures of commonly used initiators in ATRP.

species catalyzed by the copper complex. Exchange between the diethyldithiocarbamate and the Br atom of CuBr was found to be almost negligible from chain-end analysis.

Ruthenium complexes have also been studied extensively as catalysts for ATRP being one of the first catalysts reported along with the copper-based systems. The ruthenium catalyst is a divalent Ru(II) complex, carrying two anionic ligands, as well as neutral ligands. The tolerance to functional groups of Ru(II) complexes makes it possible to design large number of complexes that can be used in RDRP. Some examples of ruthenium complexes will be discussed here. [RuCl$_2$(PPh$_3$)$_3$] was the first ruthenium complex to catalyze a RDRP.[17] Cocatalysts such as aluminum isopropoxide and amines are required for controlled molecular weights and higher quantities ([Ru]$_0$/[initiator]$_0$ = 1/2) are needed for sufficient catalytic activity. Half-metallocene-type ruthenium complexes with indenyl and pentamethylcyclopentadiene (Cp*) ligands are active catalysts that give narrow molecular weight distributions and well-defined block copolymers.[18,19] Lower redox potential coupled with fast halogen exchange as evidenced in model reactions accounted for the higher activity. Introduction of dimethylamino group onto the indenyl ring resulted in faster polymerization along with narrower molecular weight distributions (M$_w$/M$_n$ = 1.07 vs 1.13 for unsubstituted derivative).[20] Steric/electronic effects of phosphine ligands studied for polymerization of MMA using ([RuCp*(μ$_3$-Cl)]$_4$) showed that for controlled polymerization, bulkiness of the ligand is more crucial than its electron-donating ability.[21] For example, ruthenium complex with tri(m-tolyl)phosphine as ligand (cone angle, θ = 165°) gave lower polydispersity (M$_w$/M$_n$ = 1.07) than triphenylphosphine derivative (θ = 145°) under the same conditions (M$_w$/M$_n$ = 1.20). Using mono- or diamine as cocatalyst further helped to decrease the catalyst loading without loss of control. It was observed from ^{31}P NMR analysis that the amine undergoes an exchange with one of the phosphine ligands *in situ* to form a phosphine/amine-coordinated complex, which is probably more reactive and would promote the activation and deactivation steps. [RuCl$_2$(p-cymene)(PR$_3$)] complexes with bulky ligands such as tricyclopentylphosphine (PCp$_3$) and *tert*-butyldicyclohexylphosphine (PCp$_2$*t*-Bu) were also found to give narrower molecular weight distribution (MWDs) (M$_w$/M$_n$ < 1.15) in the polymerization of MMA.[22] Replacement of phosphine ligands with *N*-heterocyclic ligands was also found to be effective.[23] Along this line, Grubbs first- and second-generation catalysts have also been found to catalyze controlled polymerizations of MMA.

Copper complexes have been found to be superior catalysts compared to other metal complexes and the detailed mechanistic understanding has been now established. Other attractive features are their low cost and easy handling, that is, complex is obtained by mixing a copper halide with a ligand in the reaction flask itself. Some of the highly effective and commonly used ligands are shown in Figure 5.3. The range of activity of ATRP catalysts spans six orders of magnitude. The general order of copper complex activity for ligands is tetradentate (cyclic-bridged) > tetradentate (branched) > tetradentate (cyclic) > tridentate > tetradentate (linear) > bidentate ligands.[8] Thus, bridged cyclam, tris(2-dimethylaminoethyl)-amine (Me$_6$TREN), and tris(2-pyridylmethyl)amine (TPMA) afford the most active catalysts and pyridineimine and 2,2′-bipyridine afford the least active. The nature of

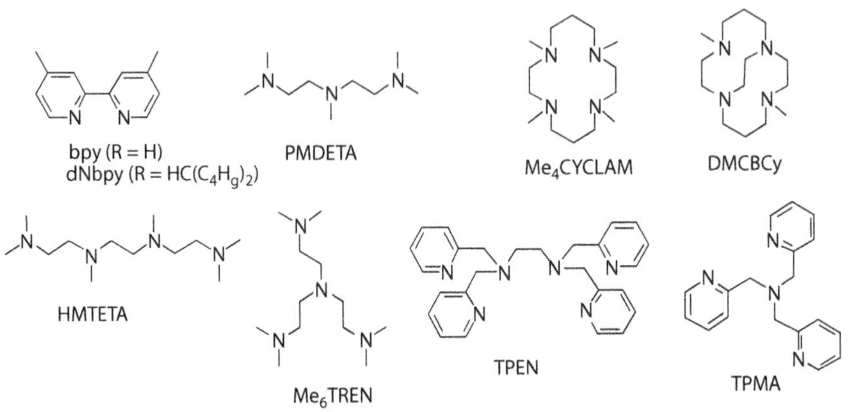

FIGURE 5.3 Chemical structures of commonly used ligands in ATRP.

nitrogen atoms in ligands also plays a role in the activity of the complexes and follows the order pyridine $\approx$ aliphatic amine > imine < aromatic amines. Most of these ligands were developed before 2000, however exploration for more effective ligands continued. A dimethyl cross-bridged cyclam derivative (DMCBCy) with CuCl was reported to show 30-fold larger equilibrium constant than that for CuCl/ Me_6TREN, one of the most active catalysts.[24] Another ligand N,N,N',N'-tetrakis(2-pyridylmethyl)ethylendiamin (TPEN) gave highly active catalyst for methyl acrylate (MA), MMA, and styrene (St) polymerization such that controlled molecular weights and low polydispersity were obtained at a very low concentration of catalyst ($[CuBr]_0/[Initiator]_0 = 0.005$).[25] The binuclear and the mononuclear states of the complex were found to be in equilibrium with each other and the mononuclear complex was proposed to be the actual activating species. Due to the dynamic structural change in the complex and the flexible coordination by the multidendate TPEN ligand, the complex is stabilized during polymerization, which prevents an extra coordination of solvents and monomers that allows for low concentration of the catalyst. Highly efficient copper-catalyzed systems have been developed in ATRP that require very low concentration of catalyst (~10 ppm) without losing the control over polymerization. An ideal catalyst required for polymerization of less reactive monomers or used at low concentration should have high K_{eqm} (i.e., larger k_{act}) while maintaining large value of k_{deact}. Activation rate constants were shown to increase with temperature.[26] Activation and deactivation rate constants of ethyl 2-bromoisobutyrate with $CuBr/Me_6TREN$ system were evaluated using electrochemistry in acetonitrile and dimethyl sulfoxide (DMSO), wherein, extremely fast activation ($k_{act} = 4 \times 10^4$ and 9×10^4 M^{-1} s^{-1}, respectively), and fast deactivation ($k_{deact} = 10^7$ M^{-1} s^{-1}) were observed.[27]

Equilibrium constants in ATRP generally increase with solvent polarity, due to stabilization of more polar (cationic) Cu(II) species and with temperature. The effect of solvent on K_{eqm} for Cu(I)Br/HMTETA with methyl 2-bromoisobutyrate as initiator was investigated for several solvents and then extrapolated to predict catalyst activity in 17 additional solvents.[28] The plot of experimental $log(K_{eqm})$ against values predicted by Kamlet–Taft relationship was linear and could be used to predict catalyst activity over seven orders of magnitude from 10^{-11} in fluoroalcohols to $\sim 10^{-4}$ in water.

5.4 PROCESSES FOR EFFICIENT ATOM TRANSFER RADICAL POLYMERIZATION

Since ATRP follows the persistent radical effect, high concentrations of catalysts (>1000 ppm) were required to maintain high polymerization rate. However, catalyst-free polymers or polymers with catalyst concentration less than 1 ppm are desirable for many high end applications such as biomedical applications where toxicity of copper is an issue and electronic applications where copper affects the photophysical and electronic properties of conjugated polymers. ATRP processes have evolved over time to make the polymerization more efficient by decreasing the amount of Cu to catalytic or even *ppm* levels. The approach is to avoid catalyst deactivation due to termination reactions and oxidation of the activator Cu(I) species by regenerating the Cu(I) complex. This can be achieved by reducing the deactivator species or through supplemental activation reactions. Initially, *reverse* ATRP was developed so that polymerization can be started by oxidatively stable but more reactive complexes containing the transition metal in higher oxidation state that could be reduced to the activating species in lower oxidation state by a standard free radical initiator; a regular alkyl halide initiator was not used. However, it did not help to reduce the amount of catalyst.[29,30] *Reverse* ATRP could not be used for preparation of block copolymers or more complex architectures since the use of free radical initiator led to homopolymerization of second monomer during block copolymer synthesis. This problem was addressed to some extent by using an ATRP initiator such as alkyl halide along with the conventional free radical initiator. This process was called *simultaneous reverse and normal initiation* (SR&NI). Both the initiators take part in the ATRP equilibrium and so the amount of catalyst required is drastically reduced making the synthesis of block copolymers possible. However, a small fraction of homopolymer of the second monomer was invariably formed due to the presence of free radical initiator.

To circumvent this problem, *activators generated by electron transfer* (AGET) process was developed, which involves the activation of the deactivator (transition metal in higher oxidation state such as Cu(II) complex) by a reducing agent instead of an organic radical initiator (Figure 5.4). This allows for the polymerization to be carried out in the presence of limited amount of air while maintaining control

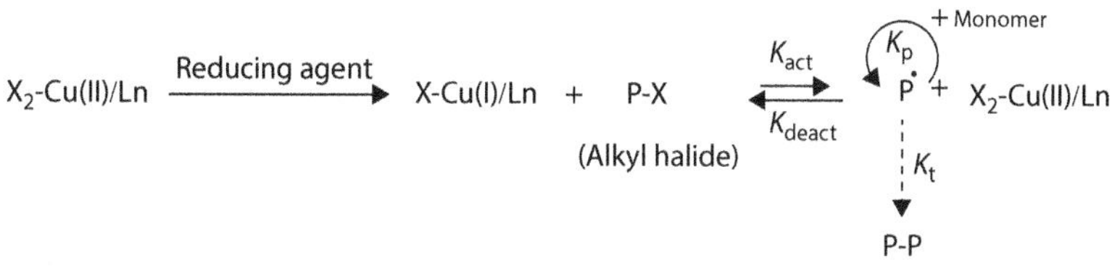

FIGURE 5.4 Mechanism of AGET ATRP process.

over polymerization. The usual alkyl halide ATRP initiator is used.[31,32] The reducing agents are used in stoichiometric amount and are such that they do not generate any initiators but only activate the catalyst by electron transfer, examples are tin(II) 2-ethylhexanoate ($Sn(EH)_2$) or Cu(0) and environmentally friendly reductants such as glucose, ascorbic acid and several others. Typical monomers used in AGET ATRP are acrylonitrile, (meth)acrylates, meth(acrylamides), and styrenes, whereas the initiators include α-halonitriles, α-haloesters, benzylic halides, and phenyl-ester halides.

Under specific circumstances, certain monomers (e.g., N,N-dimethylaminoethyl methacrylate [DMAEMA]) or ligands (TMEDA, HMTETA, PMDETA) serve as reducing agents and the use of an extra reducing agent is avoided.[33,34] A bimetallic catalyst system $FeCl_3 \cdot 6H_2O/CuCl$ was used for iron-catalyzed AGET ATRP of MMA, and rate of polymerization and control over molecular weight was found to be improved in comparison to monometallic systems.[35]

A large decrease in the amount of catalyst in AGET became possible with the development of *activator regenerated by electron transfer* (ARGET) ATRP process.[8,36] ARGET process is essentially similar to the AGET process except that: (1) much smaller concentration of activator, (2) much larger amount of the ligands, and (3) large excess of reducing agent relative to deactivator are used. The use of excess reducing agents allows for continuous regeneration of small amounts of the active catalyst to compensate for the inevitable yet appreciable levels of chain termination so that polymerization can be carried out with ppm concentrations of the catalyst—a decrease by 10^3 times from prior levels. For example, ATRP of DMAEMA initiated by ethyl 2-bromoisobutyrate (EBiB) could be carried out at low concentration of $CuCl_2$/tris[(2-pyridyl)methyl]amine (TPMA) complex (100 ppm vs monomer and 5 mol% vs initiator).[33] Advantages of ARGET ATRP include ability to carry out the polymerization without removal of inhibitors or deoxygenation and catalyst-induced side reactions are significantly reduced thereby high chain end fidelity can be ensured making preparation of higher molecular weight copolymers possible.

A *reverse* ARGET ATRP process termed as *initiators for continuous activator regeneration* ATRP (ICAR ATRP) employs a source of free radicals to regenerate the Cu(I) species.[37] The difference between SR&NI process and ICAR process is that a large excess of free radical initiator, which acts as reducing agent, is used and the radicals are slowly generated over the course of the reaction. Advantage of this process is the reduction in the amount of catalyst to ppm levels (10–50 ppm), similar to ARGET ATRP. Excellent control over molecular weight and molecular weight distribution is obtained, however synthesis of block copolymers is not possible due to the presence of free radicals. Thus, both ICAR and ARGET processes are industrially important because polymerization can be carried out using ppm level of catalyst. Mechanism of both the processes is shown in Figure 5.5.

As mentioned earlier, Cu(0) is used in some of the processes to activate the catalyst by reducing Cu(II) species. Employing Cu(0) allows for the synthesis of complex polymer architectures such as hyperbranched, decablock copolymers and pentablock star copolymers. Use of Cu(0) confers advantages on the ATRP process such as a significant decrease in the amount of soluble copper species, easy removal and reuse of excess Cu(0), and control of the polymerization rate by the amount of ligand and

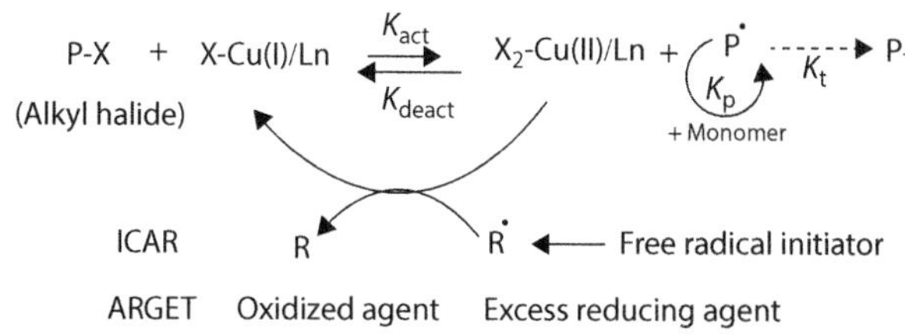

FIGURE 5.5 Mechanism of ARGET and ICAR processes.

the surface area of Cu(0).[38] It is therefore important to elucidate the mechanism of polymerization in the presence of Cu(0). Two models were proposed for this mechanism, the *supplemental activator and reducing agent* ATRP (SARA ATRP) and the single electron transfer living radical polymerization (SET-LRP).[39,40] In SARA ATRP, Cu(0) acts as the supplemental activator of alkyl halides and reducing agent for Cu(II) species through comproportionation. There is minimal contribution of disproportionation since Cu(I) primarily activates alkyl halides and all halide activation takes place by inner sphere electron transfer (ISET). In SET-LRP, Cu(0) is the exclusive activator of alkyl halides, whereas Cu(II) is the major deactivator. Cu(I) instantaneously disproportionates to Cu(0) and Cu(II) instead of activating alkyl halides and there is minimal comproportionation. Activation of alkyl halides in SET-LRP occurs by outer sphere electron transfer (OSET) mechanism. Thus, the deactivation processes in two mechanisms are similar. However, there is no clarity over: (1) which is the major activator of alkyl halides? (2) whether kinetics determine that Cu(I) activates alkyl halides or undergoes disproportionation; (3) whether Cu(0) is the supplemental or major activator; (4) what is the mode of activation of alkyl halides, that is, ISET or OSET? (5) what is the extent of termination? (6) whether disproportionation dominates or comproportionation. There are many literature reports arguing in favor of one or the other mechanism owing to the different assumptions regarding kinetics of polymerization. Recently, Matyjaszewski and coworkers have shown that SARA ATRP is the operating mechanism for RDRP in the presence of Cu(0) in organic as well as aqueous media.[39] The key evidence came from the measurement of kinetic parameters in isolated model reactions performed in DMSO and aqueous systems to determine the contributions and roles of Cu(0), Cu(I), and Cu(II) species.[41–43]

Later, in place of reducing agents or radical initiators, electrochemical potential was employed to reduce the initially added Cu(II) species by one-electron transfer and it was shown that polymerization kinetics is tunable in real time by varying the magnitude of applied potential.[44] This approach avoids the side products formed due to use of reducing chemicals in the several processes described earlier. Multistep intermittent potentials were applied to initiate the polymerization as well as to maintain its living nature by switching between dormant and active species. Controlled polymerization was afforded over a range of potentials at ppm concentrations of catalyst. Parameters, such as applied current, potential, and total charge passed, can be controlled to maintain the desired concentration of the catalytic species and thereby rate of polymerization.

5.5 AQUEOUS-PHASE ATOM TRANSFER RADICAL POLYMERIZATION

Water is a highly attractive medium for polymerization since it addresses the environmental concerns and is inexpensive. FRP of water-soluble as well as hydrophobic monomers is presently carried out on an industrial scale. It is challenging to conduct a controlled ATRP in aqueous media since in water, the deactivator may lose the halide ligand, the activator may disproportionate and the initiator may hydrolyze or react with the monomer if it contains basic or nucleophilic groups. For ATRP to be carried out in aqueous systems, conditions have to be adjusted for lower complex stability and suitable ligands should be selected. In heterogeneous systems, the complex should be sufficiently hydrophobic to be located in the organic phase and surfactant choice is important. The typical polymerization processes employed in aqueous systems such as emulsion, miniemulsion, microemulsion, dispersion, precipitation, and inverse miniemulsion polymerization have been carried out by ATRP. AGET ATRP was used under miniemulsion technique to prepare linear and star block copolymers using ascorbic acid as reducing agent due to its solubility in water and a highly hydrophobic ligand BPMODA (bis(2-pyridylmethyl)octadecylamine).[45,46] The activator species is driven back into the monomer droplets after the reduction of deactivator since the corresponding activator complex is more hydrophobic. With the ratio of ascorbic acid to Cu(II) complex being 0.4:1, the polymerization of n-BA (n-butyl acrylate) was sufficiently fast and did not show significant coupling reactions. Later, inverse miniemulsion AGET ATRP was used to obtain well-defined water-soluble polymeric particles and microgels/nanogels with narrow size distribution, the properties of which (swelling ratio, degradation behavior and colloidal stability) were found to be superior to microgels prepared by conventional FRP.[47,48] Microemulsion system could also be used to prepare stable translucent microlatexes by carrying out AGET ATRP of MMA and styrene by utilizing Brij 98 as surfactant and BPMODA as ligand. Particle size was controlled between 10 and 100 nm owing to controlled polymerization.[49]

5.6 POLYMER ARCHITECTURES AND ADVANCED MATERIALS BY ATOM TRANSFER RADICAL POLYMERIZATION

Controlled nature of polymerization and the processes that tolerate the presence of oxygen such as AGET/ARGET ATRP have allowed the synthesis of functional polymers with different topologies by ATRP. Well-defined poly(2-hydroxyethyl methacrylate) (PHEMA) and its block copolymer with MMA was prepared by AGET ATRP using $CuCl_2$/bipy catalyst system in a mixture of methyl ethyl ketone and methanol as solvent and reducing agent $Sn(EH)_2$, which is soluble in this solvent mixture.[50] Copolymerization of less reactive alkenes such as 1-octene with n-butyl acrylate could be achieved with ARGET ATRP using 10 ppm of $CuCl_2$/TPMA versus monomer and 10 mol% of $Sn(EH)_2$ versus initiator. The copolymer was prepared in higher yield and with lower polydispersity index (PDI < 1.4) than with normal ATRP due to formation of unreactive 1-octene terminated dormant species.[51] Well-controlled polymerization of polar monomers such as acrylamide and N,N-dimethylaminoethyl methacrylate was also carried out using AGET/ARGET

ATRP. An amphiphilic diblock copolymer PEO-b-PS-Br was used as a stabilizer as well as a macroinitiator in miniemulsion in AGET ATRP so that use of free surfactant could be avoided.[52]

Complex macromolecular architectures have also been synthesized using ATRP in combination with efficient coupling techniques such as click chemistry.[8,53] Polymer properties are known to depend on the topology, and polymers with specific properties for targeted applications can be synthesized by precision synthesis of the desired polymer architecture. The architectures include graft, star, dendritic, bottle-brush, cyclic, and gradient copolymers prepared using inimers (initiator-monomer) and multifunctional initiators. An example of synthesis of such an architecture is shown in Figure 5.6.[54]

Advanced materials such as protein–polymer conjugates and virus-like particles coated with polymers—materials termed as bioconjugates—and inorganic-organic hybrid materials such as silica particles grafted with polymers have also been prepared using ATRP.[53] An important part of the preparation of hybrid materials is the surface-initiated ATRP (SI-ATRP) on surfaces of various shapes, which is a more common approach than the surface-grafting technique.[55,56] Strategies for synthesis of these materials are shown in Figure 5.7.

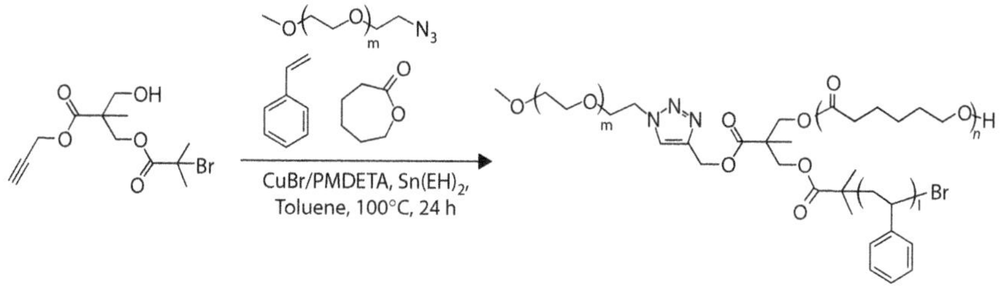

FIGURE 5.6 Synthesis of PEG-PS-PCL star polymer using ATRP and Cu-catalyzed azide-alkyne cycloaddition click chemistry.

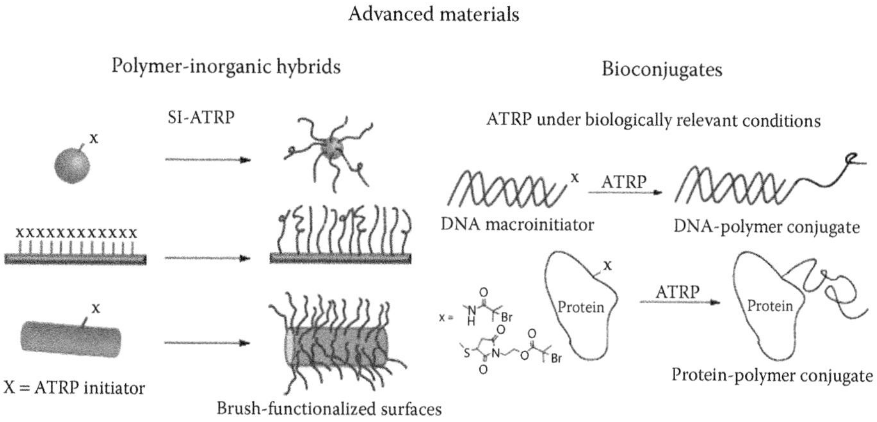

FIGURE 5.7 Approaches for synthesis of advanced materials using ATRP.

5.7   REMOVAL OF COPPER FROM POLYMERS SYNTHESIZED BY ATOM TRANSFER RADICAL POLYMERIZATION

In the conventional ATRP, the amount of catalyst used was unacceptable for most applications since high level of copper residues gave a greenish color to the polymer apart from being toxic. Although various processes such as ARGET and ICAR were developed later, exploration of strategies to remove the copper from synthesized polymers had begun quite early. The efforts can be classified into the following directions apart from the efficient ATRP processes: (1) postpolymerization purification; (2) various soluble, insoluble, immobilized/soluble, and reversible supported catalyst systems; and (3) temperature-controlled liquid–liquid biphasic-catalyzed systems.[57,58] The first choice is the postpolymerization purification methods. Copper(II) complexes are highly soluble in polar solvents such as methanol and water so that precipitation of polymers into these solvents can reduce the amount of copper although several dissolution–precipitation cycles are required.[59] Washing the solution of polymer in organic solvent with aqueous solution of ethylene diamine tetraacetic acid (EDTA) is also an efficient method to remove the copper catalyst. Copper complexes can also be removed to a large extent by adsorption onto surface of silica or alumina possibly due to bonding with Si-OH or Al-OH groups. This is achieved by passing the polymerization mixture through a column of silica gel or alumina, however the efficiency of the process depends on polarity of the eluting solvent and polymer. For example, CuBr/HMTETA complex could be removed from poly(methyl methacrylate) (PMMA) by eluting through a silica gel column using toluene as solvent but less efficiently when eluted with acetone, however removal of the same complex from poly(2-(N,N-dimethylamino)ethyl methacrylate) required a longer column or multiple elutions. Ion-exchange resins have also been used to remove the catalyst.[60] It was shown that in the presence of large excess of resin with -SO$_3$H groups, the rate of removal depended on solvent polarity, temperature, and size of the catalyst complexes such that removal of complexes with aliphatic amine ligands was faster than of those with bipyridyl ligand. Limitations of these postpolymerization processes are use of large amount of silica gel and solvents, loss of polymer, and difficulty in large-scale separation (>10 g) and inability to reuse the separated catalyst. It is also difficult to separate the catalyst from functional polymers that interact with copper complexes and from water-soluble polymers. Ionic liquids can be used to separate the catalyst after polymerization by conducting a liquid–liquid biphasic polymerization where the catalyst remains in ionic liquid phase. However, polymerization is much less controlled due to limited contact of catalyst with monomer in organic phase. Polymerization of MMA using CuBr-N-propyl-2-pyridylmethanimine complex was carried out in a mixture of ionic liquid, 1-butyl-3-methyl- imidazolium hexafluorophosphate ([bmim][PF6]), and toluene.[61] The mixture became homogeneous at 90°C, however polymerization could be conducted at lower temperature that gave lower polydispersity. The catalyst was retained in ionic liquid layer and could be easily separated from the polymer that contained only 0.17% of the original polymer used. Acrylate monomers with longer alkyl chains were less soluble in ionic liquids and polymerization was less controlled.[62] Limitations of conducting ATRP in ionic liquids are—use of inert atmosphere,

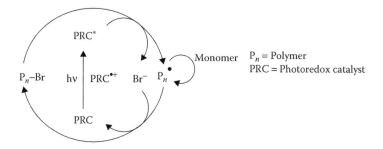

FIGURE 5.8   Proposed mechanism of metal-free ATRP catalyzed by 10-phenylphenothiazine as photoredox catalyst. Adapted from ref. 65.

limited kind of monomers, catalyst needs to be regenerated under inert atmosphere. To address these issues, a temperature-controlled phase-separable catalysis-based AGET ATRP system was designed that uses $[CH_3(OCH_2CH_2)_nN^+Et_3]$ $[CH_3SO_3^-]$ ($MPEG_{750}NIL$, $n = 16$), as ionic liquid. Polymerization was carried out at higher temperature, and catalyst was separated at room temperature by decantation and reused.[63] Extra ligand had to be added to compensate for the loss of ligand during separation. Similar strategy was also used for aqueous/organic solvent biphasic system with copper-mediated AGET ATRP using a thermoresponsive polyethylene glycol (PEG)-supported pyridyl ligand and an alkyl pseudo-halogen 2-cyanoprop-2-yl 1-dithionaphthalate (CPDN) as the initiator.[64] Polymerization takes place under homogeneous conditions at 90°C and catalyst-containing aqueous phase separates upon cooling to room temperature after the polymerization. The catalyst could be recycled three times however with less control over polymerization.

Recently, metal-free ATRP has been developed that involves the use of photoredox catalysts 10-phenylphenothiazine and perylene under irradiation at 380 nm or using white LED light.[65,66] A proposed mechanism for photomediated ATRP is shown in Figure 5.8. However, this method still lags behind metal-catalyzed ATRP in terms of control over molecular weight and range of polymerizable monomers, and the polymers appear mildly fluorescent. Nevertheless, there is scope for development of better small molecule-based photocatalysts.

5.8   POLYMER SYNTHESIS BY ATOM TRANSFER RADICAL POLYMERIZATION IN COMMERCIAL APPLICATIONS

Several patent applications have appeared on the use of RDRP polymers as macromolecular emulsifiers, dispersants, surface modifiers, rheology modifiers, thermoplastic elastomers, encapsulants for actives, binders, and cross-linkers.[67] Matyjaszewski at Carnegie Mellon University (CMU) founded the ATRP Consortium[68] in 1996 that led to issuance of several U.S. patents and signing of licenses with industrial partners. A company called ATRP Solutions was spun out of CMU in 2006.[69] Since 2003, ATRP has been licensed to 8 of the over 40 corporations funding the research at CMU. With the advent of ARGET ATRP and the possibility of lowering the catalyst amount to 10 ppm, use of ATRP for commercial applications has gathered pace. Licensees have begun commercial production of specialty polymers. One of the licensees explored use of ATRP to prepare low polydispersity, functional polymer additives of various polymer architectures such as block, gradient, star, and graft copolymers for coating applications. The Kaneka Texas Corporation uses ATRP for large-scale production of telechelic polyarcylates for use in various products such as sealants, adhesives, gaskets, and others.[70] A company in the UK (PolyTherics Limited) synthesizes comb polymers comprising short PEG chains attached to poly(methacrylate) backbone for biomedical applications. However, the number of commercial products based on RDRP polymers is limited and its development has been slower than expected. Some of the major factors could be the lack of access to initiators in large quantities at reasonable costs and the extra cost involved in purification of polymers, which is unavoidable in most cases. Development of simple methods for catalyst residue removal and determination of macromolecular architecture–property relationship would hasten the widespread use of ATRP-based polymers in commercial products.

REFERENCES

1. Ouchi, M., Terashima, T., Sawamoto, M. 2009. Transition metal-catalyzed living radical polymerization: Toward perfection in catalysis and precision polymer synthesis. *Chem. Rev.* 109:4963–5050.
2. Fischer, H. 2001. The persistent radical effect: A principle for selective radical reactions and living radical polymerizations. *Chem. Rev.* 101:3581–3610.
3. Hawker, C. J., Bosman, A. W., Harth, E. 2001. New polymer synthesis by nitroxide mediated living radical polymerizations. *Chem. Rev.* 101:3661–3688.
4. Allan, L. E. N., Perry, M. R., Shaver, M. P. 2012. Organometallic-mediated radical polymerization. *Prog. Polym. Sci.* 37:127–156.
5. Hill, M. R., Carmean, R. N., Sumerlin, B. S. 2015. Expanding the scope of RAFT polymerization: Recent advances and new horizons. *Macromolecules* 48:5459–5469.
6. Destarac, M., Brochon, C., Catala, J.-M., Wilczewska, A., Zard, S. Z. 2002. Macromolecular design via the interchange of xanthates (MADIX): Polymerization of styrene with O-ethyl xanthates as controlling agents. *Macromol. Chem. Phys.* 203:2281–2289.
7. Lopez, G., Thenappan, A., Ameduri, B. 2015. Synthesis of chlorotrifluoroethylene-based block copolymers by iodine transfer polymerization. *ACS Macro Lett.* 4:16–20.
8. Matyjaszewski, K. 2012. Atom transfer radical polymerization (ATRP): Current status and future perspectives. *Macromolecules* 45:4015–4039.
9. Matyjaszewski, K. and Xia, J. 2001. Atom transfer radical polymerization. *Chem. Rev.* 101:2921–2990.
10. Lin, C. Y., Coote, M. L., Gennaro, A., Matyjaszewski, K. 2008. Ab initio evaluation of the thermodynamic and electrochemical properties of alkyl halides and radicals and their mechanistic implications for atom transfer radical polymerization. *J. Am. Chem. Soc.* 130:12762–12774.
11. Lin, C.-Y., Marque, S. R. A., Matyjaszewski, K., Coote, M. L. 2011. Linear-free energy relationships for modeling structure-reactivity trends in controlled radical polymerization. *Macromolecules* 44:7568–7583.
12. Peng, C.-H., Kong, J., Seeliger, F., Matyjaszewski, K. 2011. Mechanism of halogen exchange in ATRP. *Macromolecules* 44:7546–7557.

13. Gillies, M. B. et al. 2003. A DFT study of R-X bond dissociation enthalpies of relevance to the initiation process of atom transfer radical polymerization. *Macromolecules* 36:8551–8559.

14. Percec, V., Barboiu, B., Kim, H.-J. 1998. Arenesulfonyl halides: A Universal class of functional initiators for metal-catalyzed "living" radical polymerization of styrene(s), methacrylates, and acrylate. *J. Am. Chem. Soc.* 120:305–316.

15. Grigoras, C. and Percec, V. 2005. Arenesulfonyl bromides: The second universal class of functional initiators for the metal-catalyzed living radical polymerization of methacrylates, acrylates, and styrenes. *J. Polym. Sci. Part A: Polym. Chem.* 43:319–330.

16. Kwak, Y. and Matyjaszewski, K. 2008. Effect of initiator and ligand structures on ATRP of styrene and methyl methacrylate initiated by alkyl dithiocarbamate. *Macromolecules* 41:6627–6635.

17. Kato, M., Kamigaito, M., Sawamoto, M., Higashimura, T. 1995. Polymerization of methyl methacrylate with the carbon tetrachloride/dichlorotris-(triphenylphosphine)ruthenium(II)/methylaluminum bis(2,6-di-tert-butylphenoxide) initiating system: Possibility of living radical polymerization. *Macromolecules* 28:1721–1723.

18. Takahashi, H., Ando, T., Kamigaito, M., Sawamoto, M. 1999. Half-metallocene-type ruthenium complexes as active catalysts for living radical polymerization of methyl methacrylate and styrene. *Macromolecules* 32:3820–3823.

19. Ando, T., Kamigaito, M., Sawamoto, M. 2000. Catalytic activities of ruthenium(II) complexes in transition-metal-mediated living radical polymerization: Polymerization, model reaction, and cyclic voltammetry. *Macromolecules* 33:5825–5829.

20. Kamigaito, M., Watanabe, Y., Ando, T., Sawamoto, M. 2002. A New ruthenium complex with an electron-donating aminoindenyl ligand for fast metal-mediated living radical polymerizations. *J. Am. Chem. Soc.* 124: 9994–9995.

21. Ouchi, M., Ito, M., Kamemoto, S., Sawamoto, M. 2008. Highly active and removable ruthenium catalysts for transition-metal-catalyzed living radical polymerization: Design of ligands and cocatalysts. *Chem. Asian J.* 3:1358–1364.

22. Simal, F., Jan, D., Delaude, L., Demonceau, A., Spirlet, M.-R., Noels, A. F. 2001. Evaluation of ruthenium-based complexes for the controlled radical polymerization of vinyl monomers. *Can. J. Chem.* 79:529–535.

23. Delaude, L., Delfosse, S., Ricel, A., Demonceau, A., Noels, A. F. 2003. Tuning of ruthenium N-heterocyclic carbene catalysts for ATRP. *Chem. Commun.* 13:1526–1527.

24. Braunecker, W. A. and Matyjaszewski, K. 2006. Recent mechanistic developments in atom transfer radical polymerization. *J. Mol. Cat. A: Chem.* 254:155–164.

25. Tang, W. et al. 2008. Understanding atom transfer radical polymerization: Effect of ligand and initiator structures on the equilibrium constants. *J. Am. Chem. Soc.* 130:10702–10713.

26. Seeliger, F. and Matyjaszewski, K. 2009. Temperature effect on activation rate constants in ATRP: New mechanistic insights into the activation process. *Macromolecules* 42:6050–6055.

27. Bell, C. A., Bernhardt, P. V., Monteiro, M. J. 2011. A rapid electrochemical method for determining rate coefficients for copper-catalyzed polymerizations. *J. Am. Chem. Soc.* 133:11944–11947.

28. Braunecker, W. A., Tsarevsky, N. V., Gennaro, A., Matyjaszewski, K. 2009. Thermodynamic components of the atom transfer radical polymerization equilibrium: Quantifying solvent effects. *Macromolecules* 42:6348–6360.

29. Wang, J.-S. and Matyjaszewski, K. 1995. "Living"/controlled radical polymerization. Transition-metal-catalyzed atom transfer radical polymerization in the presence of a conventional radical initiator. *Macromolecules* 28:7572–7573.

30. Wang, J.-S. and Matyjaszewski, K. 1995. Controlled/"living" radical polymerization. Halogen atom transfer radical polymerization promoted by a Cu(I)/Cu(II) redox process. *Macromolecules* 28:7901–7910.

31. Jakubowski, W. and Matyjaszewski, K. 2005. Activator generated by electron transfer for atom transfer radical polymerization. *Macromolecules* 38:4139–4146.

32. Bai, L., Zhang, L., Cheng, Z., Zhu, X. 2012. Activators generated by electron transfer for atom transfer radical polymerization: Recent advances in catalyst and polymer chemistry. *Polym. Chem.* 3:2685–2697.

33. Dong, H. C. and Matyjaszewski, K. 2008. ARGET ATRP of 2-(dimethylamino)ethyl methacrylate as an intrinsic reducing agent. *Macromolecules* 41:6868–6870.

34. Hu, Z. et al. 2009. AGET ATRP of methyl methacrylate with poly(ethylene glycol) (PEG) as solvent and TMEDA as both ligand and reducing agent. *Eur. Polym. J.* 45: 2313–2318.

35. Miao, J. et al. 2012. AGET ATRP of methyl methacrylate via a bimetallic catalyst. *RSC Adv.* 2:840–847.

36. Pintauer, T. and Matyjaszewski, K. 2008. Atom transfer radical addition and polymerization reactions catalyzed by ppm amounts of copper complexes. *Chem. Soc. Rev.* 37:1087–1097.

37. Matyjaszewski, K. et al. 2006. Diminishing catalyst concentration in atom transfer radical polymerization with reducing agents. *Proc. Natl. Acad. Sci. USA* 103:15309–15314.

38. Rosen, B. M. and Percec, V. 2009. Single-electron transfer and single-electron transfer degenerative chain transfer living radical polymerization. *Chem. Rev.* 109:5069–5119.

39. Konkolewicz, D. et al. 2013. Reversible-deactivation radical polymerization in the presence of metallic copper. A critical assessment of the SARA ATRP and SET-LRP mechanisms. *Macromolecules* 46:8749–8772.

40. Konkolewicz, D. et al. 2014. SARA ATRP or SET-LRP. end of controversy? *Polym. Chem.* 5:4396–4417.

41. Peng, C. H. et al. 2013. Reversible-deactivation radical polymerization in the presence of metallic copper. Activation of alkyl halides by Cu0. *Macromolecules* 46:3803–3815.

42. Konkolewicz, D. et al. 2014. Aqueous RDRP in the presence of Cu0: The exceptional activity of CuI confirms the SARA ATRP mechanism. *Macromolecules* 47:560–570.

43. Wang, Y. et al. 2013. Improving the "livingness" of ATRP by reducing Cu catalyst concentration. *Macromolecules* 46:683–691.

44. Magenau, A. J. D., Strandwitz, N. C., Gennaro, A., Matyjaszewski, K. 2011. Electrochemically mediated atom transfer radical polymerization. *Science* 332:81–84.

45. Min, K., Gao, H. F., Matyjaszewski, K. 2005. Preparation of homopolymers and block copolymers in miniemulsion by ATRP using activators generated by electron transfer (AGET). *J. Am. Chem. Soc.* 127:3825–3830.

46. Min, K. and Matyjaszewski, K. 2005. Atom transfer radical polymerization in microemulsion. *Macromolecules* 38:8131–8134.

47. Oh, J. K. et al. 2006. Inverse miniemulsion ATRP: A new method for synthesis and functionalization of well-defined water-soluble/cross-linked polymeric particles. *J. Am. Chem. Soc.* 128:5578–5584.

48. Averick, S. E. et al. 2011. Covalently incorporated protein-nanogels using AGET ATRP in an inverse miniemulsion. *Polym. Chem.* 2:1476–1478.

49. Min, K. and Matyjaszewski, K. 2005. Atom transfer radical polymerization in microemulsion. *Macromolecules* 38:8131–8134.

50. Oh, J. K. and Matyjaszewski, K. 2006. Synthesis of poly(2-hydroxyethyl methacrylate) in protic media through atom transfer radical polymerization using activators generated by electron transfer. *J. Polym. Sci. Part A: Polym. Chem.* 44:3787–3796.

51. Tanaka, K. and Matyjaszewski, K. 2007. Controlled copolymerization of n-butyl acrylate with nonpolar 1-alkenes using activators regenerated by electron transfer for atom-transfer radical polymerization. *Macromolecules* 40:5255–5260.

52. Stoffelbach, F. et al. 2007. Use of an amphiphilic block copolymer as a stabilizer and a macroinitiator in miniemulsion polymerization under AGET ATRP conditions. *Macromolecules* 40:8813–8816.

53. Matyjaszewski, K. and Tsarevsky, N. V. 2014. Macromolecular engineering by atom transfer radical polymerization. *J. Am. Chem. Soc.* 136:6513–6533.

54. Liu, H. et al. 2012. Facile synthesis of ABCDE-type H-shaped quintopolymers by combination of ATRP, ROP, and click chemistry and their potential applications as drug carriers. *J. Polym. Sci. Part A: Polym. Chem.* 50:4705–4716.

55. Banerjee, S., Paira, T. K., Mandal, T. K. 2014. Surface confined atom transfer radical polymerization: Access to custom library of polymer-based hybrid materials for specialty applications. *Polym. Chem.* 5:4153–4167.

56. Du, T. et al. 2016. Bio-inspired renewable surface-initiated polymerization from permanently embedded initiators. *Angew. Chem. Int. Ed.* 55:4260–4264.

57. Shen, Y., Tang, H., Ding, S. 2004. Catalyst separation in atom transfer radical polymerization. *Prog. Polym. Sci.* 29:1053–1078.

58. Ding, M. et al. 2015. Recent progress on transition metal catalyst separation and recycling in ATRP. *Macromol. Rapid Commun.* 36:1702–1721.

59. Kasko, A. M., Heintz, A. M., Pugh, C. The effect of molecular architecture on the thermotropic behavior of poly[11-(40-cyanophenyl-400-phenoxy)undecyl acrylate] and its relation to polydispersity. *Macromolecules* 31:256–271.

60. Matyjaszewski, K., Pintauer, T., Gaynor, S. 2000. Removal of copper-based catalyst in atom transfer radical polymerization using ion exchange resins. *Macromolecules* 33:1476–1478.

61. Carmichael, A. J., Haddleton, D. M., Bon, S. A. F., Seddon K. R. 2000. Copper (I) mediated living radical polymerisation in an ionic liquid. *Chem. Commun.* 1237–1238.

62. Biedron, T. and Kubisa, P. 2001. Atom transfer radical polymerization of acrylates in an ionic liquid. *Macromol. Rapid Commun.* 22:1237–1242.

63. Du, X. et al. 2014. Thermo-regulated phase separable catalysis (TPSC)-based atom transfer radical polymerization in a thermo-regulated ionic liquid. *Chem. Commun.* 50:9266–9269.

64. Bai, L. et al. 2013. Developing a synthetic approach with thermoregulated phase-transfer catalysis: Facile access to metal-mediated living radical polymerization of methyl methacrylate in aqueous/organic biphasic system. *Macromolecules* 46:2060–2066.

65. Treat, N. J. et al. 2014. Metal-free atom transfer radical polymerization. *J. Am. Chem. Soc.* 136:16096–16101.

66. Miyake, G. M. and Theriot, J. C. 2014. Perylene as an organic photocatalyst for the radical polymerization of functionalized vinyl monomers through oxidative quenching with alkyl bromides and visible light. *Macromolecules* 47:8255–8261.

67. Destarac, M. 2010. Controlled radical polymerization: Industrial stakes, obstacles and achievements. *Macromol. React. Eng.* 4:165–179.

68. CRP Consortium. http://www.chem.cmu.edu/ groups/maty/crp/

69. ATRP solutions. http://www.atrpsolutions.com

70. Homepage. http://www.kanekatexascorporation.com

6 Organometallic-Mediated Radical Polymerization

Daniel L. Coward, Benjamin R. M. Lake,
and Michael Shaver

CONTENTS

6.1 INTRODUCTION

Organic radicals, such as those present in a controlled radical polymerization (CRP) reaction, are able to react or interact with a transition metal center and its ligand sphere *via* a number of chemically distinct pathways.[1,2] This includes, but is not limited to, reversible transfer of a halogen atom between the radical terminus of a propagating polymer chain and the metal center (ATRP), as discussed extensively in Chapter 5. Less common but of particular importance in CRP reactions is the ability of certain metal centers to react directly with a propagating radical, (reversibly) forming a metal-carbon bond. Other processes, including atom abstraction and radical addition to the ligand scaffold, may also intervene, compete, or interplay during a reaction, with the degree of participation and (ir)reversibility of such processes markedly influencing the outcome of a given CRP reaction.[1,3–8]

Organometallic-mediated radical polymerization (OMRP) involves the reversible homolytic cleavage of a metal-carbon bond formed between the mediating metal complex and propagating polymer chain. OMRP is an attractive technique for further exploration, both in academic and industrial settings, due to the inherently broad scope of the system through careful manipulation of the electronic and steric properties of the ligand and selection of an appropriate metal center. This tuning can adjust metal-carbon bond strengths, allowing for the OMRP of a broad scope of monomers, along with the formation of copolymers inaccessible using ATRP.

The purpose of this chapter is to provide the reader with an understanding of the fundamental principles of OMRP including how different equilibria can participate in and influence the course of a CRP reaction. OMRP systems based on various transition metals will be presented in this chapter, including advantages of the use of each and specific mechanistic details. Finally, recent trends in the synthesis of co- and functionalized polymers will be described using these OMRP systems.

6.2 MECHANISMS

Mechanistically, OMRP may proceed through either a reversible termination (RT-OMRP) or degenerative transfer (DT-OMRP) pathway depending on reaction conditions, the metal center, and its ligand environment.[9,10] As its name suggests, RT-OMRP involves a metal complex acting as a reversible spin trap, temporarily deactivating a growing polymer chain *via* formation of a metal-carbon bond, producing a metal-capped dormant polymer chain. The metal-capped polymer chain does not undergo further reaction with an additional propagating chain but can undergo reactivation, producing the chemically unmodified metal complex and growing polymer chain (Figure 6.1). The nature of the metal-polymer interaction in the metal-capped dormant species is dependent on the nature of the monomer being polymerized, with the polymer chain acting as either a single σ-bonded ligand or a chelate *via* an additional metal-heteroatom dative bond.[1,11] RT-OMRP reactions can be initiated in one of two ways; either *via* homolytic decomposition of a conventional radical initiator in the presence of the reduced metal complex (M^x) or *via* homolytic decomposition of an oxidized metal complex containing a preformed metal-carbon bond (M^{x+1}–P_n), with the former method being the most widely used in the literature. In a reaction performed under RT-OMRP conditions, it is important that the rate of deactivation (k_d) is much faster than the rate of propagation (k_p) so that the concentration of active radicals is kept low, significantly reducing the rate of irreversible bimolecular radical termination reactions, and thus increasing control over the reaction.

Degenerative transfer OMRP (DT-OMRP) is the other mechanistically distinct OMRP pathway and can be thought of as conceptually analogous to a reversible addition-fragmentation chain-transfer (RAFT) polymerization. In a DT-OMRP reaction, the rate of polymerization tends to be extremely slow at the start of the reaction

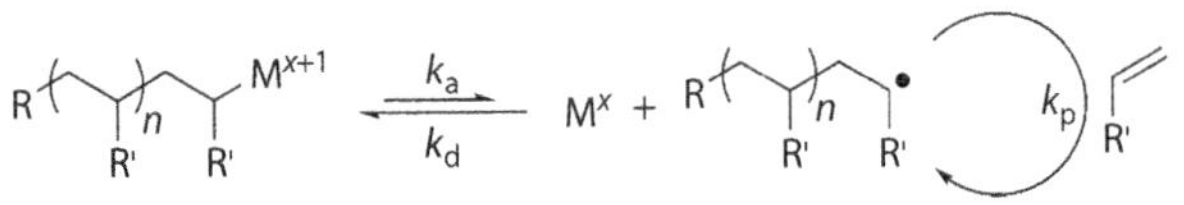

FIGURE 6.1   Mechanism of reversible termination OMRP (RT-OMRP).

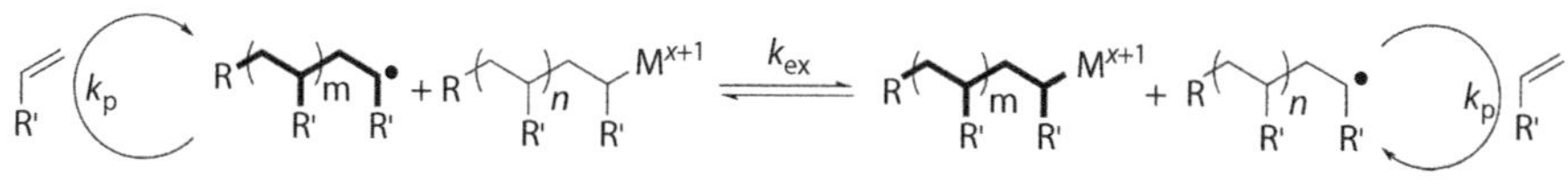

FIGURE 6.2 Mechanism of degenerative transfer OMRP (DT-OMRP).

since the radicals generated by decomposition of the initiator react immediately with the metal mediator, forming the DT-OMRP dormant species ($M^{x+1}-P_n$). It is only once the total concentration of radicals produced is greater than the concentration of metal mediator that polymerization begins to occur since additional radicals entering the system can either react with monomer or displace (associative exchange) the metal-bound radical, which then goes on to react with monomer (Figure 6.2). Control over a polymerization using a DT-OMRP mechanism is achievable since the rate of termination (k_t) is strongly dependent on chain length, with shorter chains terminating much more quickly. Therefore, since radicals are generated from the initiator continuously, associative exchange allows rapid and reversible release of longer chain radicals, which terminate far more slowly. Control over the dispersity and molecular weight in a DT-OMRP is thus achievable when the rate of associative exchange (k_{ex}) is much faster than the rate of propagation (k_p).

For a reaction established under OMRP conditions (and sometimes for reactions set-up under ATRP conditions), there are two other significant side reactions to consider, which may have a significant impact on the progress of the reaction, namely, catalytic chain transfer (CCT) and catalyzed radical termination (CRT). CCT is a process that has long been realized to occur in metal-mediated radical polymerization reactions,[12] whereas the participation of CRT has been reported far more recently.[13–15] CCT occurs as a result of H-atom transfer from the radical terminus of the growing polymer, producing an alkene-terminated dead polymer chain and a metal hydride ($M^{x+1}-H$). The transfer of a H-atom may occur either *via* direct H-atom abstraction from the radical terminus by the reduced metal complex (M^x) or *via* β-H elimination from an OMRP dormant species ($M^{x+1}-P_n$). In this case, the presence of a vacant coordination site *cis* with respect to the coordinated polymer chain is essential. The so-formed metal hydride ($M^{x+1}-H$) can then initiate the growth of a new chain by reaction with an equivalent of monomer (Figure 6.3). The polymers produced where CCT plays a significant role in the reaction tend to possess short chain lengths, an alkene terminus, and molecular weights independent of monomer conversion.

While CCT would be expected to have no effect on the overall number of radicals in a CRP reaction, CRT leads to a significant reduction in the number of radicals, with the potential to quench the entire reaction. Instead of reaction of the metal hydride ($M^{x+1}-H$) with an equivalent of monomer or β-H elimination from an OMRP dormant chain ($M^{x+1}-P_n$), either of these species may undergo reaction with an additional radical chain ($^{\bullet}P_m$) leading to a dead chain with a saturated terminus ($H-P_m$) or a coupled chain (P_n-P_m). In addition, the reduced metal complex is regenerated, allowing it to re-enter the catalytic cycle (Figure 6.4).[16]

The possibility of interplay between the mechanisms described earlier (and ATRP) became apparent soon after the early reports on ATRP, with Poli and coworkers describing a remarkable series of half-sandwich Mo-based complexes that are able to control the

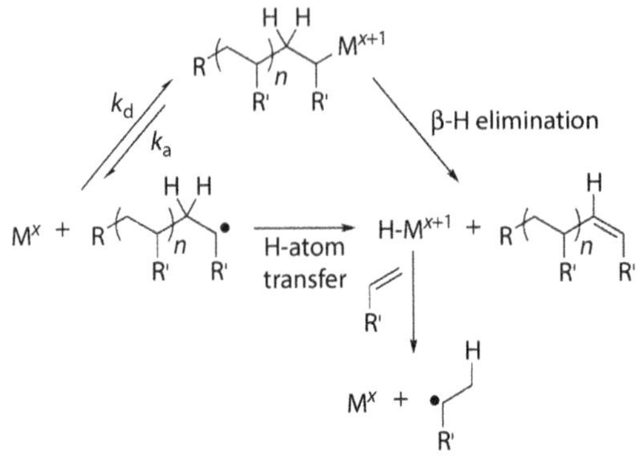

FIGURE 6.3   Mechanism of catalytic chain transfer (CCT).

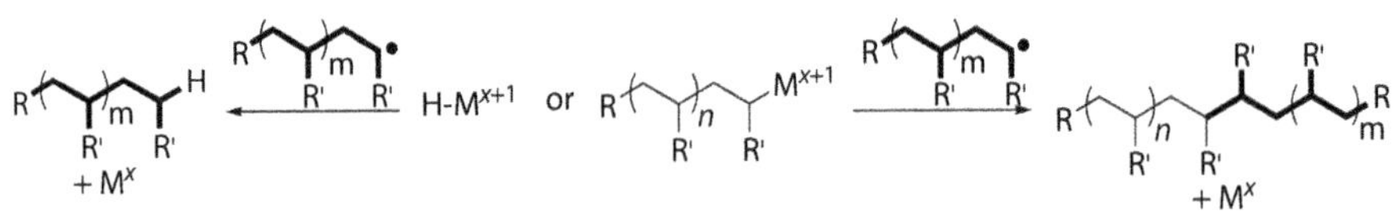

FIGURE 6.4   Mechanism of catalyzed radical termination (CRT).

polymerization of styrene *via* simultaneous RT-OMRP and ATRP mechanisms when the reactions are performed under (R-)ATRP conditions, but which may also mediate CCT under slightly different conditions.[17–19] Although further examples of possible mechanistic interplay have been described in Ti-[20] and Os-mediated[21] systems, Fe-based systems remain by far the most common and well-understood systems displaying interplay,[3] including α-diimine,[22–27] amine-*bis*(phenolate),[28–32] and β-ketiminate-ligated[33] Fe complexes. Interplay is possible between ATRP and OMRP (and thus CCT/CRT) mechanisms since the equilibrium in an ATRP reaction is weighted toward the reduced metal complex (M^x) (and halogen-capped dormant polymer). This high concentration of the reduced metal complex (M^x) can allow it to act as a spin trap for propagating radicals, reversibly forming metal-carbon bonds, thus allowing an OMRP equilibrium to potentially impart significant control over a reaction performed under (R-)ATRP conditions. However, since the formation of the OMRP dormant species (M^{x+1}–P_n) may also be a prerequisite for the occurrence of CCT and CRT reactions, the presence of an OMRP equilibrium in an ATRP reaction may not necessarily lead to an increase in the control (dispersity and molecular weight). In addition, given the high concentration of the reduced metal complex (M^x) in an ATRP reaction, CCT *via* direct H-atom transfer may occur, even in the absence of a simultaneous OMRP equilibrium leading to short-chained oligomers with alkene end-groups.

6.3   METALS USED IN OMRP

In the following section, a selection of some of the seminal examples of metals used as mediators in OMRP is provided.

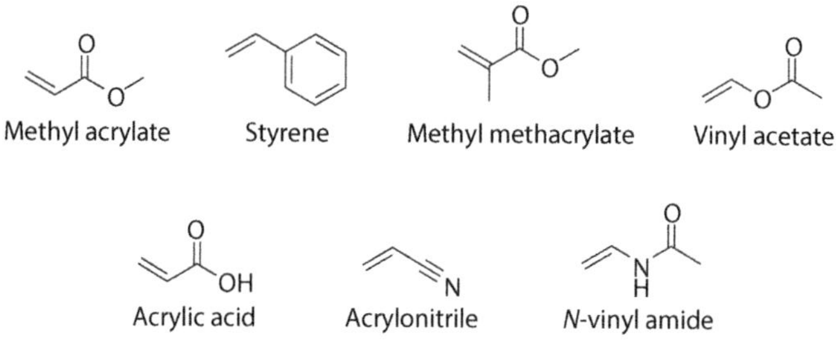

Methyl acrylate   Styrene   Methyl methacrylate   Vinyl acetate

Acrylic acid   Acrylonitrile   N-vinyl amide

FIGURE 6.5   Some of the monomers referred to within this chapter.

6.3.1 RHODIUM

The first example of OMRP was published by Wayland and coworkers in 1992 and involved tetramesitylporphyrinato rhodium(II) (**1**).[34] Two equivalents of **1** were reacted with acrylates, in the dark, to form an alkyl-chain-bridged dirhodium complex, [(TMP)RhCH$_2$CH(CO$_2$X)CH(CO$_2$X)CH$_2$Rh(TMP)] (**2**) (X=H, Me, Et). Exposure to visible light homolytically cleaves the complex resulting in [(TMP)Rh(II)]• (**3**) and [(TMP)RhCH$_2$CH(CO$_2$X)CH(CO$_2$X)CH$_2$]• (**4**). The former was unable to initiate acrylate polymerization, whereas the latter initiated the polymerization of acrylic acid, methyl acrylate, and ethyl acrylate (Figure 6.5). Complex **3** behaved as the reversible trapping species, achieving a low degree of control. However, this trapping species was also able to reinitiate polymerization, through further formation of the bridged dirhodium initiator. As a result, broad dispersities of 1.75–2.75 were attained.

6.3.2 COBALT

Cobalt is the most widely used metal for OMRP, with a significant body of work exploring a range of complexes and monomers.[7] Early work by Wayland and coworkers used cobalt porphyrin complexes to polymerize acrylates.[35] Organometallic derivatives of cobalt tetramesitylporphyrin (TMP) were used to both initiate and control the polymerization of methyl acrylate. Living characteristics were evidenced by dispersities between 1.1 and 1.3. This was further demonstrated by the synthesis of a poly(methyl acrylate)-block-poly(butyl acrylate) copolymer, with a dispersity of 1.3, achieved through the sequential addition of butyl acrylate after methyl acrylate had fully reacted. The steric bulk of the TMP ring prevents β-H abstraction.

(TMP)Co(II)• (**5**) (Figure 6.6), alongside an organic radical source such as V-70 (a low temperature azo initiator with $t_{1/2}$ = 10 hours at 30°C) generates the alkylcobalt(III) species *in situ*.[36] The observed induction period corresponds with the conversion from Co(II) to Co(III). This is achieved through β-H abstraction from the organic radical to form (TMP)Co-H, which then adds reversibly to the olefin monomer. Using this complex as the initiating species in the polymerization of methyl acrylate gave polymers with molecular weights, which increased linearly with conversion and dispersities less than

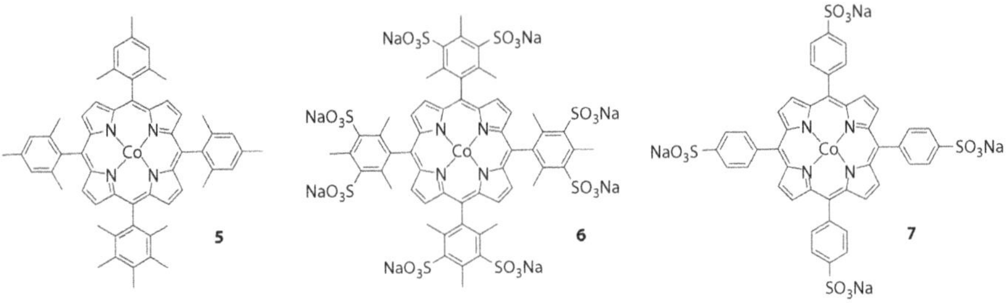

FIGURE 6.6   (TMP)Co (**5**), (TMPS)Co (**6**), and (TSPP)Co (**7**).

1.1. Poly(methyl acrylate)-block-poly(butyl acrylate) was synthesized by sequential monomer addition with a low dispersity of 1.07.

Robust cobalt porphyrin derivatives were also used in the polymerization of acrylic acid, a highly challenging monomer to control.[37] Water-soluble cobalt(II) porphyrins, (TMPS)Co (**6**) and (TSPP)Co (**7**) (Figure 6.6), were used as mediators along with V-70. A DT-OMRP mechanism was implicated, and high molecular weights (M_n = 232,000) and low dispersities (1.20) were achieved within only 30 minutes.

Vinyl acetate has been a monomer of particular interest for use in cobalt-mediated radical polymerizations, due to difficulty achieving control through ATRP, with the first example of an efficient OMRP appearing in 2005.[38] Jérôme and coworkers used [Co(acac)$_2$] (**8**) along with V-70 to polymerize vinyl acetate in the bulk at 30°C. An induction period of 12 hours was required for the Co(II) species to be oxidized to a Co(III) organometallic complex. Following this, a controlled polymerization was observed, with molecular weights linearly increasing with conversion and dispersities between 1.1 and 1.2. Using AIBN instead of V-70 and higher reaction temperatures gave polymers with considerably broader dispersities (2.0–3.5), implying irreversible chain termination reactions. The same system has also been used to polymerize vinyl acetate in both aqueous suspension[39] and miniemulsion.[40] In both cases, very high conversions were achieved.

Although V-70 is a very effective low-temperature initiator, there are persistent issues relating to storage and handling. Attempts have been made to replace this initiating method with one based on a redox system.[41] Jérôme and coworkers combined [Co(acac)$_2$], as a reducing agent, with lauroyl or benzoyl peroxide, as oxidants, at 30°C to polymerize vinyl acetate. The best system used a combination of citric acid and lauroyl peroxide. Although inefficient initiation gave molecular weights significantly higher than theoretical values, a high conversion of 80% was achieved after 3 hours and a dispersity of 1.4.

More detailed investigations, using a combination of reactivity, polymerization, and computational studies, were conducted to probe the mechanism of control.[42,43] In the presence of an additional ligand, such as pyridine or triethylamine, the mechanism of control switches from DT-OMRP to RT-OMRP. In the absence of the ligand, an induction period (30 hours) was observed before a rapid well-controlled polymerization. RT-control is promoted by addition of a ligand, as no such induction period is observed.

Aside from vinyl acetate, other monomers have also been polymerized with good control using this [Co(acac)₂]/V-70 system. Acrylonitrile is a notoriously difficult monomer to control, owing to its high reactivity and low solubility of the polymer product. Using DMF as solvent at 30°C, molecular weights increased with conversion, albeit with dispersities of 1.6–1.9 and higher molecular weights than the theoretical values, due to poor initiation and low polymer solubility.[44] Using a low molecular weight cobalt(III) adduct in DMSO at 0°C, acrylonitrile was polymerized with excellent control and dispersities as low as 1.15. This cobalt(III) adduct was R-Co(acac)₂ (**9**), where R was a poly(vinyl acetate) oligomer, which behaved as a macroinitiator. DFT calculations and X-ray analysis were used to confirm that coordination of solvent molecules to the cobalt complex changes its reactivity, weakening the Co-C bond and enabling fast initiation. The same method of using an alkylcobalt(III) initiator was applied to control the polymerization of n-butyl acrylate.[45] Optimizing the initiation step, temperature, and presence of additives led to a system where well-controlled polyacrylates were synthesized. Molecular weights increased linearly with conversion and dispersities were as low as 1.2.

A range of N-vinyl amides have also been polymerized using the alkylcobalt(III) initiator **9**.[46] In particular, N-vinylcaprolactam (NVCL) was very well controlled, with dispersities less than 1.1 and a conversion of 40% reached after just 3 hours at 40°C. N-Methyl-N-vinylacetoamide (NMVA) was also well controlled with a low dispersity (1.1–1.2). It was shown that the amide moiety in the last monomer unit of the metal-terminated polymer chain chelates with the cobalt, thus strengthening the Co-C bond. The nature of this intramolecular chelation affects the kinetics and the control.

There has been work to examine why the [Co(acac)₂] mediator does not suffer the same problem of sequence errors as other OMRP-based systems.[47] It was postulated that [Co(acac)₂] either reduces the number of head-to-head (HH) additions or the dormant Co-PVAc species allows for easy reactivation, irrespective of head-to-tail (HT) or HH addition. Evidence showed that [Co(acac)₂] neither suppresses HH additions, nor does it change the relative fraction of HH additions in the PVAc chain. DFT calculations into the difference between bond-dissociation energy (BDE) of the five-membered dormant species (formed by HH addition) and the six-membered dormant species (formed by HT addition) showed that the difference is virtually zero (Figure 6.7a and b). Although chelation of the carbonyl group has a greater effect in the five-membered ring, the six-membered ring is more reactive and thus the two properties cancel each other out. This lack of any difference between the BDE of the two dormant species is responsible for suppressing the impact of sequence errors.

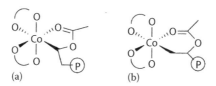

(a) (b)

FIGURE 6.7 (a) Five-membered dormant species. (b) Six-membered dormant species.

More recent work has focussed on using cobalt-mediated radical polymerization to synthesize poly(ionic liquid)s, a type of polyelectrolyte. Two different methods were used to polymerize various *N*-vinyl-3-alkylimidazolium-type monomers (Figure 6.8).[48] The first method used **9** as a macroinitiator. The second method used [Co(acac)₂] itself, along with *tert*-butylhydroperoxide (*t*-BuOOH) as initiator. In both cases, relatively low temperatures (30°C–50°C) were used with water as the solvent.

The first method showed excellent control, with molecular weights linearly increasing with conversion, reaching full conversion after only 6 hours using an initial monomer:complex ratio of 75. As the concentration of monomer increased the degree of control was maintained, with the expected increase in the reaction time. In all cases dispersities were low, between 1.1 and 1.4. The second method also gave excellent control; with dispersities always lower than 1.25, albeit with slower rates.

Small alterations to the *acac* framework offer little advantage over the initial *acac* system. Adding a bulky *tert*-butyl group to *acac* yields tetramethylheptadionato (tmhd), and the complex [Co(tmhd)₂] (**10**, Figure 6.9) was used alongside V-70 at 30°C.[49] Using different ratios of V-70 to **10**, polymerizations gave well-controlled polymers with dispersities between 1.1 and 1.5. After six half-lives of V-70, the mechanism of control switches from DT-OMRP to RT-OMRP, albeit with poorer control than [Co(acac)₂]. Use of a solvent extends the reaction times, thus slowing down the DT mechanism and promoting the RT mechanism. At long reaction times, deviation from the first-order rate law is observed. This is due to the breakdown of the persistent radical effect and the presence of competing catalytic chain transfer. Further promotion of RT was achieved by decreasing the amount of V-70 used and preventing complete conversion of Co(II) to Co(III), the species required for DT-control. Furthermore, use of an external base, such as water or pyridine, blocks the vacant coordination site on the Co(III)-R species and prevents DT-control. Comparisons between **10** and [Co(acac)₂] show that in the absence of an external base the polymerization rate is faster for the former, whereas the latter is five times

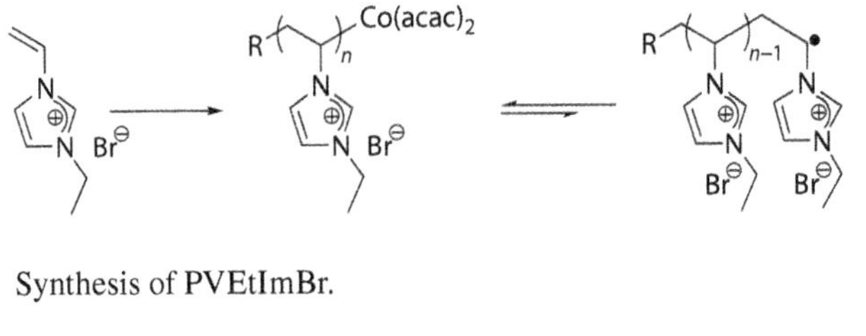

FIGURE 6.8   Synthesis of PVEtImBr.

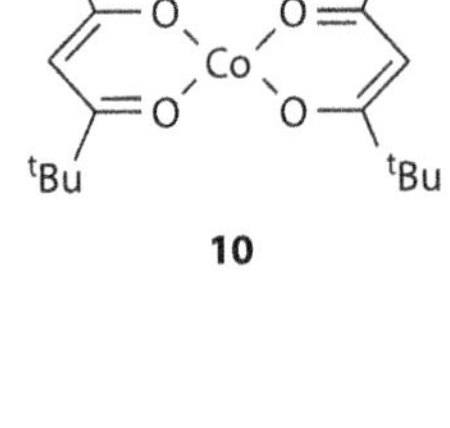

10

FIGURE 6.9   Co(tmhd)₂.

faster in the presence of an external base. This is due to the competitive steric effects on the Co(III)-R and Co(II)-L bond strengths (where L is the external base) and was supported by ^{1}H NMR and DFT studies.

Aside from [Co(acac)$_2$] and its derivatives, other cobalt complexes have been explored for their efficacy in OMRP. A range of complexes supported by the 1,3-bis(2-pyridylimino)isoindolate (bpi) ligand (**11**, Figure 6.10) were synthesized and screened for the OMRP of methyl acrylate.[50] Using V-70 initiator in benzene at 60°C yielded well-controlled polymers. Molecular weights increased linearly with conversion and were close to theoretical values. Dispersities as low as 1.1 were achieved, with conversions as high as 70%. Various constituents of the ligand were altered (R$_1$, R$_2$ and R$_3$), using both electron-withdrawing and electron-donating groups, with very little effect on the polymerization results. DFT calculations verified this result, showing only a small change in the bond dissociation energy of the Co-C bond. This implies that the Co-C bond is electronically decoupled from the substituents on the bpi ligand.

β-Ketoiminate cobalt(II) complexes (**12**, Figure 6.11) have been used in the OMRP of vinyl acetate, initiated by V-70 at 30°C.[51] The ligand framework allows for steric and electronic modifications to tune polymerization results. With 0.8 equivalents of initiator per cobalt, slow but well-controlled polymerizations were observed, with dispersities between 1.1 and 1.5, and the reaction continued well beyond the time taken for all the initiator to decompose, implying RT-control. Higher molecular weights than theoretical values suggest the Co(II) complexes used are not efficient radical traps. This is rationalized by larger steric crowding around the metal, compared to [Co(acac)$_2$]. Using six equivalents of initiator,

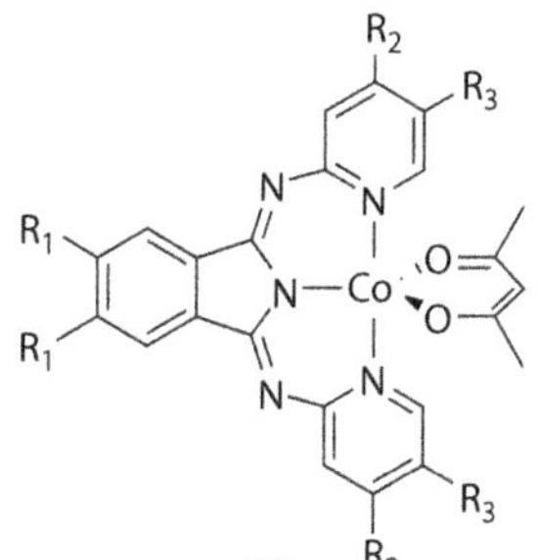

FIGURE 6.10   [Co(acac)(bpi)].

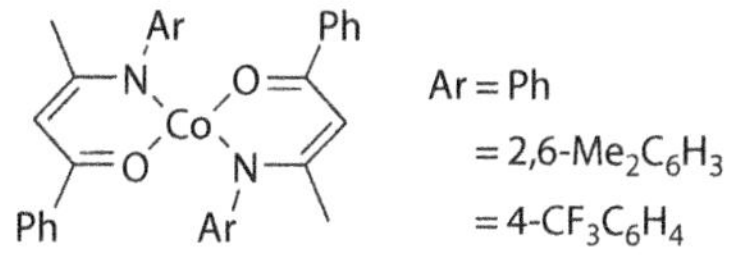

FIGURE 6.11   β-Ketoiminate cobalt(II) complexes.

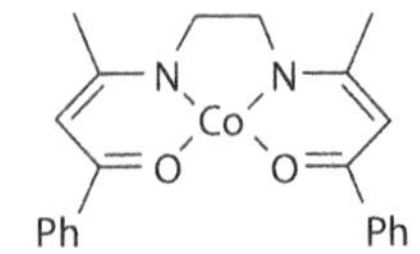

FIGURE 6.12 Bis(benzoylacetone)ethylenediaminate cobalt(II).

a large excess, promotes control through DT and yields a well-controlled faster polymerization, with dispersities around 1.2. The increased polymerization rate is due to associative radical exchange occurring faster than addition of the propagating radical to the Co(II) species. Addition of pyridine had little effect on the polymerization mediated by the Ph-substituted complex. In contrast the polymerization rate quadrupled when pyridine was used with the CF_3-substituted complex. Dispersities remain low, albeit with a slight loss of control. ^{1}H NMR spectroscopic analysis showed weak binding between pyridine and the Ph-substituted cobalt complex, whereas stronger binding was observed with the CF_3-substituted cobalt complex. This shifted the OMRP equilibrium toward the active radical species and increased the polymerization rate.

An extension to this work involved using a ketoaminato cobalt(II) complex in the OMRP of methyl acrylate.[52] Bis(benzoylacetone)ethylenediaminate cobalt(II) (**13**, Figure 6.12) was synthesized and used alongside 0.6 equivalents of V-70 in toluene at 50°C. A linear first-order consumption of monomer was observed. This continued beyond six half-lives of the initiator, suggesting an RT mechanism is in play. Increasing to 0.9 equivalents of V-70 introduced a 2.3-hour induction time, during which time the polymerization rate was comparable to the rate with 0.6 equivalents. After the induction time, the rate rapidly accelerated as the DT-mechanism becomes more dominant. Dispersities increased throughout the reaction, from 1.1 to 1.5, indicating a gradual loss of control, suggesting an increasing presence of irreversible terminations such as catalytic chain transfer. The alkylcobalt(III) derivative, Co(III)-Et, was synthesized and used as both initiator and mediator. Initial rates were fast with nonlinear kinetics. After one hour the rate slows and first-order kinetics are observed. This suggests that radicals from the alkylcobalt initiator were generated slower than those from a polymer chain bound to the Co(II) complex, giving a lower initial concentration of the Co(II)-trapping species and a faster polymerization. After consumption of all initiator the concentration of Co(II) is now higher, control occurs through the RT mechanism and the original first-order rate returns.

6.3.3 IRON

α-Diimine iron complexes (**14**, Figure 6.13) have been studied in detail for use in both ATRP[22–24,53] and OMRP.[25] Investigations into the relationship between the metal spin state of the oxidized species and the tendency to undergo ATRP over catalytic chain transfer (CCT) were conducted.[23] High-spin Fe(III) complexes were effective for ATRP, whereas lower spin Fe(III) complexes were effective for organometallic pathways, paving the way to study these complexes for OMRP.

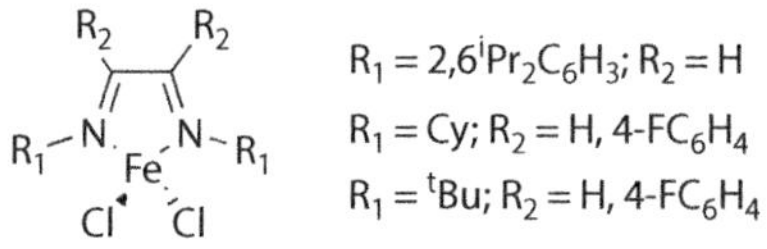

$R_1 = 2,6^iPr_2C_6H_3; R_2 = H$
$R_1 = Cy; R_2 = H, 4\text{-}FC_6H_4$
$R_1 = {}^tBu; R_2 = H, 4\text{-}FC_6H_4$

14

FIGURE 6.13 α-Diimine iron complexes.

The OMRP of styrene was investigated,[25] with dispersities as low as 1.4 and molecular weights reasonably close to the theoretical values. However, this required eight equivalents of catalyst per initiator radical and reached only 7.5% conversion after 24 hours at 120°C. This excessive catalyst loading is due to the instability of the Fe(III)-R species at the high reaction temperature. Using a lower temperature initiator, such as V-70, allows for lower reaction temperatures. As a result, lower catalyst loadings, only two equivalents, were required to achieve dispersities of 1.3 and molecular weights in good agreement with theoretical values.

Amine-*bis*(phenolate)s (ABPs) are a well-researched ligand framework, able to stabilize high oxidation state metals due to their strong σ- and π-donating character. In particular, iron(III) amine-bis(phenolate) complexes (**15**) (Figure 6.14) have been synthesized and previously used as catalysts in cross-coupling[54] and hydro-functionalization reactions.[55–57] Shaver and coworkers have extensively studied both iron(II) and iron(III) amine-bis(phenolate) complexes as mediators in ATRP and OMRP.

The first report, in 2012, synthesized a range of iron(III) complexes with different substituents on the phenolate ring and different donor arms.[28] These complexes were screened in the R-ATRP of styrene and methyl methacrylate. Dispersities as low as 1.11 were achieved with molecular weights in good agreement with theoretical values.

The next study probed deeper into the mechanism of control.[29] A comparison between sterics and electronics (through screening of both chloro- and methyl-substituted complexes, which have a similar size but markedly different electronic properties) showed that electronic effects are primarily responsible for the degree of control. Furthermore it was shown that the donor arm has little effect on the degree of control. In addition, the concentration of AIBN initiator was changed to investigate its effect on the efficiency of polymerization. With styrene, higher concentrations of initiator gave faster reaction rates and higher conversion but a loss of control. With methyl methacrylate, higher initiator concentrations also gave higher conversions, albeit with poorer agreement of molecular weights to theoretical values,

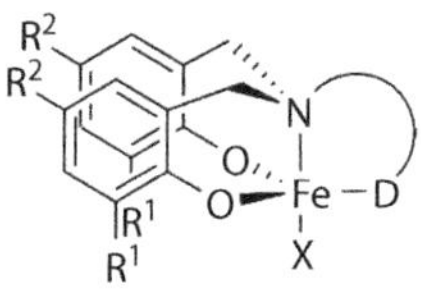

15

FIGURE 6.14 Amine-bis(phenolate) iron(III).

but also offered a marginally better degree of control. This supports the belief that OMRP may be more important for methyl methacrylate, since this increased initiator concentration promotes a degenerative transfer mechanism of control.

To explore the mechanistic interplay between ATRP and OMRP in controlling styrene and methyl methacrylate polymerizations, an analogous iron(II) complex (**16**) was synthesized using [Fe(N(SiMe$_3$)$_2$)$_2$] (**17**) as a precursor.[31] The advantage of such iron(II) complexes is that OMRP can be isolated from ATRP. Using an iron(II) mediator and an azo initiator affords no free halide atoms, and so it is impossible for the polymerization to be controlled through ATRP. Mechanistic studies show that the styrene polymerizations operate predominantly through ATRP, with no evidence of reversible termination *via* an organometallic iron(III) species. Polymerizations under ATRP conditions yielded well-controlled polymers, whereas under solely OMRP conditions it gave polymers with considerably broader dispersities. Conversely, studies into the methyl methacrylate (MMA) polymerization showed a dual control mechanism involving both ATRP and OMRP. This is true even when an excess of ATRP initiator, R-Cl, is used. Under solely OMRP conditions, dispersities as low as 1.29 were achieved.

Iron(II) acetylacetonate, [Fe(acac)$_2$] (**18**), has been shown to give good control over the polymerization of vinyl acetate.[58] Using V-70 initiator and no solvent at 30°C, dispersities as low as 1.29 were achieved, albeit with considerably higher molecular weights than theoretical values. The higher oxidation state derivative, [Fe(acac)$_3$] (**19**), has also been shown to impart control over the polymerization.[59] A reducing agent was used to generate *in situ* the iron(II) species required for OMRP.

6.3.4 TITANIUM

The earliest example of using titanium in radical polymerization appeared in 2003.[60] Titanocene dichloride, Cp$_2$TiCl$_2$ (**20**, Cp=cyclopentadienyl) (Figure 6.15), was used as a chain growth mediator in the polymerization of MMA, initiated by AIBN. Dispersities were broad (Đ = 1.9–2.9) implying only a low degree of control. Further studies showed that the method of control was due to reduction of **20** to the titanium(III) analogue (**21**).[20]

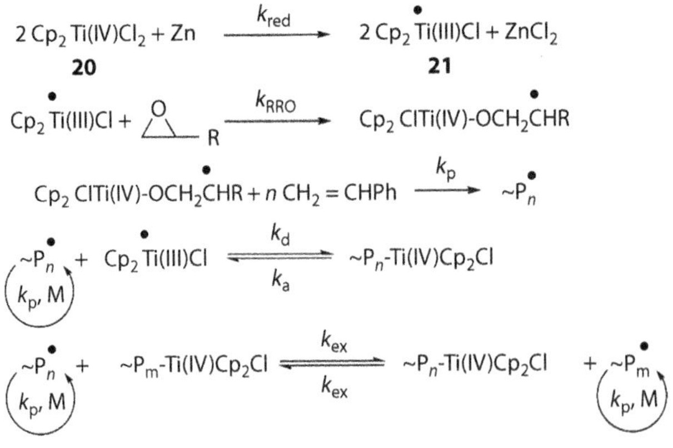

FIGURE 6.15 Mechanism of polymerization of styrene using Cp$_2$TiCl$_2$.

Virtually all work on titanium-mediated radical polymerizations has focussed on titanocene dichloride and its derivatives, predominantly coming from Asandei and coworkers. The first report, in 2004, showed that a system involving **20**, Zn, and an epoxide was able to initiate and control the polymerization of styrene.[61] The system showed the typical characteristics of a controlled polymerization, with a linear dependence of molecular weight on monomer conversion and narrow dispersities (Đ = 1.1–1.3). The mechanism of control was proposed to occur in a number of stages (Figure 6.15):

1. *In situ* formation of the Ti(III) analog from **20** through reduction using zinc metal
2. Radical ring opening (RRO) of the epoxide, yielding a Ti-alkoxide radical species
3. Radical initiation of the polymerization, using the Ti-alkoxide radical
4. Reversible trapping of the propagating radical species using a second equivalent of **21**, through both reversible-deactivation and degenerative chain transfer

The best initiator was found to be the monosubstituted, oxygen-rich diepoxide 1,4-butanediol diglycidyl ether (BDGE). In addition to styrene, *para*-substituted styrenes were also polymerized and copolymerized with good control.

Asandei and coworkers optimized the polymerization by first exploring the ligand effects, showing that the titanocene framework was a better mediator than complexes based on alkoxides, bisketonates, scorpionates, or half-sandwiches. Bisketonates were shown to be ineffective mediators of styrene polymerizations,[62] as were scorpionates.[63] However alkoxides were shown to exert some degree of control over the polymerization, using $(^{i}PrO)_3TiCl$ (**22**) as the mediator, albeit with dispersities between 1.4 and 1.5.[62] Half-sandwich complexes, such as $CpTiCl_3$ (**23**) and Cp^*TiCl_3 (**24**), also demonstrated poorer performance than **20**.[63] Cp* (pentamethylcyclopentadiene) is more electron donating than Cp and thus gives a stronger C-Ti bond. The consequence is that temperature in excess of 110°C is required for *living* behavior.

In addition to the earlier, a range of substituted metallocenes were synthesized to investigate the effect of increased steric bulk on the Cp ring.[64] The work concluded that there is only a weak dependence of the polymerization on the nature of the substituent. This was attributed to a balance between the electronic and steric effects, as the size of the substituent increased. The effect of the halide was more pronounced, with **20** a better mediator than Cp_2TiBr_2 (**25**).

Other parameters were also investigated, including the effect of reducing agents, temperature, and reagent ratios.[65] Cu, Ni, and Cr were unable to reduce **20**, whereas Al and Fe gave free radical or poorly controlled polymerizations (Đ > 1.5). Zinc alloy, powder, or nanoparticles were the most effective at reducing **20** and achieving a well-controlled polymerization, with an optimum ratio of **20**:Zn = 1:2 and optimum temperature of 70°C–90°C. The effect of solvents and additives was also explored.[66] With respect to solubility of **21** and dispersity of the resulting polymer, dioxane was shown to be the best solvent. Finally the nature of the initiator was explored, with a range of peroxide and aldehyde initiators tested.[67] Two equivalents of Ti(III) are still required per equivalent of peroxide—one used in the redox initiation and one used in the reversible termination process. In addition, similar rates of polymerization to epoxide initiation show that initiation of the peroxide occurred through the redox reaction, rather than thermal decomposition. Benzoyl peroxide was shown to be the superior initiator of those tested.

To date, there has been only a little work to investigate the mechanism of this titanium-mediated polymerization. The mechanism proposed by Asandei and coworkers (Figure 6.15), while plausible, has yet to be conclusively demonstrated to be the mechanism of control. Grishin and coworkers investigated how **20** controlled the polymerization of styrene and methyl methacrylate, through the use of quantum chemical calculations and electron spin resonance (ESR) spectroscopy.[20] ESR measurements confirmed the presence of **21** during the polymerization of styrene and was suggested to mediate the polymerization through reversible binding to the propagating radical. Furthermore, the formation of **21** had no significant thermodynamic restriction. However, the lack of a complete mechanistic understanding warrants further study.

6.3.5 CHROMIUM

In the 1970s, Minoura and coworkers first used a chromium complex to control a radical polymerization, where [Cr(acac)$_2$] (**26**) and benzoyl peroxide were used to initiate a range of vinyl monomers.[68–71] The rate of polymerization for methyl methacrylate, methyl acrylate, acrylonitrile, and acrylamide was fast, with an increase in the molecular weight with respect to conversion observed for methyl acrylate and acrylonitrile. On the other hand, styrene, vinyl chloride, and vinyl acetate polymerized with a much slower rate, with a *living* polymerization only achieved below 30°C. Matyjaszewski further explored this system, using **26** and benzoyl peroxide with a range of polyamine ligands to polymerize vinyl acetate and methyl methacrylate.[72] Although dispersities as low as 1.5 were achieved, molecular weight did not increase linearly with conversion.

Further advances in using chromium for OMRP came in 2008, where Poli and coworkers explored the controlled polymerization of vinyl acetate using chromium β-diketiminate complexes (**27**, Figure 6.16).[73] The strength of the formed chromium-carbon bond is greatly dependent on the steric nature of the aryl substituents and thus can be finely tuned for optimum control.

The bulkiest substituent, R = 2,6-di*iso*propylphenyl, was used to mediate the polymerization of vinyl acetate with V-70 as thermal initiator. Conversion of 70% after 46 hours at 30°C was achieved, with dispersities increasing from 1.4 to 1.8 over the course of the reaction. Using an excess of initiator yielded an uncontrolled polymerization (Ð > 2.4); these complexes are unable to control the polymerization

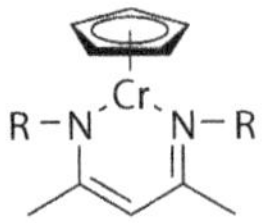

27

FIGURE 6.16 Chromium β-diketiminate complex.

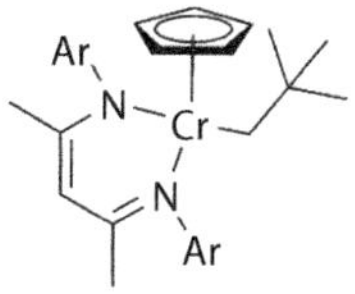

28

FIGURE 6.17 CpCrL(CH$_2$CMe$_3$), where L=2,6-dimethylphenyl β-diketiminato.

through an associative DT process due to the lack of a vacant coordination site on the Cr atom. Using less bulky substituents gave lower polymerization rates, as verified by DFT calculations. Bulky substituents yield longer and weak Cr-C bonds, favoring the growing radicals and thus increased polymerization rates.

Further work using the bulky 2,6-di*iso*propylphenyl substituent found that increasing the polymerization temperature to 90°C lowered the polymerization rate, with only 34% conversion after 84 hours.[74] A final dispersity of 1.2 was achieved, with molecular weights linearly dependant on conversion. Attempts were also made to develop a chromium(III) alkyl complex for use as a single-component OMRP reagent, which would behave both as the initiator and mediator. The complex CpCrL(CH$_2$CMe$_3$), where L=2,6-dimethylphenyl β-diketiminato (**28**, Figure 6.17), was synthesized and used in the room temperature polymerization of vinyl acetate. The molecular weight increased linearly with conversion, albeit to only 14% conversion after an excessively long 400 hours. The rate constant of polymerization also decreased with time, which points toward the partial deactivation of the growing chains. This resulted in a slightly broader dispersity of 1.46.

Similarly with iron amine-bis(phenolate) complexes, chromium β-diketiminate complexes are involved in both ATRP and OMRP equilibria, with an interplay between the two. This interplay has been explored using cyclopentadienyl half-sandwich complexes, finding that the OMRP equilibria dominate over ATRP.[75]

6.3.6 VANADIUM

In 2010, Shaver and coworkers first used a vanadium complex to mediate a controlled radical polymerization. Bis(imino)pyridine [BIMPY]VCl$_3$, where [BIMPY]=2,6-(ArN=CMe)$_2$C$_5$H$_3$N and Ar=2,6-^{i}Pr$_2$C$_6$H$_3$) (**29**, Figure 6.18), was used for the OMRP of vinyl acetate.[76] Using AIBN initiator and a reaction temperature of 120°C, dispersities as low as 1.3 were achieved, with molecular weights linearly increasing with conversion. Lower reaction temperatures yield polymers with broader dispersities and no dependence of conversion on molecular weight, which is attributed to inferior initiation and chain exchange. The lack of halogen end groups and the ability for the metal-terminated polymer to behave as a macroinitiator are further evidence for the dominance of an OMRP equilibrium and lack of an ATRP equilibrium.

The active catalyst was suggested to be [BIMPY]VCl$_2$ (**30**), a V(II) species formed *in situ* by reduction of the initial V(III) complex, through chlorine abstraction by AIBN-derived radicals. Further work complimented experimental studies with computational analysis of the potential reaction pathways.[77] For both styrene

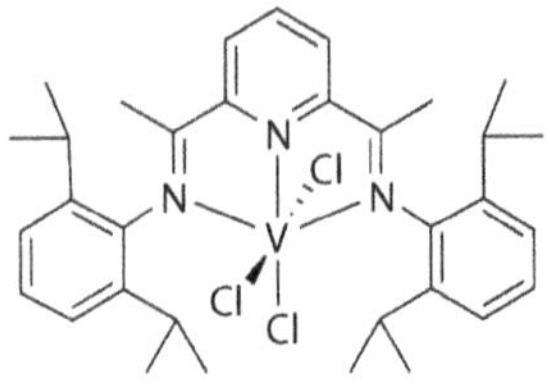

29

FIGURE 6.18　[BIMPY]VCl$_3$.

and vinyl acetate, halogen abstraction was irreversible, and so control through ATRP was not possible. Other monomers were also tried with the system under OMRP conditions. Although methyl methacrylate and acrylonitrile were poorly controlled, other vinyl esters, such as vinyl propionate and vinyl pivalate, were better controlled with dispersities around 1.4.

A range of [BIMPY]VCl$_3$ complexes were synthesized and screened for efficacy as mediators of the OMRP of vinyl acetate (Figure 6.19).[78] It was shown that complex **A**, the original [BIMPY]^{dipp}VCl$_3$ derivative, was the best mediator. Optimization, by changing the temperature and initiator, made little difference. In addition, challenges in catalyst death, potentially from trapping head-to-head radical addition products, were still persistent.

	R^1	R^2
A	Me	2,6-^{i}Pr$_2$C$_6$H$_3$
B	Me	Ph
C	Me	4-MeC$_6$H$_4$
D	Me	4-MeOC$_6$H$_4$
E	Me	4-FC$_6$H$_4$
F	Me	2,6-Me$_2$C$_6$H$_3$
G	Me	2,6-Et$_2$C$_6$H$_3$
H	Me	2,4,6-Me$_3$C$_6$H$_2$
I	Me	Cy
J	H	2,6-^{i}Pr$_2$C$_6$H$_3$
K	Ph	2,6-^{i}Pr$_2$C$_6$H$_3$
L	Et	2,6-^{i}Pr$_2$C$_6$H$_3$
M	^{i}Pr	2,6-^{i}Pr$_2$C$_6$H$_3$
N	Ph	Ph

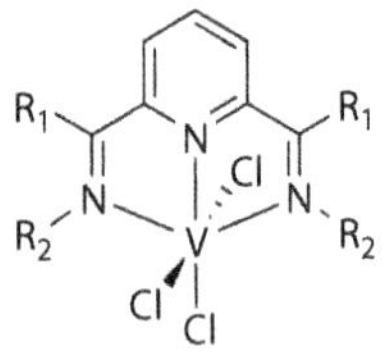

FIGURE 6.19　[BIMPY]VCl$_3$.

6.4 COPOLYMERS AND FUNCTIONAL POLYMERS

A major advantage of CRP is the ability to develop a range of new materials, with different properties and functionality. Although this has been extensively explored in ATRP, it remains underexplored for OMRP.

[Co(acac)$_2$]-controlled polymerizations of vinyl acetate have been extended to substitute the metal-capped polymer product with a reactive functionality, through the use of radical scavengers.[79] Thio compounds and nitroxides are able to displace the cobalt complex and replace it with a reactive group: 1-Propanethiol or TEMPO were used to terminate the polymer with a proton or nitroxide, respectively, yielding an almost colorless polymer without any increase in the dispersity. Furthermore, a nitroxide containing a functional group was also shown to efficiently displace the cobalt complex, while introducing a functionality to the polymer chain end.

A wide range of block copolymers incorporating poly(vinyl acetate) have been made. An early example synthesized cobalt-terminated poly(vinyl acetate) before replacing the metal with an α-bromide, thus producing a macroinitiator for the ATRP of other monomers, such as styrene, ethyl acrylate, and methyl methacrylate.[80] Nitroxides containing either an α-bromoester or an α-bromoketone were used to displace [Co(acac)$_2$] from the polymer, giving a colorless bromo-functionalized macroinitiator. Styrene and ethyl acrylate were then polymerized using CuBr and hexamethyltriethylenetetramine (HMTETA) in toluene, whereas methyl methacrylate was polymerized using CuCl, CuCl$_2$, and HMTETA in toluene. The block copolymer poly(vinyl acetate-block-poly(styrene) had a dispersity as low as 1.15, albeit with a conversion of only 24% for the styrene block. Poly(vinyl acetate)-block-poly(methyl methacrylate) gave an even lower dispersity of 1.10, with an improved conversion of 48% for the methyl methacrylate block.

Poly(vinyl acetate)-block-poly(styrene) was also synthesized using purely OMRP, with styrene added to the [Co(acac)$_2$]-PVAc macroinitiator without any modification or functionalization.[81] Using this method, dispersities increased from 1.20 to 1.70 after polymerization of the styrene block implying that the OMRP of styrene mediated by [Co(acac)$_2$] is poorly controlled. Styrene conversion was also limited to 30%–45%, suggesting irreversible termination of the polymer chains.

A more powerful copolymerization is that of vinyl acetate with 1-alkenes, such as octene and ethylene.[82] First poly(vinyl acetate)-co-poly(octene) was synthesized, with molecular weights linearly increasing with conversion and dispersities between 1.1 and 1.3. ^{1}H NMR of the polymer suggested a gradient sequence distribution. Using an isolated [Co(acac)$_2$]-PVAc macroinitiator gave a lower dispersity (1.1–1.2), although the conversion of octene was limited to 14%. This was rationalized by chain transfer to the 1-alkene, forming stable allylic radicals that were slow to reinitiate. This was also verified by an unsuccessful homopolymerization of 1-octene using [Co(acac)$_2$] and V-70. Poly(vinyl acetate)-co-poly(ethylene) was also synthesized using [Co(acac)$_2$] and V-70 at 30°C. Compared to octene the polymerization was fast, giving high molecular weights and a dispersity of 2.4. This poor control was due to irreversible termination, including chain transfer to the monomer and polymer. Using the [Co(acac)$_2$]-PVAc macroinitiator afforded better control, with a dispersity of 1.4, albeit with only a 10% incorporation of ethylene into the gradient copolymer.

More recent work looked at using cobalt to mediate the copolymerization of ethylene with polar monomers such as vinyl acetate, acrylonitrile, and N-methyl vinyl acetamide (NMVA).[83] The [Co(acac)$_2$]-PVAc macroinitiator was used with a reaction temperature of 40°C. Ethylene-vinyl acetate copolymers were made at a range of ethylene pressures from 10 bar to 50 bar. In all cases, dispersity was less than 1.2 except at very high conversion. ^{1}H NMR showed that the composition of polymer, and as a result the thermal properties, was dependent on the ethylene pressure. As the pressure was increased from 10 to 50 bar the ethylene content in the polymer increased from 13 to 54 mol%. NMR also suggested the copolymer had a random sequence structure with negligible branching. The composition of the ethylene-NMVA copolymer was also dependent on the ethylene pressure used. However this was not true for the ethylene-acrylonitrile copolymer, which contained fewer ethylene units. This was due to the reactivity ratio of ethylene-acrylonitrile supporting preferential acrylonitrile addition, whereas the reactivity ratios for ethylene-vinyl acetate and ethylene-NMVA are roughly the same. To emphasize the versatility of the system a *block-like* ethylene-vinyl acetate copolymer was made, with the first segment containing a larger ethylene content than the second block.[82]

The only method discussed thus far about synthesizing an alkylcobalt(III) macro-initiator is through the isolation of [Co(acac)$_2$]-PVAc, where PVAc is an oligomer of approximately four vinyl acetate units. A recent study by Detrembleur and cowork-ers was explored using [Co(acac)$_2$] to trap halomethyl radicals (XCH$_2$˙), which then acts as the initiating species.[84] This also has the added benefit of end-chain functional-ity, termed a telechelic polymer. The alkylcobalt(III) initiator, [Co(acac)$_2$-CH$_2$X], was synthesized using V-70 or AIBN, tris(trimethylsilyl)silane (TTMSS), and an excess of CH$_2$X$_2$ (X=Cl or Br). TTMSS is proposed to behave as a radical reducing agent that, in combination with V-70/AIBN, yields tris(trimethylsilyl) radicals ((Me$_3$Si)$_3$Si˙), which in turn abstract a halogen atom from CH$_2$X$_2$ giving XCH$_2$˙. This species is then trapped by [Co(acac)$_2$]. The presence of TTMSS is crucial for the synthesis, and no product is attained in its absence. [Co(acac)$_2$-CH$_2$X] was isolated and fully characterized by X-ray crystallography, after addition of pyridine to successfully obtain crystals. [Co(acac)$_2$-CH$_2$Br] was used to initiate and control the polymerization of vinyl acetate at 40°C. No induction period was observed and molecular weights were increased with conversion. Dispersities were low (Đ < 1.2) at low conversion but increased as molecular weights reached 100,000 g mol^{-1}. To highlight the new functionality afforded by the α-chain end halomethyl group, novel α-azido functional poly(vinyl acetate) was made using the copper-catalyzed azide-alkyne cycloaddition (CuAAC) reaction. This method paves the way for new end-functional poly(vinyl acetate) to be developed (Figure 6.20).

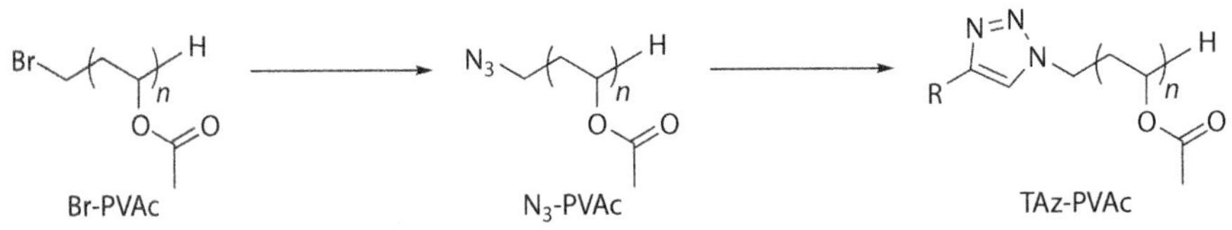

FIGURE 6.20   Synthesis of α-azido poly(vinyl acetate).

6.4 CONCLUSION

This chapter presents an overview of OMRP, including the fundamental mechanisms involved in synthesizing well-defined polymers. A selection of work that highlights the tremendous and complementary capability of OMRP is also reviewed.

A wide variety of metal complexes have been used for OMRP including rhodium, cobalt, iron, titanium, chromium, and vanadium, with a myriad of monomers now well controlled as a result. Despite this, there are still several avenues of potential new research in the field. Although some metals, in particular cobalt, have been studied in considerable detail, most are relatively unexplored. In addition, few ligand frameworks have been used and innovative ligand design will yield novel complexes. The above-mentioned two factors, in combination with an increased understanding of metal-carbon bond strengths in the systems, will further enhance the applicability of OMRP.

CRP has opened doors to synthesize a range of materials that were simply impossible to achieve with conventional free radical polymerization. The notion of a well-defined block copolymer before techniques such as ATRP and OMRP were discovered was unthinkable. Although this chapter has covered a selection of the co- and functionalized polymers synthesized through OMRP, the vast majority of work to date has focussed on homopolymers. There is a considerable potential to expand the monomer scope, the monomers used within copolymers to synthesize new materials, and the architectures adopted by said polymers, in line with other CRP techniques.

REFERENCES

1. Poli, R. 2011. Radical coordination chemistry and its relevance to metal-mediated radical polymerization. *Eur. J. Inorg. Chem.* 2011: 1513–1530.
2. Poli, R. 2006. Relationship between one-electron transition-metal reactivity and radical polymerization processes. *Angew. Chem. Int. Ed.* 45: 5058–5070.
3. Lake, B. R. M., Shaver, M. P. 2015. The interplay of ATRP, OMRP and CCT in iron-mediated controlled radical polymerization. In *Controlled Radical Polymerization: Mechanisms*, Matyjaszewski, K., Sumerlin, B.S., Tarevsky, N.V., Chiefari, J., eds. Washington, DC: American Chemical Society, pp. 311–326.
4. Poli, R., Allan, L. E. N., Shaver, M. P. 2014. Iron-mediated reversible deactivation controlled radical polymerization. *Prog. Polym. Sci.* 39: 1827–1845.
5. Braunecker, W. A., Matyjaszewski, K. 2007. Controlled/living radical polymerization: Features, developments and perspectives. *Prog. Polym. Sci.* 32: 93–146.
6. di Lena, F., Matyjaszewski, K. 2010. Transition metal catalysts for controlled radical polymerization. *Prog. Polym. Sci.* 35: 959–1021.
7. Debuigne, A., Poli, R., Jérôme, C., Jérôme, R., Detrembleur, C. 2009. Overview of cobalt-mediated radical polymerization: Roots, state of the art and future prospectus. *Prog. Polym. Sci.* 34: 211–239.
8. Smith, K. M., McNeil, W. S., Abd-El-Aziz, A. S. 2010. Organometallic-mediated radical polymerization: Developing well-defined complexes for reversible transition metal-alkyl bond homolysis. *Macromol. Chem. Phys.* 211: 10–16.
9. Allan, L. E. N., Perry, M. R., Shaver, M. P. 2012. Organometallic mediated radical polymerization. *Prog. Polym. Sci.* 37: 127–156.

10. Poli, R. 2012. Organometallic-mediated radical polymerization. In *Polymer Science: A Comprehensive Reference*. Moeller, M., Matyjaszewski, K., eds. Amsterdam, the Netherlands: Elsevier.

11. Poli, R. 2015. New phenomena in organometallic-mediated radical polymerization (OMRP) and perspectives for control of less active monomers. *Chem. Eur. J.* 21: 6988–7001.

12. Gridnev, A. A., Ittel, S. D. 2001. Catalytic chain transfer in free-radical polymerization. *Chem. Rev.* 101: 3611–3659.

13. Schröder, K., Konkolewicz, D., Poli, R., Matyjaszewski, K. 2012. Formation and possible reactions of organometallic intermediates with active Cu(I) catalysts in ATRP. *Organometallics* 31: 7994–7999.

14. Wang, Y., Soerensen, N., Zhong, M., Schroeder, H., Buback, M., Matyjaszewski, K. 2013. Improving the "livingness" of ATRP by reducing Cu catalyst concentration. *Macromolecules* 46: 683–691.

15. Schroeder, H., Buback, M. 2014. SP-PLP-EPR measurement of iron-mediated radical termination in ATRP. *Macromolecules* 47: 6645–6651.

16. Wahidur Rahaman, S. M., Matyjaszewski, K., Poli, R. 2016. Cobalt(III) and copper(II) hydrides at the crossroad of catalysed chain transfer and catalysed radical termination: A DFT study. *Polym. Chem.* 7: 1079–1087.

17. Le Grognec, E., Claverie, J., Poli, R. 2001. Radical polymerization of styrene controlled by half-sandwich Mo(III)/Mo(IV) couples: All basic mechanisms are possible. *J. Am. Chem. Soc.* 123: 9513–9524.

18. Stoffelbach, F., Poli, R., Richard, P. 2002. Half-sandwich molybdenum(III) compounds containing diazadiene ligands and their use in the controlled radical polymerization of styrene. *J. Organomet. Chem.* 663: 269–276.

19. Stoffelbach, F., Poli, R., Maria, S., Richard, P. 2007. How the interplay of different control mechanisms affects the initiator efficiency factor in controlled radical polymerization: An investigation using organometallic MoIII -based catalysts. *J. Organomet. Chem.* 692: 3133–3143.

20. Grishin, D. F., Ignatov, S. K., Shchepalov, A. A., Razuvaev, A. G. 2004. Mechanism of controlled radical polymerization of styrene and methyl methacrylate in the presence of dicyclopentadienyl titanium dichloride. *Appl. Organomet. Chem.* 18: 271–276.

21. Braunecker, W. A., Brown, W. C., Morelli, B. C., Tang, W., Poli, R., Matyjaszewski, K. 2007. Origin of activity in Cu-, Ru-, and Os-mediated radical polymerization. *Macromolecules* 40: 8576–8585.

22. Gibson, V. C., O'Reilly, R. K., Reed, W., Wass, D. F., White, A. J. P., Williams, D. J. 2002. Four coordinate iron complexes bearing α-diimine ligands: Efficient catalysts for atom transfer radical polymerisation (ATRP). *Chem. Commun.* 2: 1850–1851.

23. Shaver, M. P., Allan, L. E. N., Rzepa, H. S., Gibson, V. C. 2006. Correlation of metal spin states with catalytic: Polymerizations mediated by α-diimine-iron complexes. *Angew. Chem. Int. Ed.* 45: 1241–1244.

24. Allan, L. E. N., Shaver, M. P., White, A. J. P., Gibson, V. C. 2007. Correlation of metal spin-state in α-diimine iron catalyst with polymerization mechanism. *Inorg. Chem.* 46: 8963–8970.

25. Shaver, M. P., Allan, L. E. N., Gibson, V. C. 2007. Organometallic intermediates in the controlled radical polymerization of styrene by α-diimine iron catalyst. *Organometallics* 26: 4725–4730.

26. Johansson, M. P., Swart, M. 2011. Subttle effects control the polymerisation mechanism in α-diimine iron catalysts. *Dalton Trans.* 40: 8419–8428.

27. Poli, R., Shaver, M. P. 2014. ATRP/OMRP/CCT interplay in styrene polymerization mediated by iron(II) complexes: A DFT study of the α-diimine system. *Chem. Eur. J.* 20: 17530–17540.

28. Allan, L. E. N., MacDonald, J. P., Reckling, A. M., Kozak, C. M., Shaver, M. P. 2012. Controlled radical polymerization mediated by amine-bis(phenolate) iron(III) complexes. *Macromol. Rapid Commun.* 33: 414–418.

29. Allan, L. E. N., Macdonald, J. P., Nichol, G. S., Shaver, M. P. 2014. Single component iron catalysts for atom transfer and organometallic mediated radical polymerization: Mechanistic studies and reaction scope. *Macromolecules* 47: 1249–1257.

30. Poli, R., Shaver, M. P. 2014. Atom transfer radical polymerization (ATRP) and organometallic mediated radical polymerization (OMRP) of styrene mediated by diaminobis(phenolato)iron(II) complexes: A DFT study. *Inorg. Chem.* 53: 7580–7590.

31. Schroeder, H., Lake, B. R. M., Demeshko, S., Shaver, M. P., Buback, M. 2015. A synthetic and multispectroscopic speciation analysis of controlled radical polymerization mediated by amine-bis(phenolate)iron complexes. *Macromolecules* 48: 4329–4338.

32. Buback, M., Schroeder, H., Kattner, H. 2016. Detailed kinetic and mechanistic insight into radical polymerization by spectroscopic techniques. *Macromolecules* 49: 3193–3213.

33. Lake, B. R. M., Shaver, M. P. 2016. Iron(II) β-ketiminate complexes as mediators of controlled radical polymerisation. *Dalton Trans.* 45: 15840–15849.

34. Wayland, B. B., Poszmik, G., Fryd, M. 1992. Metalloradical reactions of rhodium(II) porphyrins with acrylates: Reduction, coupling, and photopromoted polymerization. *Organometallics* 11(11): 3534–3542.

35. Wayland, B. B., Poszmik, G., Mukerjee, S. L., Fryd, M. 1994. Living radical polymerization of acrylates by organocobalt phorphyrin complexes. *J. Am. Chem. Soc.* 116: 7943–7944.

36. Lu, Z., Fryd, M., Wayland, B. B. 2004. New life for living radical polymerization mediated by cobalt(II) metalloradicals. *Macromolecules* 37: 2686–2687.

37. Peng, C.-H., Fryd, M., Wayland, B. B. 2007. Organocobalt mediated radical polymerization of acrylic acid in water. *Macromolecules* 40: 6814–6819.

38. Debuigne, A., Caille, J.-R., Jérôme, R. 2005. Highly efficient cobalt-mediated radical polymerization of vinyl acetate. *Angew. Chem. Int. Ed.* 44: 1101–1104.

39. Debuigne, A., Caille, J.-R., Detrembleur, C., Jérôme, R. 2005. Effective cobalt mediation of the radical polymerization of vinyl acetate in suspension. *Angew. Chem. Int. Ed.* 44: 3439–3442.

40. Detrembleur, C., Debuigne, A., Bryaskova, R., Charleux, B., Jérôme, R. 2006. Cobalt mediated radical polymerization of vinyl acetate in miniemulsion: Very fast formation of stable poly(vinyl acetate) latexes at low temperature. *Macromol. Rapid Commun.* 27: 37–41.

41. Bryaskova, R., Detrembleur, C., Debuigne, A., Jérôme, R. 2006. Cobalt-mediated radical polymerization (CMRP) of vinyl acetate initiated by redox systems: Toward the scale-up of CMRP. *Macromolecules* 39: 8263–8268.

42. Maria, S., Kaneyoshi, H., Matyjaszewski, K., Poli, R. 2007. Effect of electron donors on the radical polymerization of vinyl acetate mediated by [Co(acac)2]: Degenerative transfer versus reversible homolytic cleavage of an organocobalt(III) complex. *Chem. Eur. J.* 13: 2480–2492.

43. Debuigne, A., Champouret, Y., Jérôme, R., Poli, R., Detrembleur, C. 2008. Mechanistic insights into the cobalt-mediated radical polymerization (CMRP) of vinyl acetate with cobalt(III) adducts as initiators. *Chem. Eur. J.* 14: 4046–4059.

44. Debuigne, A., Michaux, C., Jérôme, C., Jérôme, R., Poli, R., Detrembleur, C. 2008. Cobalt-mediated radical polymerization of acrylonitirle: Kinetics investigations and DFT calculations. *Chem. Eur. J.* 14: 7623–7637.

45. Hurtgen, M., Debuigne, A., Jérôme, C., Detrembleur, C. 2010. Solving the problem of bis(acetylacetonato)cobalt(II)-mediated radical polymerization (CMRP) of acrylic esters. *Macromolecules* 43: 886–894.

46. Debuigne, A., Morin, A. N., Kermagoret, A., Piette, Y., Detrembleur, C., Jérôme, C., Poli, R. 2012. Key role of intramolecular metal chelation and hydrogen bonding in the cobalt-mediated radical polymerization of N-vinyl amides. *Chem. Eur. J.* 18: 12834–12844.

47. Morin, A. N., Detrembleur, C., Jérôme, C., De Tullio, P., Poli, R., Debuigne, A. 2013. Effect of head-to-head addition in vinyl acetate controlled radical polymerization: Why is Co(acac)2-mediated polymerization so much better? *Macromolecules* 46: 4303–4312.

48. Cordella, D., Kermagoret, A., Debuigne, A., Riva, R., German, I., Isik, M., Jérôme, C., Mecerreyes, D., Taton, D., Detrembleur, C. 2014. Direct route to well-defined poly(ionic liquid)s by controlled radical polymerization in water. *ACS Macro Lett.* 3: 1276–1280.

49. Santhosh, K. K. S., Gnanou, Y., Champouret, Y., Daran, J.-C., Poli, R. 2009. Radical polymerization of vinyl acetate with bis(tetramethylheptadionato)cobalt(II): Coexistance of three different mechanisms. *Chem. Eur. J.* 15: 4874–4885.

50. Langlotz, B. K., Fillol, J. L., Gross, J. H., Wadepohl, H., Gade, L. II. 2008. Living radical polymerization of acrylates mediated by 1, 3-bis(2-pyridylimino) isoindolatocobalt(II) complexes: Monitoring the chain growth at the metal. *Chem. Eur. J.* 14: 10267–10279.

51. Santhosh-Kumar, K. S., Li, Y., Gnanou, Y., Baisch, U., Champouret, Y., Poli, R., Robson, K. C. D., McNeil, W. S. 2009. Electronic and steric ligand effects in the radical polymreization of vinyl acetate mediate by β-ketoiminate complexes of cobalt(II). *Chem. Asian J.* 4: 1257–1265.

52. Sherwood, R. K., Kent, C. L., Patrick, B. O., McNeil, W. S. 2010. Controlled radical polymerization of methyl acrylate initiated by a well-defined cobalt alkyl complex. *Chem. Commun.* 46: 2456–2458.

53. Gibson, V. C., O'Reilly, R. K., Wass, D. F., White, A. J. P., Williams, D. J. 2003. Polymerization of methyl methacrylate using four-coordinate (α-diimine)iron catalysts: Atom transfer radical polymerization vs catalytic chain transfer. *Macromolecules* 36: 2591–2593.

54. Chowdhury, R. R., Crane, A. K., Fowler, C., Kwong, P., Kozak, C. M. 2008. Iron(III) amine-bis(phenolate) complexes as catalysts for the coupling of alkyl halides with aryl Grinard reagents. *Chem. Commun.* 1: 94–96.

55. Zhu, K., Shaver, M. P., Thomas, S. P. 2015. Stable and easily handled FeIII catalyst for hydrosilylation of ketones and aldehydes. *Eur. J.Org. Chem.* 2015: 2119–2123.

56. Zhu, K., Shaver, M. P., Thomas, S. P. 2016. Chemoselective nitro reduction and hydroamination using a single iron catalyst. *Chem. Sci.* 7: 3031–3035.

57. Zhu, K., Shaver, M. P., Thomas, S. P. 2016. Amine-bis(phenolate)iron(III)-catalyzed formal hydroamination of olefins. *Chem. Asian J.* 11: 977–980.

58. Xue, Z., Poli, R. 2013. Organometallic mediated radical polymerization of vinyl acetate with Fe(acac)2. *J. Polym. Sci. Part A: Polym. Chem.* 51: 3494–3504.

59. Wang, J., Zhou, J., Sharif, H. S. E. M., He, D., Ye, Y. S., Xue, Z., Xie, X. 2015. Living radical polymerization of vinyl acetate mediated by iron(III) acetylacetonate in the presence of a reducing agent. *RSC Adv.* 5: 96345–96352.

60. Grishin, D. F., Semyonycheva, L. L., Telegina, E. V., Smirnov, A. S., Nevodchikov, V. I. 2003. Dicyclopentadienyl complexes of titanium, niobium, and tungsten in the controlled synthesis of poly(methyl methacrylate). *Russ. Chem. Bull.* 52: 505–507.

61. Asandei, A. D., Moran, I. W. 2004. TiCp2Cl catalyzed living radical polymerization of styrene initiated by oxirane radical ring opening. *J. Am. Chem. Soc.* 126: 15932–15933.

62. Asandei, A. D., Moran, I. W. 2005. The ligand effect in Ti-mediated living radical styrene polymerizations initiated by epoxide radical ring opening. 1. Alkoxide and bisketonate Ti complexes. *J. Polym. Sci. Part A: Polym. Chem.* 43: 6028–6038.

63. Asandei, A. D., Moran, I. W. 2005. The ligand effect in Ti-mediated living radical styrene polymerizations initiated by epoxide radical ring opening. 2. Scorpionate and half-sandwich LTiCl3 complexes. *J. Polym. Sci. Part A: Polym. Chem.* 43: 6039–6047.

64. Asandei, A. D., Moran, I. W. 2006. The ligand effect in Ti-mediated living radical styrene polymerizations initiated by epoxide radical ring opening. III. substituted sandwich metallocenes. *J. Polym. Sci. Part A: Polym. Chem.* 44: 1060–1070.

65. Asandei, A. D., Moran, I. W., Saha, G., Chen, Y. 2006. Titanium-mediated living radical styrene polymerizations. VI. Cp2TiCl-catalyzed initiation by epoxide radical ring opening: Effect of the reducing agents, temperature, and titanium/epoxide and titanium/zinc ratios. *J. Polym. Sci. Part A: Polym. Chem.* 44: 2156–2165.

66. Asandei, A. D., Moran, I. W., Saha, G., Chen, Y. 2006. Titanium-mediated living radical styrene polymerizations. V. Cp2TiCl-catalyzed initiation by epoxide radical ring opening: Effect of solvents and additives. *J. Polym. Sci. Part A: Polym. Chem.* 44: 2015–2026.

67. Asandei, A. D., Saha, G. 2006. Cp2TiCl-catalyzed living radical polymerization of styrene initiated from peroxides. *J. Polym. Sci. Part A: Polym. Chem.* 44: 1106–1116.

68. Lee, M., Minoura, Y. 1978. Polymerization of vinyl monomers initiated by cromium(II) accetate + organic peroxides. *J. Chem. Soc., Faraday Trans.1* 74: 1726–1737.

69. Lee, M., Morigami, T., Minoura, Y. 1978. "Living" radical polymerization of vinyl monomers initiated by aged "Cr+2 + BPO" in homogeneous solution. *J. Chem. Soc. Faraday Trans. 1* 74: 1738–1749.

70. Lee, M., Utsumi, K., Minoura, Y. 1979. Effect of hexamethylphosphoric triamide on the living radical polymerization initiated by aged chromium ion + BPO. *J. Chem. Soc. Faraday Trans. 1* 75: 1821–1829.

71. Lee, M., Ishida, Y., Minoura, Y. 1982. Effects of ligands on the "living" radical polymerization initiated by the aged Cr2+ plus benzoyl peroxide system. *J. Polym. Sci. Part A: Polym. Chem.* 20: 457–465.

72. Mardare, D., Gaynor, S. G., Matyjaszewski, K. 1994. Radical polymerization of vinyl acetate and methyl methacrylate using organo-chromium initiators complexed with macrocyclic polyamines. *Polym. Prepr. (Am. Chem. Soc. Div. Polym. Chem.)* 35: 700–701.

73. Champouret, Y., Baisch, U., Poli R., Tang, L., Conway, J. L., Smith, K. M. 2008. Homolytic bond strengths and formation rates in half-sandwich chromium alkyl complexes: Relevance for controlled radical polymerization. *Angew. Chem. Int. Ed.* 47: 6069–6072.

74. Champouret, Y., Macleod, K. C., Baisch, U., Patrick, B. O., Smith, K. M., Poli, R. 2010. Cyclopentadienyl chromium β-diketiminate complexes: Initiators, ligand steric effects, and deactivation processes in the controlled radical polymerization of vinyl acetate. *Organometallics* 29: 167–176.

75. Champouret, Y., Macleod, K. C., Smith, K. M., Patrick, B. O., Poli, R. 2010. Controlled radical polymerization of vinyl acetate with cyclopentadienyl chromium β-diketiminate complexes: ATRP vs OMRP. *Organometallics* 29: 3125–3132.

76. Shaver, M. P., Hanhan, M. E., Jones, M. R. 2010. Controlled radical polymerization of vinyl acetate mediated by a vanadium complex. *Chem. Commun.* 46: 2127–2129.

77. Allan, L. E. N., Cross, E. D., Francis-Pranger, T. W., Hanhan, M. E., Jones, M. R., Pearson, J. K., Perry, M. R., Storr, T., Shaver, M. P. 2011. Controlled radical polymerization of vinyl acetate mediated by a bis(imino)pyridine vanadium complex. *Macromolecules* 44: 4072–4081.

78. Perry, M. R., Allan, L. E. N., Decken, A., Shaver, M. P. 2013. Organometallic mediated radical polymerization of vinyl acetate using bis(imino)pyridine vanadium tricholoride complexes. *Dalton Trans.* 42: 9157–9165.

79. Debuigne, A., Caille, J.-R., Jérôme, R. 2005. Synthesis of end-funtional polyvinyl acetate by cobalt-mediated radical polymerization. *Macromolecules* 38: 5452–5458.
80. Debuigne, A., Caille, J.-R., Willet, N., Jérôme, R. 2005. Synthesis of poly (vinyl acetate) and poly(vinyl alchol) containing blcok copolymers by combination of cobalt mediated radical polymerization and ATRP. *Macromolecules* 38: 9488–9496.
81. Bryaskova, R., Willet, N., Debuigne, A., Jérôme, R., Detrembleur, C. 2007. Synthesis of poly(vinyl acetate)-b-polystyrene copolymers by cobalt-mediated radical polymerization. *J. Polym. Sci. Part A: Polym. Chem.* 45: 81–89.
82. Bryaskova, R., Willet, N., Degée, P., Dubois, P., Jérôme, R., Detrembleur, C. 2007. Copolymerization of vinyl acetate with 1-octene and ethylene by cobalt-mediated radical polymerization. *J. Polym. Sci. Part A: Polym. Chem.* 45: 2532–2542.
83. Kermagoret, A., Debuigne, A., Jérôme, C., Detrembleur, C. 2014. Precision design of ethylene- and polar-monomer-based copolymer by organometallic-mediated radical polymerization. *Nat. Chem.* 6: 179–187.
84. Demarteau, J., Kermagoret, A., German, I., Cordella, D., Robeyns, K., De Winter, J., Gerbaux, P., Jérôme, C., Debuigne, A., Detrembleur, C. 2015. Halomethyl-cobalt(bis-acetylacetonate) for the controlled synthesis of functional polymers. *Chem. Commun.* 51: 14334–14337.

7 Metal-Catalyzed Condensation Polymerization

Ashootosh V. Ambade

CONTENTS

Conjugated polymers are semiconducting materials and an integral component of solar cells and other optoelectronic gadgets, which are important as environment-friendly devices. Conjugated polymers are less expensive than inorganic materials and are easy to process, hence several synthetic methods have been developed based on transition metal-catalyzed polycondensation involving sp^2-sp^2 C–C bond formation for their production. The coupling reaction between monomers involves the following steps: (a) metal-catalyzed oxidative addition (as presented in Chapter 1) across the C–X bond of the monomer; (b) transmetallation with a main group organometallic compound; and (c) reductive elimination (as presented in Chapter 1) yielding the C–C bond formation and regeneration of the active catalyst, which are typical in a C–C bond formation catalyzed by organometallic complex (Figure 7.1).[1] The reactions are characterized by mild conditions, tolerance to many functional groups, and possibility of synthesizing regioregular polymers.

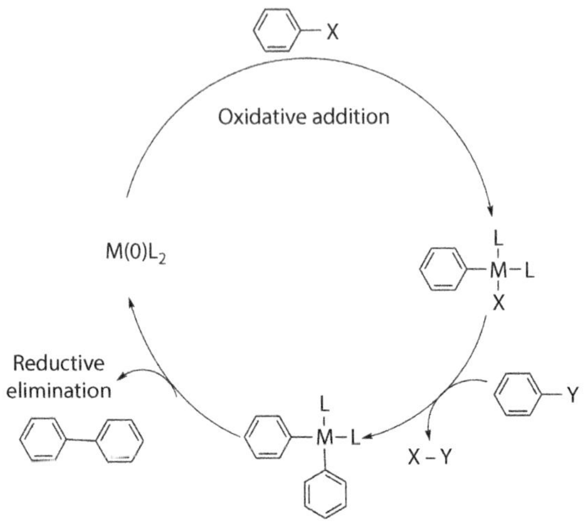

FIGURE 7.1   General mechanism of metal-catalyzed C–C bond formation.

7.1   SUZUKI–MIYAURA COUPLING

Suzuki–Miyaura reaction involves coupling of an organoboron reagent with a halide or pseudohalide in the presence of palladium or nickel catalyst and a base.[2] Organoboron reagents are inert to water and oxygen, are thermally stable, and are readily available. The reaction requires mild conditions and is tolerant to various functional groups. It is one of the most widely used methods for conjugated polymer synthesis and follows a step-growth mechanism. Conjugated polymers containing various types of monomers have been synthesized by Suzuki–Miyaura coupling. Typical conditions involve [Pd(PPh$_3$)$_4$] as catalyst, K$_2$CO$_3$ as base, and a phase-transfer catalyst in a solvent mixture (toluene/degassed water), and polymers with high molecular weights and low polydispersity index (PDI) are obtained.[1]

Polyfluorene-based conjugated polymers are important candidates for applications in optoelectronic devices due to their high quantum yields and high stability. Fluorene unit is copolymerized with an electron acceptor unit to decrease the band gap of the conjugated polymer so as to make it amenable for photo absorption in solar cells. Considering requirements for reactivity the 9,9-dialkylfluorene unit is functionalized at 2 and 7 positions with boronic esters, and dibromide of another aromatic unit is used.

An alternating copolymer comprising 9,9-dihexylfluorene and the 1,1-dimethyl-3,4-diphenyl-2,5-bis(2′-thienyl)-silole unit has been synthesized by Suzuki coupling (Figure 7.2).[3] 2,7-Linked carbazole is used as a donor unit along with acceptor units to obtain low band gap-conjugated polymers. An alternating copolymer of carbazole and 4,7-dithien-5-yl-2,1,3-benzodiathiazole was synthesized using [Pd$_2$(dba)$_3$/P(o-tol)$_3$] as a catalyst system to achieve a band gap of 1.88 eV, which is one of the lowest values for carbazole polymers. Long alkyl chains were attached to the nitrogen of carbazole to improve the solubility.[4]

Indolo[3,2-b]carbazole is used as a strong donor unit in donor-acceptor conjugated polymers. It is a large coplanar π-conjugated system with arylamine structure and

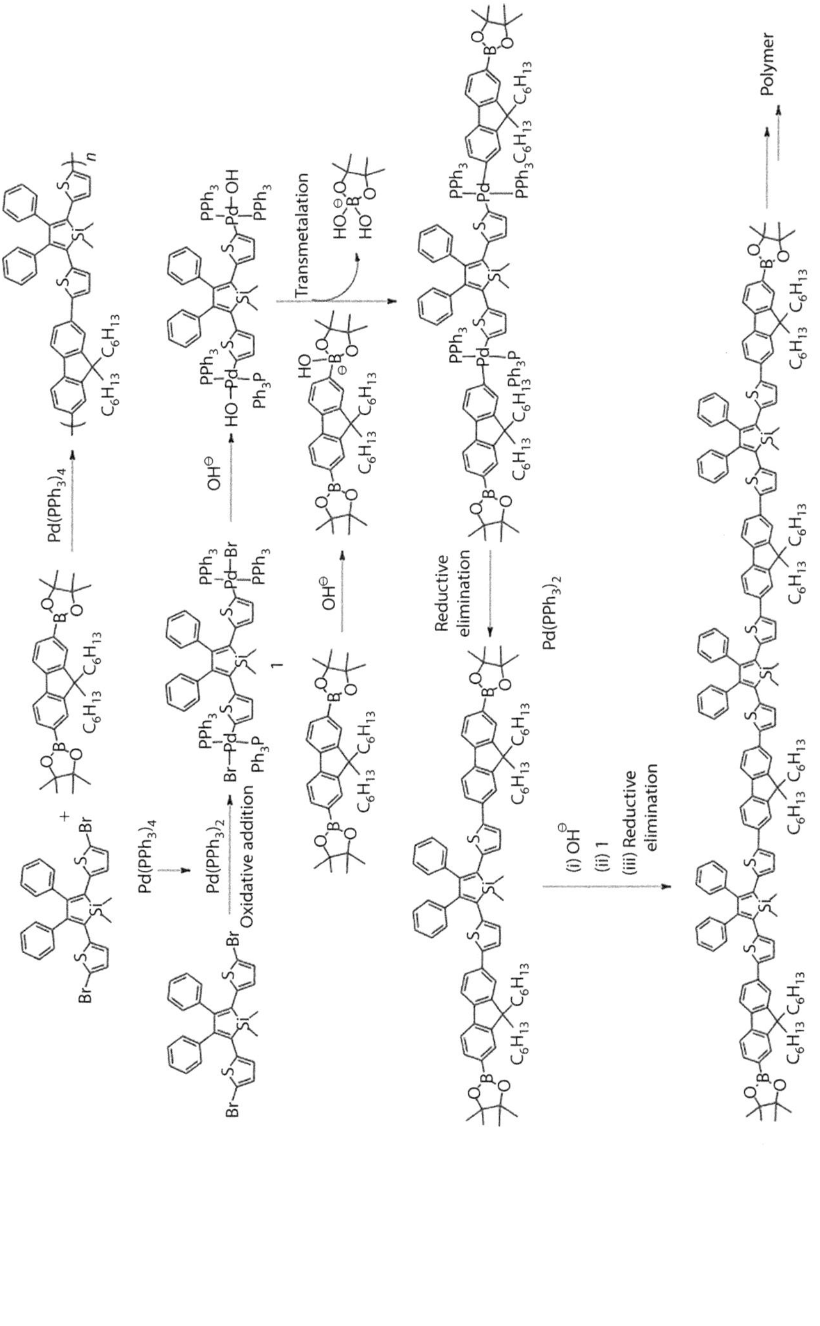

FIGURE 7.2 Suzuki coupling polymerization and the possible mechanism involved.

low-lying highest occupied molecular orbital (HOMO) that makes it a good hole transporter. A copolymer of indolocarbazole with pinacol boronate ester units at 3,9 positions and electron-poor benzothiadiazole oligothiophene was synthesized.[5] Poly(3-hexylthiophene) (P3HT), the most widely studied conjugated polymer, has been shown to possess improved optoelectronic properties in the regioregular form than in the regioirregular/regiorandom form. This is typically synthesized by the Grignard metathesis (GRIM) polymerization (see Section 7.7). Recently, it has been synthesized by Suzuki–Miyaura polymerization reaction also. To achieve regioregularity by this route, N-methyliminodiacetic acid (MIDA) boronate ester thienyl monomer is required. This yields the corresponding active boronic acid upon slow hydrolysis in low concentration that gives P3HT with >98% head-to-tail couplings in >94% yield in a palladium-catalyzed reaction. Corresponding pinacol boronate ester gave a low molecular weight polymer in low yield under identical conditions.[6] Conjugated polymers comprising heterocycles of main group elements such as five-membered chalcogenophene rings based on S, Se, and Te are inaccessible by conventional routes however could be synthesized using Suzuki–Miyaura coupling. Regioregular hybrid thiophene-selenophene-tellurophene and selenophene-fluorene copolymers were also synthesized. The reaction involved replacement of Zr from the metallacycle boronate ester by S, Se, or Te followed by cross-coupling polymerization.[7]

Polymer micro/nanoparticles in an emulsion are useful for coating applications. Stable emulsions of spherical and rod-like conjugated polymer nanoparticles were synthesized via Suzuki–Miyaura polymerization. Boronate ester of 9,9-dioctylfluorene was copolymerized with different dibromoarene monomers such as benzothiadiazole, 4-sec-butylphenyldiphenylamine and bithiophene in xylene/water mixture using a nonionic surfactant in the presence of [(N-heterocyclic carbene)PdCl$_2$(triethylamine)] and tetraethylammonium hydroxide, at room temperature.[8] Hyperbranched poly(m-phenylene)s was synthesized by polycondensation of a m-terphenyl-derived branched AB$_2$ monomer containing a boronate ester and two triflate groups using [Pd(OAc)$_2$] and S-Phos catalyst system.[9] Molecular weight and polydispersity could be controlled by varying the catalyst loading and the monomer concentration. A pseudo-chain-growth pathway involving intramolecular catalyst transfer was proposed. The hyperbranched polymer has triflate end groups on the periphery that can be converted to other functionalities after the monomer is fully consumed using Suzuki coupling.

Limitations of Suzuki–Miyaura coupling polymerization are: (a) the requirement for basic conditions may prevent use of several monomers that are susceptible to basic conditions. This may be circumvented by using complex protection–deprotection approach. (b) Need for a two-phase system may not provide polymers with high molecular weight and in good yields due to decreasing solubility. (c) Water is necessary for Suzuki–Miyaura coupling, and side reactions such as deboronation and dehalogenation occur for certain substrates.

7.2   STILLE COUPLING

Thiophene-containing polymers have emerged as the frontrunners among conjugated polymers. Stille reaction involves palladium-catalyzed coupling of organostannanes with aryl halides, triflates, and acyl chlorides.[10] Stannyl groups attached to the benzene ring of monomer were shown to give poor reactivity with aryl halides under Stille coupling conditions.[11] However, this coupling gives excellent yields in preparation of polythiophenes and hence is the method of choice for their synthesis.[12] It is a stereospecific and regioselective reaction. Organotin compounds can be prepared without the need for protection–deprotection chemistry. They are less oxygen- and moisture-sensitive than other organometallic species such as Grignard reagents and organolithium reagents.

The accepted mechanism of Stille reaction is similar to the general mechanism described earlier involving Pd(0) species as the active catalyst formed due to reduction of Pd(II) catalysts by the organostannane monomers (Figure 7.3). Transmetalation of Pd(0) species with the organostannane even in the presence of excess of stannane has been demonstrated to be slow and rate-determining step. This is followed by *trans* to *cis* isomerization, a fast step that leads to the product via reductive elimination of the Pd(0) catalyst. The actual process is thought to be complex and alternate pathways exist depending on reaction conditions such as solvents, ligands, and additives.[13] Solvent is an important parameter since apart from the catalyst and monomer, solubility of the growing polymer chain is of concern for successful polymerization. Reactivity of the two monomers is another factor; typically electron-rich organotin compounds and electron-deficient halide or triflate are used to obtain high molecular weight polymers. Therefore, in contrast to small molecule synthesis, optimization of reaction conditions is time-consuming for polymerization reactions using Stille coupling.

Functional group tolerance of Stille coupling is highly advantageous for synthesis of functional polymers. Selected examples are discussed here. Polythiophenes containing nitro, amine, ketone, and quinone groups were synthesized using Pd(0) and CuI precatalyst system.[14] Polythiophenes containing rhenium complexes were synthesized to obtain polymers with low band gap for solar cell application.[15] Photorefractive polymers comprising nonlinear optical (NLO) chromophore attached to conjugated poly(p-phenylene–thiophene)s were synthesized using Stille reaction since NLO chromophores are labile under many other conditions.[16] Influence of reaction conditions such as solvent and catalyst on synthesis of polythiophenes under microwave heating was studied.[17] Higher molecular weights ($M_n > 15,000$) and a lower polydispersity (PDI $\approx$ 2) were obtained using [Pd$_2$(dba)$_3$/P(o-tol)$_3$] as catalyst, LiCl as additive, and chlorobenzene as solvent than when conventional heating was used with same catalyst and solvent. Chlorobenzene gave higher molecular weights than those obtained using tetrahydrofuran (THF) due to better solubility of the polymer. Alternating copolymers of perylene diimide and dithienothiophene were synthesized for electron-transporting layer in organic field effect transistors (OFET) using [Pd(PPh$_3$)$_4$] as catalyst in toluene at 90°C.[18] Small molecules and oligomers have also been explored for photovoltaic applications due to well-defined structure, no batch-to-batch variation, and ease of purification compared to polymers. However, oligomers require multistep repetitive and tedious synthesis. A one-step method to synthesize quinoxaline–thiophene oligomers was developed using Stille reaction wherein stoichiometry of the two monomers was controlled.[19] The tin monomer was added to the dibromo monomer in slow dropwise manner to keep its concentration low in order to avoid the formation of high molecular weight polymers. The series of oligomers containing 5, 7, and 9 repeat units shown in Figure 7.4 were obtained that were separated using column chromatography.

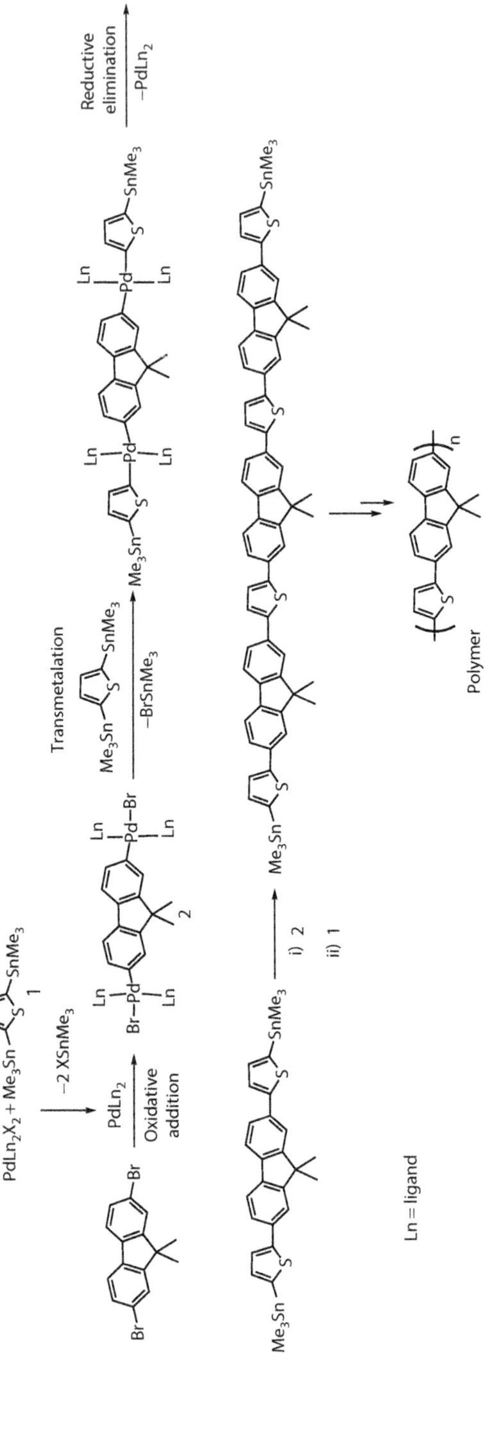

FIGURE 7.3 Mechanism of Stille coupling polymerization based on the general mechanism of Stille reaction.

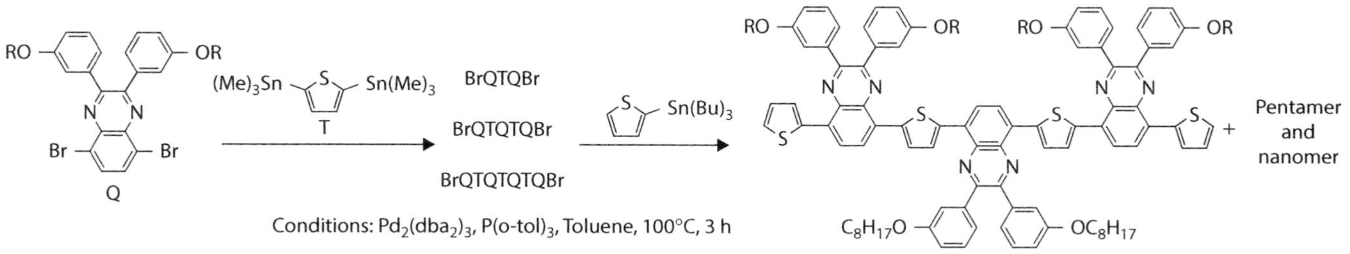

FIGURE 7.4 One-step synthesis of oligomers using Stille coupling reaction.

In the aforementioned polymers for photovoltaic applications, electron-donor and acceptor units are used to improve charge mobility. New units are always explored to improve the performance of the device. Electron-deficient pyrimidine units functionalized with nitrile and fluoro groups were copolymerized with benzodithiophene as electron-rich unit to obtain donor-acceptor conjugated polymers using Stille coupling conditions. Tin compound of thiophene derivative was coupled with trichloropyrimidine derivative.[20] Polyoxometalates, atomic clusters of transition metal oxyanions, have also been incorporated into conjugated polymers. Hexamolybdate clusters containing iodoaryl groups were directly copolymerized with thiophene-based stannanes under Stille coupling conditions using $[Pd_2(dba)_3]$ to obtain hybrid conjugated polymers.[21]

Stille coupling has been used for the synthesis of polymers for light-emitting diodes. A red light-emitting polymer comprising N-dodecylpyrrole and 2,1,3-benzo-thiadiazole repeat units was synthesized using $[Pd(PPh_3)_2Cl_2]$ as a catalyst in THF. The molecular weights (M_n) were between 7.1 and 14.6 kDa, with a PDI of 2.06.[22] For comparison of properties, regioregular and regioirregular poly(thienylenevinylenes) were synthesized using Stille coupling and Heck coupling reactions, respectively. Stille reaction was carried out using $[Pd_2(dba)_3]$ as catalyst and triphenylarsenic as additive to obtain polymers with M_n between 8–15 kDa. Efforts to prepare regioregular polymers by Heck coupling were not successful, whereas attempts to prepare regiorandom polymers by Stille coupling gave minimum 90% regioregularity. Regioregular polymers showed a shift in absorbance maximum of 40 nm from solution to film, whereas regioirregular polymers showed a shift of only 12 nm that was attributed to α-coupling and cis and $trans$ additions in Heck reaction resulting in polymers with lower effective conjugation length.[23] Conjugated polymer-fullerene bulk heterojunction (BHJ) devices have shown high-power conversion efficiencies in solar cells and can also be used as photodetectors for light-sensing applications due to ultrafast electron transfer from polymer to fullerene. In this direction, two low band gap polymers based on the boron dipyrromethane (BODIPY) core as electron-acceptor and either bis(3,4-ethylenedioxythiophene) (bis-EDOT) or its all-sulfur analog bis(3,4-ethylenedithiathiophene) (bis-EDTT) as electron donor were synthesized using Stille reaction. $[Pd(PPh_3)_4]$ was used as a catalyst for coupling of stannylated thiophene derivatives and dichloro-BODIPY derivative. Polymer containing bis-EDOT derivative showed a band gap of 1.18 eV.[24]

7.3   HECK COUPLING

Heck coupling, also called Mizoroki–Heck reaction, involves reaction of vinyl/aryl/benzyl halide or triflate with an alkene (electron deficient) in the presence of a base and a Pd catalyst to form a substituted alkene. The reaction follows the same catalytic cycle for palladium(0)-catalyzed reactions as described earlier. Heck coupling is used to synthesize poly(phenylene vinylene) (PPV)-based conjugated polymers that contain alternating phenyl rings and C=C bonds.

Triarylamine-based polymers are excellent hole-transporting materials and are studied for applications in organic light-emitting diodes (OLEDs). A donor-acceptor triarylamine-vinylene-benzothiadiazole alternating copolymer was synthesized using Heck coupling.[25] To explore the possibility of Heck coupling polymerization following chain-growth mechanism (see Section 7.7), formation of PPV under Heck conditions was studied. Number-average molecular weight remained low until 90% monomer conversion and then increased sharply indicating the step-growth mechanism that was supported by the broad molecular weight distribution. This observation was attributed to lower coordination ability of $[H-Pd(II)-X({}^tBu_3P)]$ (X = Br or I), formed in the catalytic cycle of the reaction.[26] Luminescent nanoporous organic–inorganic hybrid materials were synthesized using $[Pd(PPh_3)_4]$ and K_2CO_3 in dimethylformamide (DMF) at 120°C. Octavinylsilsesquioxane and various aromatic bromides such as dibromo-substituted biphenyl, naphthyl, and tetraphenylethene were used as monomers. The materials were used as picric acid sensors.[27] A limitation with Heck coupling is that it gives regioirregular polymers due to significant α-arylation as mentioned earlier, for example, polymerization of 2-bromo-3-dodecyl-5-vinylthiophene under Heck conditions gave a regiorandom polymer.[23]

7.4   NEGISHI COUPLING

Negishi coupling is similar to Heck coupling with the difference that it uses organozinc compound of an alkene or arene. It is catalyzed by Pd complexes that give higher yields and display better functional group tolerance, although Ni complexes are also used. Organozinc reagents are more sensitive to air and moisture however they are also more reactive and afford faster reaction times. A hyperbranched polymer poly(triphenylamine) was synthesized by Negishi coupling reaction from tris(4-bromophenyl)amine with the terminal bromo group on the hyperbranched polymer bromothiophene. This functionalized hyperbranched polymer was used as a core for the synthesis of a star-structured regioregular poly(3-hexylthiophene). The star polymer had a fully conjugated system with conjugated arms of poly(3-hexylthiophene) attached to a conjugated core of poly(triphenylamine).[28] A chiral binaphthalene-oligothiophene copolymer was synthesized by Negishi coupling between 2,2′-bis(2-thiophene)-1,1′-binaphthalene and 2,2′-bis(5-bromo-2-thiophene)-1,1′-binaphthalene. The monomers were prepared by Stille coupling and Negishi coupling.[29] An efficient synthesis of 2,5-functionalized silole derivatives was accomplished using Negishi coupling. Bis(phenylethynyl)dimethylsilane underwent intramolecular reductive cyclization upon treatment with lithium naphthalenide to form 2,5-dilithiosilole, which was converted to dizincated silole by reaction with $ZnCl_2$/tetramethylethylenediamine. Negishi coupling of this zincate intermediate with 2-bromothiophene gave 2,5-bis(2′-thienyl)silole in good yield.[30]

7.5   SONOGASHIRA COUPLING

Carbon–carbon bond formation between a terminal alkyne and aryl/vinyl halide in the presence of a Pd(0) and Cu(I) halide as catalyst takes place in Sonogashira coupling, sometimes also referred to as Sonogashira–Hagihara coupling. The catalytic cycle involves reaction of Pd(0) complex with aryl or vinyl halide in oxidative addition to form Pd(II) complex, which undergoes transmetallation with copper acetylide formed in the presence of a base so that the acetylenic ligand is

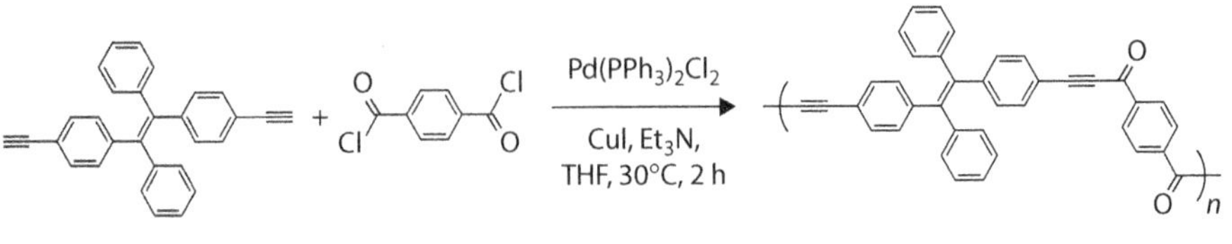

FIGURE 7.5   An example of Sonogashira coupling polymerization.

introduced into Pd(II) complex. When applied to conjugated polymer synthesis it gives poly(aryleneethynylene)s, where the aryl group is mainly thiophene or phenyl. Alternating porphyrin-diethynyldithienothiophene copolymers were synthesized by Sonogashira coupling using Pd(0) and CuI.[31] Organometallic platinum alkyne units have been incorporated into conjugated polymers because d-orbital of the Pt can overlap with the p-orbital of the alkyne unit, leading to an increase in the π-electron delocalization along the polymer chain, and the probability of efficient charge separation is enhanced due to formation of triplet excited states. Copolymers of platinum diacetylene and electron-poor bithiazole with different numbers of electron-donating thiophene units have been synthesized.[32] Multicomponent tandem polymerization involving multicomponent reactions was developed using Sonogashira coupling between alkynes and carbonyl chlorides to give a polymer containing alkynone groups. The alkynyl groups in the polymer backbone underwent hydrothiolation to yield unsaturated polymers with high sulfur content. The polymerization was conducted at 30°C in the presence of [Pd(PPh₃)₂Cl₂], [CuI] and triethylamine (Figure 7.5). Polymers with high molecular weights (38–44 kDa) could be obtained under optimized conditions only since the ratio between [Pd(PPh₃)₂Cl₂] and [CuI] seemed to influence the molecular weight.[33]

Sonogashira coupling has been used to prepare conjugated microporous polymers (CMP) that are useful as composite materials by inclusion of guest molecules or nanomaterials for photovoltaic applications. Tetraphenylethene-based CMP were prepared by first synthesizing a hyperbranched polymer by Sonogashira coupling in miniemulsion followed by crosslinking of terminal groups under solvothermal conditions to obtain microporous particles with enhanced porosity (surface area = 1214 m²/g).[34] Nile Red, a fluorescent dye, was immobilized in the micropores to obtain a light-harvesting composite that showed efficient photon absorption and energy migration characteristics. In another example, a CMP was prepared by reacting difluoro-BODIPY with 1,3,5-triethynylbenzene under Sonogashira coupling conditions in the presence of bromo-functionalized carbon nanospheres (0D), single-walled carbon nanotubes (1D), and reduced graphene oxide (2D) as templates.[35] The CMPs, which showed well-defined morphologies same as the directing templates, acted as donors whereas the nanocarbon structures functioned as acceptors in the composite p/n junction nanomaterials. Mechanism of formation of these CMPs and the effect of alkyne homocoupling on microporosity has been studied in detail recently.[36] Hybrid organic/inorganic nanoparticles have also been developed as an alternative to microporous polymers in this direction. Sonogashira polymerization of dibromoaryl and diethynylaryl compounds was carried out in miniemulsion technique and the polymer particles were doped with modified titanium dioxide and cadmium selenide nanocrystals.[37] Monomer reactivity affected the polymerization rates that influenced the microstructure, which in turn had an impact on uniform distribution of inorganic guest particles. To improve the binding of inorganic particles to polymer, a functionalized monomer was used to obtain a terpolymer followed by postpolymerization dispersion of the presynthesized conjugated polymer and inorganic particles.

7.6 DIRECT (HETERO)ARYLATION POLYMERIZATION

Suzuki and Stille polymerization involves the use of toxic starting materials such as boronic acid and stannanes. The boronic acid or stannane coupling partners need multiple-step synthesis, which limits their direct use. To circumvent this problem, a metal-catalyzed coupling of aromatic C–Br bonds with aromatic C–H bonds called *direct (hetero)arylation polymerization (DHAP)* or *direct arylation polymerization (DArP)* was developed.[38,39] This method affords higher yields as well as higher molecular weights compared to conventional polymerizations and even unfunctionalized monomers can be used. The conjugated polymers produced show properties comparable to analogous polymers synthesized by Suzuki or Stille coupling. The mechanism follows a base-assisted, concerted metalation–deprotonation pathway for most heterocycles, however electrophilic aromatic substitution and Heck-type coupling may occur depending on the substrate leading to certain defects in the resultant polymer structure. Typically, carboxylate or carbonate anions are used as additives although the reaction also works without these additives. Under carboxylate-mediated conditions, after the oxidative addition of the carbon–halogen bond to Pd(0) catalyst the halogen ligand is exchanged for the carboxylate anion. Thiophene substrate is deprotonated by carboxylate ligand while simultaneously forming a metal–carbon bond. The phosphine ligands, or the solvent, can recoordinate to the metal center or the carboxylate group can remain coordinated throughout the entire process. The product is then obtained by reductive elimination process. A major drawback of this process is the lack of C–H bond selectivity, particularly for thiophene substrates, which can result in cross-linked material during polymerization reactions.

Poly(3-hexylthiophene) (P3HT) was synthesized from 2-bromo-3-hexylthiophene using Herrmann-Beller catalyst (also called Herrmann catalyst) (*trans*-bis(acetato) bis[*o*-(di-*o*-tolylphosphino)benzyl]dipalladium(II)) and tris-(2-dimethylaminophenyl)phosphine as catalyst precursor. Highly regioregular (98% H-T) P3HT was obtained with high molecular weight (M_n = 30,600 g/mol, PDI = 1.6) in almost quantitative yield (Figure 7.6).[40] The catalyst was selected such that it could withstand high reaction temperatures (130°C). These conditions are used widely for DHAP and include trianisylphosphine as a ligand, THF or toluene as a solvent, and Cs_2CO_3 as a base with or without pivalic acid. The reaction temperature is typically 110°C–120°C, therefore the reaction is carried out in a pressure reactor. These conditions were also used to obtain a donor-acceptor copolymer of thieno[3,4-c]pyrrole-4,6-dione, an electron-deficient monomer, and an electron-rich 3-octylthiophene with high molecular weight (56,000 Da) in high yields.[41] When the same polymer was prepared using Stille reaction, low molecular weight was obtained. Similarly,

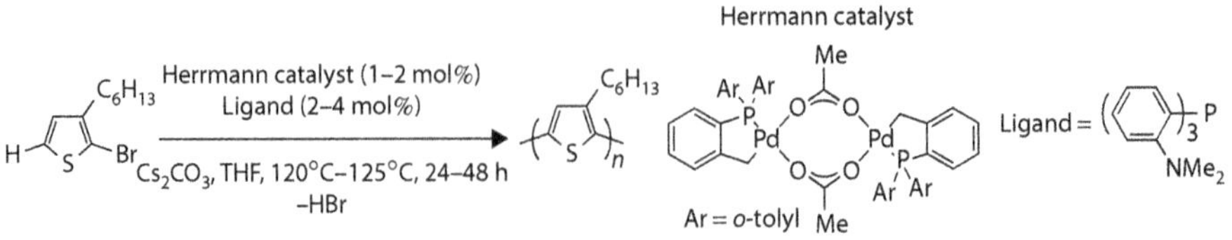

FIGURE 7.6 Synthesis of regioregular polythiophene using direct (hetero)arylation polymerization. (From Wang, Q. et al., *J. Am. Chem. Soc.*, 2010, 132, 11420–11421.)

a high molecular weight copolymer of 9,9-dioctylfluorene and 2,2′-bithiazole was obtained by DHAP[42] than under Suzuki polymerization conditions. Another set of conditions uses [Pd(OAc)$_2$] as catalyst with a phosphine as ligand, pivalic acid as proton shuttle, K$_2$CO$_3$ as base, and dimethyl acetamide as solvent at 70°C–110°C so that pressurized vessel is not required.[43] A limitation of DHAP is the requirement of optimization of reaction conditions for each monomer and a proper choice of monomer and conditions is necessary to obtain well-defined (linear) polymer material. Mainly four types of defects are possible in this polymerization: homocoupling, branching, end-group defects, and residual metal defects.[44]

7.7 CATALYST-TRANSFER POLYCONDENSATION

The synthetic methods discussed earlier are step-growth polycondensations that do not allow control over molecular weight, polydispersity, and chain-end functionality of the polymer. To address these issues, a *living* chain-growth method called *catalyst-transfer polycondensation* (*CTP*) was developed mainly for synthesis of P3HT that could be applied later to other monomers as well.[45] Conventional step-growth polymerization can be transformed into *living* chain-growth polymerization by increasing the reactivity of polymer end-group compared to the monomer reactivity. This is achieved in the polycondensation of A-XY-B monomer by activation of the polymer end-group B by intramolecular transfer of catalyst M after the addition of monomer to the polymer end as shown in Figure 7.7.[46]

The mechanism involves a metal-polymer π-complex, so that the active catalyst remains associated with the growing polymer chain and the propagation occurs *via* an intramolecular oxidative addition. Evidence for the π-complex was provided in the polymerization of 3,6-dioctylthiothieno[3,2-b]thiophene using Ni(0) and Pd(0) catalysts. The strongly bound Ni-complex could be determined from ^{31}P NMR and did not catalyze polymerization, whereas the weakly bound Pd complex did.[47] The polymerization shows typical characteristics of living nature, that is, control over molecular weights by feed ratio, ability to prepare block copolymers, and the presence of defined functional group at chain end, however deviations are observed

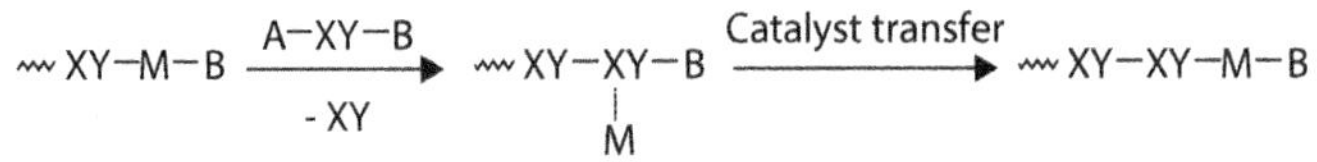

FIGURE 7.7 Mechanism for chain-growth polycondensation by catalyst transfer.

routinely. Mostly electron-rich monomers have been polymerized in a controlled manner; electron-deficient monomers afford nonliving polymerizations. $LnNiX_2$ precatalysts and Pd catalysts have been used successfully.

Grignard metathesis (GRIM) polymerization, a *living* chain-growth poly-condensation, is a method developed for preparation of regioregular P3HT on large scale at room temperature.[48] This method is also referred to as *Kumada catalyst-transfer polycondensation* by some researchers.[49] In this polymerization, 2,5-dibromo-3-hexylthiophene when reacted with *t*-butyl magnesium chloride gives a mixture of two Grignard regioisomers (at 2 or 5 position) in a ratio of 80:20. In the presence of Ni(dppp)Cl$_2$ (dppp = 1,3-bis(diphenylphosphino) propane), the regioisomer with -MgCl at 5 position is incorporated into the polymer whereas the more sterically hindered 2-substituted isomer is not polymerized. Presence of tail–tail thiophene diads at the end and middle of the polythiophene backbone due to unidirectional as well as bidirectional growth has been confirmed in literature reports.[46] Ability of introducing specific end-groups at the end of conjugated polymer chains using GRIM method has been used to prepare novel block copolymers (see Section 7.8). Mechanism for GRIM/Kumada catalyst-transfer polymerization is shown in Figure 7.8.

Catalyst-transfer polymerization has been observed under Suzuki–Miyaura conditions. Boronate ester of iodo-substituted 3-hexylthiophene was polymerized at 0°C in the presence of CsF and 18-crown-6 in THF/water. P3HT with a narrow molecular weight distribution and very high head-to-tail regioregularity and low polydispersity was obtained while the M_n values increased in proportion to the monomer/catalyst feed ratio. Presence of phenyl ring at one end and a hydrogen atom at the other was confirmed by matrix-assisted laser desorption/ionization time-of-flight (MALDI-TOF) technique. A block copolymer of polyfluorene and P3HT was also synthesized.[50] The catalyst-transfer method was used to synthesize a star polymer with terrylene diimide (TDI) core and polyfluorene arms as a light-harvesting system by Suzuki–Miyaura coupling polymerization. Four iodophenyl rings on the TDI core were used as initiating sites for polymerization of bromo-substituted fluorine boronate ester using Pd(dba)$_2$ as catalyst. It was proposed that aromatic core of TDI allowed *chain walking* of the catalyst promoting the catalyst-transfer mechanism.[51] Polyfluorenes with a single amine or phosphonic acid at the end-group were synthesized by Pd-catalyzed Suzuki–Miyaura coupling polymerization. The polymers were used as stabilizing ligands for synthesis of cadmium selenide quantum dots to obtain inorganic nanocrystals with polyfluorene attached to the surface.[52]

Limitations of the catalyst-transfer method are: (1) chain walking (as discussed in Chapter 2) of the catalyst by successively forming π-complexes along the polymer chain leading to propagation from both ends of the polymer that gives a mixture of mono- and di-telechelic polymers. This is particularly observed when monomers have different π-binding affinities. (2) Chain transfer, chain termination, and dispro-portionation reactions can occur giving rise to same end groups (Br/Br and H/H) on the polymer. (3) The rate of precatalyst initiation has been found to be slow leading to broader (PDI > 1.1) molecular weight distributions. Using Ni/dppe it was found that initiation was 20 times slower than propagation. By modifying the reactive

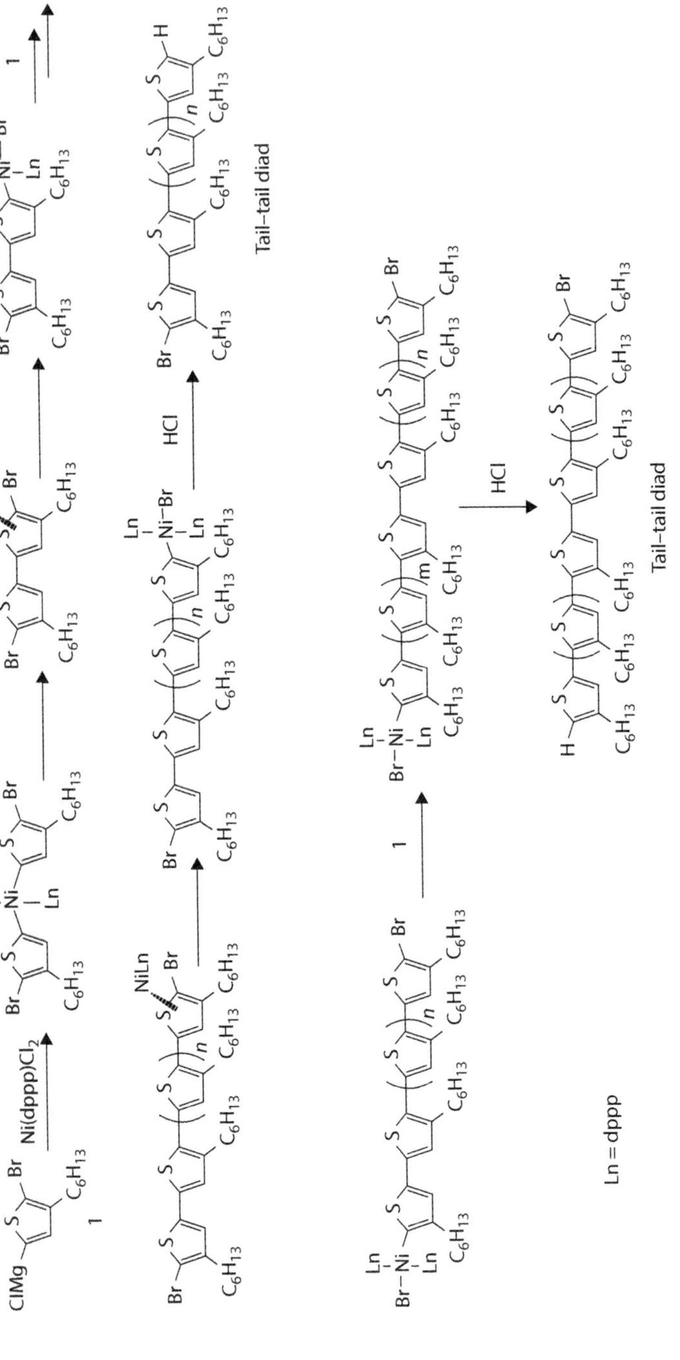

FIGURE 7.8 Mechanism of Kumada catalyst-transfer polymerization.

ligand, reductive elimination was accelerated resulting in faster initiation. Thus, each monomer and catalyst combination provides polymers with varying polydispersities and inconsistent end-groups and therefore does not necessarily follow the chain-growth mechanism.

7.8 BLOCK COPOLYMERS BY COMBINATION WITH OTHER POLYMERIZATION METHODS

Block copolymers of conjugated polymers with nonconjugated polymers are synthesized to make them amenable for certain applications such as sensing and bioimaging that require nanoscale assemblies or molecular dispersion in aqueous solutions. On the other hand, such block copolymers may also improve the processability and charge transfer properties of conjugated polymers. Some representative examples are discussed here to show that complex polymer architectures can be synthesized by combining conjugated polymer synthesis with other controlled polymerization techniques and efficient coupling methods. Conjugated polyelectrolyte brush copolymers with 2,7-linked carbazole-fluorene units in the main chain attached to poly(acrylic acid) (PAA) side chains were synthesized.[53] The main chain was obtained by Suzuki coupling copolymerization of carbazole monomers containing pentynyl groups and fluorene monomers containing polar methoxy(tetraethylene glycol) chains. The pentynyl groups were reacted with azide-terminated poly(t-butyl acrylate), which was prepared by atom transfer radical polymerization (ATRP), using Copper-catalyzed azide-alkyne cycloaddition (CuAAC) reaction. The t-butyl ester groups were hydrolyzed to obtain PAA chains, the charged state of which can be controlled by pH of the solution. An amphiphilic diblock copolymer with regioregular P3HT as hydrophobic block and poly(methyl methacrylate-r-2-hydroxyethyl methacrylate) (P(MMA-r-HEMA)) as the hydrophilic block was synthesized by sequential GRIM polymerization and ATRP. Living nature of GRIM polymerization was exploited to obtain an end-functionalized P3HT. Postpolymerization end-group modification of P3HT was done to afford the macroinitiator for ATRP. Molecular weight of 11,000 g/mol and polydispersities lower than 1.5 were obtained.[54] Similar end-group modification of P3HT was done to carry out cationic ring-opening polymerization of tetrahydrofuran (THF) and 2-ethyl-2-oxazoline (EOx) to obtain P3HT-b-PTHF.[55] In another example, alkyne-terminated P3HT was prepared using Ni-catalyzed GRIM polymerization and azide-terminated poly(N-isopropyl acrylamide)—a temperature-sensitive polymer, was synthesized by ATRP using azide-functionalized initiator. The two polymers were connected using CuAAC reaction.[56] This strategy was also used to prepare amphiphilic diblock copolymer, poly[3-(2,5,8,11-tetraoxatridecanyl) thiophene]-block-poly(ethylene glycol) (PTOTT-b-PEG). The polymer was self-assembled into various well-defined nanostructures such as vesicles, sheets, and nanoribbons in aqueous solution.[57] GRIM polymerization has also been combined with anionic polymerization to obtain block copolymers of P3HT with poly(methyl methacrylate) (PMMA). This was achieved by reacting α-phenyl acrylate-terminated P3HT with living PMMA carbanions in a Michael addition reaction.[58]

REFERENCES

1. Cheng, Y.-J., Yang, S.-H., Hsu, C.-S. 2009. Synthesis of conjugated polymers for organic solar cell applications. *Chem. Rev.* 109:5868–5923.
2. Maluenda, I. and Navarro, O. 2015. Recent developments in the Suzuki-Miyaura reaction: 2010–2014. *Molecules* 20:7528–7557.
3. Wang, F., Luo, J., Yang, K., Chen, J., Huang, F., Cao, Y. 2005. Conjugated fluorene and silole copolymers: Synthesis, characterization, electronic transition, light emission, photovoltaic cell, and field effect hole mobility. *Macromolecules* 38:2253–2260.
4. Blouin, N., Michaud, A., Leclerc, M. 2007. A low-bandgap poly(2,7-Carbazole) derivative for use in high-performance solar cells. *Adv. Mater.* 19:2295–2300.
5. Lu, J. et al. 2008. Crystalline low band-gap alternating indolocarbazole and benzothiadiazole-cored oligothiophene copolymer for organic solar cell applications. *Chem. Commun.* 42:5315–5317.
6. Carrillo, J. A., Ingleson, M. J., Turner, M. L. 2015. Thienyl MIDA boronate esters as highly effective monomers for Suzuki-Miyaura polymerization reactions. *Macromolecules* 48:979–986.
7. He, G. et al. 2013. The marriage of metallacycle transfer chemistry with Suzuki-Miyaura cross-coupling to give main group element-containing conjugated polymers. *J. Am. Chem. Soc.* 135:5360–5363.
8. Muenmart, D. et al. 2014. Conjugated polymer nanoparticles by Suzuki-Miyaura cross-coupling reactions in an emulsion at room temperature. *Macromolecules* 47:6531–6539.
9. Xue, Z., Finke, A. D., Moore, J. S. 2010. Synthesis of hyperbranched poly(m-phenylene)s via Suzuki polycondensation of a branched AB2 monomer. *Macromolecules* 43:9277–9282.
10. Stille, J. K. 1986. The palladium-catalyzed cross-coupling reactions of organotin reagents with organic electrophiles [new synthetic methods (58)]. *Angew. Chem. Int. Ed. Engl.* 25:508–524.
11. Bao, Z., Chan, W. K., Yu, L. 1995. Exploration of the Stille coupling reaction for the synthesis of functional polymers. *J. Am. Chem. Soc.* 117:12426–12435.
12. Carsten, B., He, F., Son, H.-J., Xu, T., Yu, L. 2011. Stille polycondensation for synthesis of functional materials. *Chem. Rev.* 111:1493–1528.
13. Espinet, P. and Echavarren, A. M. 2004. The mechanisms of the Stille reaction. *Angew. Chem. Int. Ed.* 43:4704–4734.
14. Devasagayaraj, A. and Tour, J. M. 1999. Synthesis of a conjugated donor/acceptor/passivator (DAP) polymer. *Macromolecules* 32:6425–6430.
15. Mak, C. S. K., Cheung, W. K., Leung, Q. Y. and Chan, W. K. 2010. Conjugated copolymers containing low bandgap rhenium(I) complexes. *Macromol. Rapid Commun.* 31:875–882.
16. You, W., Cao, S., Hou, Z., Yu, L. 2003. Fully functionalized photorefractive polymer with Infrared Sensitivity Based on Novel Chromophores. *Macromolecules* 36:7014–7019.
17. Tierney, S., Heeney, M., McCulloch, I. 2005. Microwave-assisted synthesis of polythiophenes via the Stille coupling. *Synth. Met.* 148:195–198.
18. Zhan, X. et al. 2007. A high-mobility electron-transport polymer with broad absorption and its use in field-effect transistors and all-polymer solar cells. *J. Am. Chem. Soc.* 129:7246–7247.
19. Li, W. et al. 2015. One-step synthesis of precursor oligomers for organic photovoltaics: A comparative study between polymers and small molecules. *ACS Appl. Mater. Interfaces* 7:27106–27114.
20. Kim, J. et al. 2016. Syntheses of pyrimidine-based polymers containing electron-withdrawing substituent with high open circuit voltage and applications for polymer solar cells. *J. Polym. Sci. Part A: Polym. Chem.* 54:771–784.

21. Wang, R. et al. 2015. Luminescent polythiophene-based main-chain polyoxometalate-containing conjugated polymers with improved solar-cell performance. 2015:656–663.
22. Dhanabalan, A. et al. 2001. Synthesis, characterization, and electrooptical properties of a new alternating *N*-dodecylpyrrole–benzothiadiazole copolymer. *Macromolecules* 34:2495–2501.
23. Loewe, R. S. and McCullough, R. D. 2000. Effects of structural regularity on the properties of poly(3-alkylthienylenevinylenes). *Chem. Mater.* 12:3214–3221.
24. Cortizo-Lacalle, D. et al. 2012. BODIPY-based conjugated polymers for broadband light sensing and harvesting applications. *J. Mater. Chem.* 22:14119–14126.
25. Huo, L. et al. 2007. Alternating copolymers of electron-rich arylamine and electron-deficient 2,1,3-benzothiadiazole: Synthesis, characterization and photovoltaic properties. *J. Polym. Sci. Part A: Polym. Chem.* 45:3861–3871.
26. Nojima, M., Saito, R., Ohta, Y., Yokozawa, T. 2015. Investigation of Mizoroki-Heck coupling polymerization as a catalyst-transfer condensation polymerization for synthesis of poly(p-phenylenevinylene). *J. Polym. Sci. Part A: Polym. Chem.* 53:543–551.
27. Sun, L., Liang, Z., Yu, J. 2015. Octavinylsilsesquioxane-based luminescent nanoporous inorganic-organic hybrid polymers constructed by the Heck coupling reaction. *Polym. Chem.* 6:917–924.
28. Wang, F. et al. 1999. Electroactive and conducting star-branched poly(3-hexylthiophene)s with a conjugated core. *Macromolecules* 32:4272–4278.
29. Li, J., Rajca, A., Rajca, S. 2003. Synthesis and conductivity of binaphthyl-based conjugated polymers. *Synth. Met.* 137:1507–1508.
30. Tamao, K., Yamaguchi, S., Shiro, M. 1994. Oligosiloles: First synthesis based on a novel endo-endo mode intramolecular reductive cyclization of diethynylsilanes. *J. Am. Chem. Soc.* 116:11715–11722.
31. Huang, X. et al. 2008. Porphyrin-dithienothiophene π-conjugated copolymers: Synthesis and their applications in field-effect transistors and solar cells. *Macromolecules* 41:6895–6902.
32. Wong, W. Y. et al. 2007. Tuning the absorption, charge transport properties, and solar cell efficiency with the number of thienyl rings in platinum-containing poly(aryleneethynylene)s. *J. Am. Chem. Soc.* 129:14372–14380.
33. Zheng, C. et al. 2015. Multicomponent tandem reactions and polymerizations of alkynes, carbonyl chlorides, and thiols. *Macromolecules* 48:1941–1951.
34. Zhang, P., Wu, K., Guo, J., Wang, C. 2014. From hyperbranched polymer to nanoscale CMP (NCMP): Improved microscopic porosity, enhanced light harvesting, and Enabled Solution Processing into White-Emitting Dye@NCMP Films. *ACS Macro Lett.* 3:1139–1144.
35. Zhuang, X. et al. 2015. Conjugated microporous polymers with dimensionality-controlled heterostructures for green energy devices. *Adv. Mater.* 27: 3789–3796.
36. Laybourn, A. et al. 2014. Network formation mechanisms in conjugated microporous polymers. *Polym. Chem.* 5:6325–6333.
37. Jung, C., Krumova, M., Mecking, S. 2014. Hybrid nanoparticles by step-growth Sonogashira coupling in disperse systems. *Langmuir* 30:9905–9910.
38. Mercier, L. G. and Leclerc, M. 2013. Direct (hetero)arylation: A new tool for polymer chemists. *Acc. Chem. Res.* 46:1597–1605.
39. Bura, T., Morin, P.-O., Leclerc, M. 2015. Route to defect-free polythiophene derivatives by direct heteroarylation polymerization. *Macromolecules* 48:5614–5620.
40. Wang, Q., Takita, R., Kikuzaki, Y., Ozawa, F. 2010. Palladium-catalyzed dehydrohalogenative polycondensation of 2-bromo-3-hexylthiophene: An efficient approach to head-to-tail poly(3-hexylthiophene). *J. Am. Chem. Soc.* 132:11420–11421.
41. Berrouard, P. et al. 2012. Synthesis of 5-alkyl[3,4-*c*]thienopyrrole-4,6-dione-based polymers by direct heteroarylation. *Angew. Chem. Int. Ed.* 51:2068–2071.

42. Lu, W. et al. 2012. Synthesis of π-conjugated polymers containing fluorinated arylene units via direct arylation: Efficient synthetic method of materials for OLEDs. *Macromolecules* 45:4128–4133.
43. Lafrance, M. and Fagnou, K. 2006. Palladium-catalyzed benzene arylation: Incorporation of catalytic pivalic acid as a proton shuttle and a key element in catalyst design. *J. Am. Chem. Soc.* 128:16496–16497.
44. Rudenko, A. E. and Thompson, B. C. 2015. Optimization of Direct Arylation Polymerization (DArP) through the identification and control of defects in polymer structure. *J. Polym. Sci. Part A: Polym. Chem.* 53:135–147.
45. Bryan, Z. J. and McNeil, A. J. 2013. Conjugated polymer synthesis via catalyst-transfer polycondensation (CTP): Mechanism, scope, and applications. *Macromolecules* 46:8395–8405.
46. Yokozawa, T. and Ohta, Y. 2016. Transformation of step-growth polymerization into living chain-growth polymerization. *Chem. Rev.* 116:1950–1968.
47. Willot, P. and Koeckelberghs, G. 2014. Evidence for catalyst association in the catalyst-transfer polymerization of thieno[3,2-*b*]thiophene. *Macromolecules* 47:8548–8555.
48. Stefan, M. C., Bhatt, M. P., Sista, P., Magurudeniya, H. D. 2012. Grignard metathesis (GRIM) polymerization for the synthesis of conjugated block copolymers containing regioregular poly(3-hexylthiophene). *Polym. Chem.* 3:1693–1701.
49. Kiriy, A., Senkovskyy, V., Sommer, M. 2011. Kumada catalyst-transfer polycondensation: Mechanism, opportunities, and challenges. *Macromol. Rapid Commun.* 32:1503–1517.
50. Yokozawa, T. et al. 2011. Precision synthesis of poly(3-hexylthiophene) from catalyst-transfer Suzuki-Miyaura coupling polymerization. *Macromol. Rapid Commun.* 32:801–806.
51. Fischer, C. S., Jenewein, C., Mecking, S. 2015. Conjugated star polymers from multidirectional Suzuki-Miyaura polymerization for live cell imaging. *Macromolecules* 48:483–491.
52. de Roo, T. et al. 2014. A direct approach to organic/inorganic semiconductor hybrid particles via functionalized polyfluorene ligands. *Adv. Func. Mater.* 24:2714–2719.
53. Rugen-Penkalla, N., Klapper, M., Mullen, K. 2012. Highly charged conjugated polymers with polyphenylene backbones and poly(acrylic acid) side Chains. *Macromolecules* 45:2301–2311.
54. Nguyen, H. T., Dong, B. C., Nguyen, N. H. 2014. Novel conducting amphiphilic diblock copolymer containing regioregular poly(3-hexylthiophene). *Macromol. Res.* 22:85–91.
55. Alemseghed, M. G., Gowrisanker, S., Servello, J., Stefan, M. C. 2009. Synthesis of diblock copolymers containing regioregular poly(3-hexylthiophene) and poly(tetrahydrofuran) by a combination of Grignard metathesis and cationic polymerization. *Macromol. Chem. Phys.* 210:2007–2014.
56. Kumari, P., Khawas, K., Hazra, S., Kuila, B. K. 2016. Poly(3-hexyl thiophene)-*b*-poly(*N*-isopropyl acrylamide): Synthesis and its composition dependent structural, solubility, thermoresponsive, electrochemical, and electronic properties. *J. Polym. Sci. Part A: Polym. Chem.* 54:1785–1794. doi: 10.1002/pola.28040.
57. Kamps, A. C., Cativo, M. H. M., Fryd, M., Park, S.-J. 2014. Self-assembly of amphiphilic conjugated diblock copolymers into one-dimensional nanoribbons. *Macromolecules* 47:161–164.
58. Moon, H. C., Anthonysamy, A., Kim, J. K., Hirao, A. 2011. Facile synthetic route for well-defined poly(3-hexylthiophene)-*block*-poly(methyl methacrylate) copolymer by anionic coupling reaction. *Macromolecules* 44:1894–1899.

Index

Note: Page numbers followed by f and t refer to figures and tables, respectively.

Printed and bound by CPI Group (UK) Ltd, Croydon, CR0 4YY

24/10/2024

01778301-0004